More features...

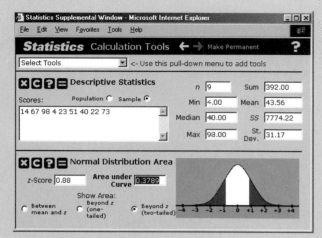

Calculation Tools

Interactive Calculation Tools are designed to take the tedium out of statistical computation while still requiring you to think through the calculations. Tools include: Data Transformations, Descriptive Statistics, and z-Scores.

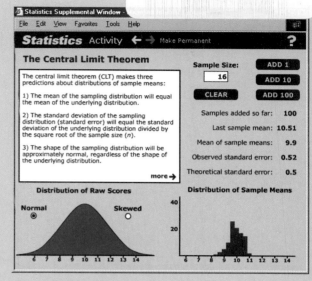

Activities

Activities explicate tricky and/or important concepts and calculations, such as the normal distribution, the Central Limit Theorem, and ANOVA. Links to Activities appear at appropriate points throughout the text.

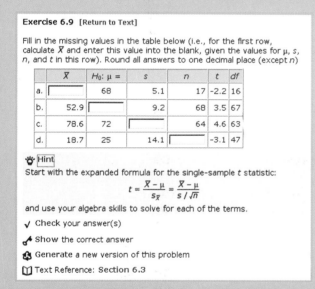

Review Exercises

Review Exercises reinforce the concepts and calculations covered in each chapter. These exercises give you instant feedback on your answers, including detailed explanations. Many Review Exercises are "regenerative," meaning you can practice them again and again with different numbers. Your professor can assign exercises, and you can submit your answers online.

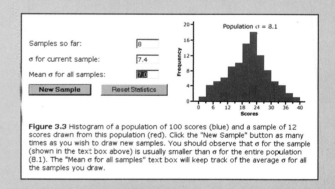

Figure 3.3 Histogram of a population of 100 scores (blue) and a sample of 12 scores drawn from this population (red). Click the "New Sample" button as many times as you wish to draw new samples. You should observe that σ for the sample (shown in the text box above) is usually smaller than σ for the entire population (8.1). The "Mean σ for all samples" text box will keep track of the average σ for all the samples you draw.

Interactive Figures

Rather than relying solely on traditional static figures, *Interactive Statistics* features many interactive figures that can help you visualize statistical principles by overlaying curves, displaying multiple results for different values, building graphs, and more.

Interactive Statistics
for the Behavioral Sciences

Pepper Williams, Ph.D.
Portland, OR

SA *Sinauer Associates, Inc. Publishers*
Sunderland, Massachusetts U.S.A.

Interactive Statistics for the Behavioral Sciences

Sinauer Associates
23 Plumtree Road
Sunderland, MA 01375 U.S.A.
FAX: 413-549-1118

Contents

Chapter 12

Inference for Categorical Data: Chi-Square Tests   331

Preface

Many traditional printed textbooks now come with a CD- or Web-based supplement. A couple of years ago, I proposed to Sinauer Associates that we turn this formula around, and produce a CD- and Web-based textbook that comes with a printed book. The result is *Interactive Statistics for the Behavioral Sciences*.

I designed and wrote the text to be read through a Web browser. Sections are broken up into pages that are the length of Web pages, and they are written so that most can be read without needing to flip back and forth to other pages for context. When another page is referred to, a hyperlink allows you to jump directly to the page (and a click of the "Back" button in your browser returns you to the page you just left). Short queries that appear every fourth page or so keep you on your toes as you read. Interactive and animated figures greatly aid explanations of tricky concepts such as the difference between population and sample standard deviations, power, and partitioning variance in ANOVA.

I've also provided ways to customize your text: You can add highlighting, bookmarks, and notes (which can include pictures and links to other websites) to any page of the text. And notes added by a professor are visible to all of her or his students.

In addition to the interactive text, *Interactive Statistics for the Behavioral Sciences* includes a set of Calculation Tools, a collection of Activities, and interactive Summaries and Review Exercises for each chapter. The Calculation Tools are designed to make statistical calculations less daunting without releasing you from the obligation to think through what statistics mean and how they are derived. Activities explicate tricky or important concepts and calculations such as the Central Limit Theorem and the Pearson correlation. Chapter Summary/Reviews provide quick checks on the concepts covered in each chapter, and Review Exercises reinforce these concepts and provide practice with statistical calculations. Many Review Exercises are regenerative, meaning that you can practice them over and over again, with a different set of data each time.

If you have a reasonably fast Internet connection, I suggest that you use the online version of the interactive textbook (register online at **www.introstats.net** using the registration number on the card bound into the printed textbook). If you're working on a computer without a good Internet connection, however, you can access most of the features via the CD included in the printed textbook. (Due to technical limitations, the customization tools noted above are not available in the CD version; in addition, a full-text search function is available only when you're using the system online.)

Of course, there are times when you need to read your text and just can't get to a computer. This is why we also provide a printed version of the text, which is nearly identical to the online version. However, I strongly encourage you to use the printed version sparingly. The printed textbook contains the main text and the Boxes, but you will find that there are many elements embedded in the interactive textbook that cannot function on the printed page, such as figures that adjust to your input and Activities that walk you through important theories and analyses. Though you may not be used to reading from a computer screen, countless studies show that you learn material more completely and more quickly when you are actively engaged, and the interactive textbook is designed to be as engaging as possible.

Structure of the Book

Chapter 1 introduces the purpose of statistics and covers the distinctions between populations and samples and between descriptive and inferential statistics. **Chapter 2** describes ways to look at datasets by constructing tables and graphs. **Chapter 3** covers the basic statistics for describing the central tendency and variability in datasets. **Chapter 4** includes a smorgasbord of topics (basic probability theory, normal distributions, and the Central Limit Theorem) necessary to provide the groundwork for our discussion of hypothesis testing, which begins in **Chapter 5** with the z-score test statistic. In **Chapter 6** we move on to testing hypotheses using t statistics. **Chapter 7** discusses what can go wrong when hypothesis testing, how to assess the likelihood of errors (power analysis), and what we can do to avoid errors. **Chapter 8** covers t tests for two related or independent samples, and confidence intervals are covered in **Chapter 9.** In **Chapter 10** we tackle analysis of variance (ANOVA), with full coverage of single-factor independent- and related-samples designs and a brief discussion of multiple-factor designs. The topics in **Chapter 11** are correlation, regression, and hypothesis testing for relationships between two variables, with short sections on multiple regression/correlation and the General Linear Model. Finally, **Chapter 12** covers nonparametric hypothesis testing using various forms of chi-square tests.

A Note about Hypothesis Testing

The biggest overarching concept in *Interactive Statistics for the Behavioral Sciences* is hypothesis testing, as in most introductory statistics books aimed toward the behavioral sciences. Hypothesis testing uses a somewhat perverse kind of logic: When looking at the results of a study, we hypothesize that the study revealed no significant effects whatsoever, and then we try to prove this straw-man hypothesis wrong.

For example, suppose you're analyzing data from an experiment testing the efficacy of a drug for treating depression. The null hypothesis in this study claims that people who are treated with the drug are no less depressed at the end of the treatment than people who don't get the drug. Given this null hypothesis and the data you observe in your experiment, you can calculate a test statistic, probably a t statistic for this experimental design. The t value that you calculate is, in turn, associated with a certain probability level (a p-value), which can be roughly described as the likelihood that you would have observed the data you observed if the null hypothesis were in fact true.

Most books present hypothesis testing as if we can unambiguously answer the yes/no question, "Is the null hypothesis correct?" The proscribed method for answering this question is to set a certain criterion, called the alpha level, which is usually $p = .05$ in behavioral science research. If the p-value you derived from your t statistic is smaller than .05, you conclude that the null hypothesis was incorrect, and you claim that the drug works—people who take it are less depressed.

In my opinion (and this is an opinion shared by many good statisticians), there are at least two important reasons why this is a bad way to conceptualize hypothesis testing. First of all, if your p-value ends up being greater than .05, this way of thinking encourages you to claim that the null hypothesis *is* correct—that the drug has absolutely no effect. But this claim cannot be justified. There is no way to justify a conclusion that the null hypothesis is correct, no matter what the p-value is. Indeed, my graduate statistics teacher, who was an eminent statistician, forcefully argued that the null hypothesis is almost *never* strictly correct.

The second reason to avoid the yes/no conceptualization of hypothesis testing is that categorizing the results of an experiment as "$p < .05$" or "$p \geq .05$" greatly oversimplifies those results. If $p = .99$, we can be pretty sure that our experiment doesn't show a significant drug effect (again, we can't say for sure that the drug has no effect, but we can say for sure that our study didn't reveal a meaningful effect). If $p = .00001$, we can be pretty darned confident that the drug did have some kind of effect. But suppose $p = .049$. Clearly, the evidence in this case is not as strong as if p had been .00001, but making thumbs-up/thumbs-down decisions about the null hypothesis leads people to draw the same conclusion in both cases.

An even bigger problem is assessing the difference between an experiment that has a p of .049 and another experiment with $p = .051$. There is really no difference whatsoever between the evidence revealed by the two experiments, so it is silly to say that one showed a significant effect and one didn't.

I believe that the right way to analyze an experiment is to calculate the p-value for your test statistic, then form an argument about what your data show based on this p-value. Again, if p is very low, you can make a strong case that there was an effect. The higher p is, the more you have to temper your argument. If p is around .05, you can claim that there was probably an effect, but you're going to need to replicate your study to make your claim convincing, regardless of whether p is .049 or .051. If p is .99, it's probably time to give up on this line of research. Probability values form a continuum, and if you simply divide the continuum up into two categories, you are throwing away important information about your experimental results.

I think there are two reasons why most books emphasize the yes/no approach, even though most statisticians would agree that the continuum approach is more reasonable. First, if all you have is a printed book with a bunch of statistical tables in the back, it is impossible to calculate the exact p-value associated with a given test statistic. Therefore, most books teach students to work backward by looking up the test statistic value that corresponds to $p = .05$, and then comparing that test statistic to this criterial test statistic value. This is a non-issue for users of *Interactive Statistics,* because I provide calculation tools for converting test statistics to exact p-values. Indeed, this is one of the best features of my system. (If a statistical table must be used in some situation, such as when taking an exam, tables are provided for estimating the p-value associated with any given test statistic.)

The second reason why other authors may prefer the yes/no approach is that it simplifies things. We all hate uncertainty. It's much more satisfying to say "the experiment worked" or "it didn't work" than to say "the experiment probably worked" or "the experiment probably didn't work" or, worst of all, "I just can't say whether or not the experiment worked."

In this book, I do cover alpha levels and discuss what it means to declare an experimental result "significant" or "not significant." We do need to draw conclusions from experiments, and the .05 level has been almost universally adopted in the behavioral sciences as the guideline for judging whether or not there is sufficient evidence to make claims. But I warn students about the dangers of treating .05 as an absolute boundary, and point out the foolishness of simply dividing experiments into two categories based on this boundary. In other words, I don't sugarcoat things: Students will read over and over that they just have to live with the uncertainty inherent in the hypothesis testing procedure.

Resources for Instructors

Apart from providing a more enjoyable experience for students, *Interactive Statistics for the Behavioral Sciences* offers benefits for the instructor. First, the text is customizable via the "Note" feature: Instructors can add notes to any page as well as links to graphics or websites of interest; their students see these notes whenever they are reading the text online. As students progress through the chapters, instructors can assign specific problems to assess their progress. When students submit their answers, the server checks them and stores graded answers. Instructors can then view the answers in their own Web browsers, assign a grade, and add individual notes to each student on each exercise. This saves class and grading time (not to mention trees).

Finally, there is an Instructor's Resource CD (ISBN 0-87893-931-8), available to qualified adopters free of charge. This cross-platform (Macintosh/Windows) resource includes all of the figures from the printed text and an electronic test bank (authored by Will Hayward, Chinese University of Hong Kong). Contact Sinauer Associates for more information (custserv@sinauer.com).

Acknowledgments

At Sinauer Associates, Pete Farley, Jason Dirks, Dean Scudder, Kerry Falvey, Graig Donini, Parker Morse, and Andy Sinauer provided the professional support necessary to bring this ambitious and novel project to fruition. Judd Volino and Patrick Rennich helped with early development of the programming and design. The probability function Calculation Tools are based on JavaScript routines provided by John C. Pezzullo of Georgetown University, and I also generated the statistical tables using these routines. Will Hayward and Kathryn Oleson class-tested the system and gave me great feedback.

Pepper Williams
Portland, OR
August 2003

Chapter 1
Introduction to Statistics

1.1 Statistics in Society and Science

Mark Twain once said that there are three kinds of lies: lies, damn lies, and statistics. But despite the mistrust in statistics reflected in this quote and shared by many people, statistics show up everywhere—in newspaper articles, baseball game box scores, presidential debates, product advertisements, electricity bills, hurricane warnings, instruction manuals, *Consumer Reports* recommendations, and so on.

Statistics as data reduction

And despite occasional abuses and most people's misgivings, statistics serve an indispensable purpose in our information-age society. Essentially, that purpose is to reduce the vast amount of data that humans produce these days into a manageable and understandable form.

For example, imagine that the school district in the small town of Springfield administers a standardized test to each of its graduating seniors every year.* Scores on the test can range from 0 to 100, with 60 considered the "passing" score. For the 116 students who graduated one spring, the scores were:

---

*Unless otherwise noted, every example used in this book (including this one) is fictional.

```
61 78 34 89 62 58 100 72 48 43 71 82 86 40 79
100 66 88 100 66 92 96 61 84 64 33 45 64 62 60
68 66 44 57 78 40 50 83 60 72 50 78 94 43 27
56 83 75 77 99 47 59 71 84 38 71 100 10 58 87
57 76 70 71 59 79 70 93 72 85 65 81 65 90 80
76 68 83 75 100 78 88 80 74 69 63 72 67 69 100
73 58 82 74 54 47 41 69 72 61 64 94 100 79 92
32 82 88 35 66 77 64 60 90 93 89
```

So how did the students do? Inspecting the scores, we can see that there is quite a bit of variability—some students did much better than some others. You may also get the sense that most students appear to have passed, although quite a few did not. But it is impossible to tell much more than this from the raw data alone. To analyze the data further, we need a **statistic:** a single number that summarizes a set of other numbers.

Glossary Term: statistic

If we add all the scores together and divide by the number of students who took the test, we get a statistic called the **mean.** In common parlance, the mean is often called the average, because it seeks to describe the average, or most typical, individual in the set of raw data. The mean score for the 116 students of Springfield High turns out to be 70.

Interactive Page 4

Clearly, a single number is much easier to work with than 116 numbers. But on its own, the number "70" is no more meaningful than the 116 raw scores. **Statistics** are only useful when they are interpreted by humans.

So what does our mean score of 70 mean? In general, to interpret a statistic we need to compare it to some other number. But what is the relevant comparison number? In a Springfield town meeting, one optimistic parent might compare the **mean** score to the passing score of 60 and conclude that students this year did pretty well. However, a pessimistic parent might instead compare the mean score this year to last year's mean score of 74 and conclude that the school district is slipping.

A statistic is only meaningful if compared to another statistic, or to some other standard.

You might think that this is exactly the kind of behavior that gives statistics a bad name: two people take the same number and use it to draw completely opposite conclusions. But the fact is that both parents are right—the average score this year was above passing, while at the same time it was also worse than the average last year. These are both important pieces of information, and we cannot glean this information directly from the raw scores of 116 children.

Moreover, consider where we would be if we didn't have statistics. People will debate issues regardless of whether or not they use statistics in their arguments. Statistics, at least, provide an objective means of evaluating competing claims. If a third parent finds the 10 point difference between this year's mean and the passing score to be more compelling than the 4 point difference between last year's and this year's mean, she will side with the more optimistic parent at the town meeting. Otherwise, she will side with the pessimistic parent. Evaluating the statistics is better than, say, simply siding with the person who shouts the loudest at the meeting!

Statistics are important in town meeting debates, but they are even more important in scientific debates. Let's consider a behavioral science example:

> Dr. L. Durley, a gerontologist, theorizes that aging leads to diminished problem-solving skills. To test this theory, he develops a simple test with 25 problems to solve and administers it to twenty 30-year-olds and twenty 60-year-olds. Half of the people in each age group are men and half are women. The numbers of correct question responses for the subjects in the experiment are:

30-year-olds		60-year-olds	
Men	**Women**	**Men**	**Women**
24	14	15	14
17	20	19	21
22	23	14	17
16	13	19	15
22	22	19	21
20	22	20	12
18	18	17	15
16	19	12	24
21	17	24	16
19	25	13	15

What can Dr. Durley conclude from these data?

As with Springfield High's test scores, just looking at the raw numbers will tell us relatively little. It appears that the older subjects generally got fewer problems correct than the younger ones. But exactly how big a difference is there between the younger and older groups? Is there a gender difference? Do women's problem-solving abilities suffer more or less than men's with age, or vice versa? To even attempt to answer these questions, we need—you guessed it—**statistics.**

More precisely, two different types of statistics are needed to fully analyze the results of a scientific study. "Descriptive" statistics describe the results of the study, and "inferential" statistics help us interpret these results. The upcoming sections introduce these two types of statistics.

1.2 Descriptive and Inferential Statistics

Here's the description of Dr. Durley's experiment again:

> Dr. L. Durley, a gerontologist, theorizes that aging leads to diminished problem-solving skills. To test this theory, he develops a simple test with 25 problems to solve and administers it to twenty 30-year-olds and twenty 60-year-olds. Half of the people in each age group are men and half are women. The numbers of correct question responses for the subjects in the experiment are:

30-year-olds		60-year-olds	
Men	**Women**	**Men**	**Women**
24	14	15	14
17	20	19	21
22	23	14	17
16	13	19	15
22	22	19	21
20	22	20	12
18	18	17	15
16	19	12	24
21	17	24	16
19	25	13	15

Descriptive statistics, as their name implies, describe the results of a study. We have already encountered the most common descriptive statistic, the **mean.** For Dr. Durley's experiment, we will calculate four means: one for young males, one for young females, one for elderly males, and one for elderly females. The mean will be covered in detail later in the course, but you can probably calculate a simple mean on your own right now.

Glossary Term: descriptive statistics

Try figuring out the mean score for the young males in Dr. Durley's experiment by adding up the scores in the leftmost column of data above, then dividing by 10 (the number of young male subjects).

On the previous page, you were asked to find the **mean** for the young males in Dr. Durley's experiment. Adding the numbers together gives a sum of 195, and dividing by 10 produces a mean of 19.5.

The means for the other three groups of subjects in Dr. Durley's experiment are 19.3 for young females, 17.2 for elderly males, and 17.0 for elderly females. Whenever more than two **descriptive statistics** are necessary to characterize the results of an experiment, a graph is useful (some would say necessary) in comprehending the results. Dr. Durley might create the graph shown in Figure 1.1 to present his results.

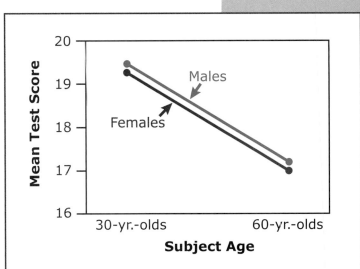

Figure 1.1 Graph of the results of Dr. Durley's experiment on problem-solving abilities in 30- and 60-year-olds.

Now we can begin to answer some of the questions posed a couple of pages ago regarding this experiment:

- How big is the difference between young and old subjects? The mean for all 20 young subjects (i.e., the 10 young males combined with the 10 young females) is 19.4, 2.3 points higher than the mean of 17.1 for the 20 elderly subjects.

- Was there a gender difference for the subjects of the experiment? Yes, but it was much smaller than the age difference. Ignoring the ages of subjects, males scored 0.2 points higher than females.

- Do women's problem-solving abilities suffer more or less than men's with age? This question is best answered by inspecting the graph. If men's problem-solving abilities declined more than women's, the line representing male performance would have a steeper slope than that representing female performance, indicating an **interaction** between gender and age. As you can see from Figure 1.1, the slopes of the two lines are identical (i.e., the two lines are parallel to each other), so these data seem to indicate that neither gender loses its problem-solving abilities more or less than the other with age.

Earlier, we noted that a single **descriptive statistic** becomes meaningful only when compared to some standard (such as the passing score on Springfield's standardized test) or to another statistic (the mean score from the previous year's crop of seniors).

The nice thing about experimental designs such as Dr. Durley's is that they have meaningful comparisons built right in. For example, the meaning of the elderly subjects' average score is easily assessed by comparing it to the average for the younger subjects. We can summarize this comparison with a single number, the difference between the two means (2.3).

This difference statistic tells us unequivocally that the 20 younger subjects scored higher than the 20 older subjects in Dr. Durley's experiment. But he would probably like to draw a broader conclusion: he would probably like to argue that 60-year-old people *in general* are worse problem solvers than 30-year-old people.

In order to make this argument effectively (that is, in a way that will convince other scientists), Dr. Durley will need the aid of the second class of statistics you'll learn in this course, called **inferential statistics.** The name comes from the notion that they help us to draw inferences about large groups on the basis of smaller groups. Let's look more closely at this relationship between the set of individuals assessed in an experiment, called a **sample,** and the larger set of all individuals in which an experimenter is interested, called a **population.**

1.2.1 Populations and Samples

Sometimes, every member of the **population** of interest can be assessed in a study. For example, the Springfield school board tests every graduating senior every year, so the parents at Springfield's town meeting need only look at the mean scores to know definitively that the seniors this year scored worse than the seniors last year. No further analyses are needed to be able to draw this conclusion. (A number that summarizes the data from every single member of a population is called a **parameter.**)

In scientific experiments, though, it is very rare that every member of a population can be tested. Dr. Durley has not tested the memory of all or even a fraction of a percent of 30-year-olds and 60-year-olds. Therefore, on the basis of his results it is impossible to say with absolute certainty whether older people in general solve problems less well than younger people.

However, if Dr. Durley's groups of subjects form representative **samples** of the relevant populations, he may be able to make a compelling argument for a **hypothesis** about the relative problem-solving abilities of older and younger people. The relationship between populations and samples is very important, and will be covered in detail in Chapter 4. For now, read the Box on **Populations and Samples** to get a preliminary idea of this relationship.

Glossary Terms: inferential statistics, population, sample

Box: Populations and Samples

Box: Populations and Samples

A simple way of stating the relationship between populations and samples is as follows:

> A **population** is a complete set of all individuals that meet some criteria or another.

> A **sample** is a subset of a population.

For example, one population that is commonly studied in the behavioral sciences is "human beings." Broadly defined, this population includes all members of the species *Homo sapiens* who are now living, lived at some time in the past, or will be born in the future.

Obviously, it is impossible to study every single member of this population, so researchers who want to draw conclusions about human behavior select a sample of subjects for their research. A sample can include any number of individuals from the population—2, 10, 100, 1000, etc. The larger the sample, the more likely it will be to fairly represent the population.

Note that one researcher's sample might be another's population, and vice versa. For example, many states give all public school students assessment tests. One researcher might be hired by the state of Massachusetts to determine whether or not 5th-graders in the state have adequate reading skills. For this researcher, the set of all 5th-graders in Massachusetts is itself the population of interest. For a second researcher, however, the Massachusetts 5th-graders might be considered a sample of the population of all 10-year-old human children.

Technically, the term *statistic* only applies to samples; when a single number is used to summarize data from all the members of a population, it is called a *parameter*. For example, the mean score on a reading skills assessment test given to a randomly selected group of 100 Massachusetts 5th-graders is a sample statistic. But if we calculate the mean score for *all* Massachusetts 5th-graders, we are determining a population parameter (if we are treating Massachusetts 5th-graders as the population of interest).

Sample statistics are usually used to provide estimates of population parameters. However, the process of generalizing from a statistic to a parameter only works properly if the sample is randomly drawn from the population as a whole. For example, 100 5th-graders from Boston do not make up a random sample of Massachusetts 5th-graders; a random sample would have to include students from around the state (and meet other more formal requirements).

In some cases, statistics are computed slightly differently than their corresponding parameters (the most notable such case is the standard deviation). Also, parameters are usually symbolized by Greek or capital letters (e.g., σ and N for the population standard deviation and the population size, respectively), whereas statistics usually go by lowercase letters (e.g., s and n).

1.2.2 Hypothesis Testing

Researchers usually collect data for one of two reasons. First, they may simply want to know more about some aspect of a population.

Parameter estimation as a research goal

For example, we might want to know at what age the average American child can do basic multiplication, or how many alcoholic drinks the average American adult consumes in a week. In Chapter 9 we will discuss procedures for assessing population parameters on the basis of samples.

Theory testing as a research goal

The second, more interesting goal of research is to evaluate the validity of a theory. Dr. Durley is not really concerned with the absolute values of the problem-solving scores for elderly and young people. Rather, he has a theory that a slide in problem-solving ability is a normal effect of aging. This theory implies that scores on a problem-solving test should be lower for older than for younger people, and his experiment is designed to test this prediction of his theory.

Dr. Durley's study uses what is called the quasi-experimental method, in which two preexisting groups of subjects are tested and the results from the two groups compared. Other research designs include true experiments, in which the researcher creates different groups by applying different manipulations to different subjects, and correlational studies, in which a single group of subjects is assessed on two different **variables** to determine if there is a relationship between the variables.

Dr. Durley's experiment provides an excellent test of his theory. But unfortunately, no study (regardless of how it is designed) can conclusively prove or disprove anything about a population on the basis of a sample of subjects. The best we can ever do is to propose and test a very simple pair of **hypotheses** (educated guesses) about the population. The topic of hypothesis testing is probably the most complex one that will be covered in this course, and we won't get to its details until Chapter 5. But we will at least give a sketch here of how the hypothesis testing procedure works.

**Interactive
Page 11**

What Dr. Durley must do is propose a **null hypothesis** that makes the opposite claim to the one he really wants to make. In this case, he would like to argue that there is a difference between young and elderly people's problem-solving abilities, so the null hypothesis will state that there is *no difference* between the abilities of the two populations.

Note that the null hypothesis makes a claim about the **populations,** not just the **samples** studied in the experiment. The **inferential statistic** will provide a bridge between the samples observed in the experiment and the populations referred to by the experimenter's hypotheses.

Inferential statistics provide a bridge between samples and populations.

After stating the null hypothesis, Dr. Durley will collect his data and compute the **descriptive statistics** we've already gone over. He will then use these descriptives to compute an inferential statistic. In this case, an appropriate inferential would be an independent-samples *t* **statistic** (covered in Chapter 8). Then on the basis of this inferential statistic, he can come up with an estimate for the **probability** that the null hypothesis is correct, given the sample data that he observed.

This probability estimate will turn out to be a bit less than .05. That is, there is less than a 1 in 20 chance that if the null hypothesis were correct (and there really was absolutely no difference between the problem-solving abilities of 30- and 60-year-olds) one would obtain a difference in sample means as large as the one Dr. Durley observed. Since 1 in 20 are pretty long odds, Dr. Durley can make a compelling claim that, in fact, the null hypothesis is *not* correct. And if there isn't *no difference* between the two populations, it logically follows that there is *some difference* between them.

If it seems to you that Dr. Durley has to make a ridiculously complicated train of inferences to get to a conclusion that isn't even all that satisfying, your assessment isn't far off. You have to jump through a lot of hoops in the hypothesis testing procedure, and in the end, all you can say for sure is "the two populations I studied probably aren't exactly the same." But as lame as this statement may sound, it forms the basis for debates without which science would never advance, as discussed in the next section.

1.2.3 Statistics and Rhetoric

Most scientists are born skeptics. They are trained to regard every proposal suspiciously until convinced that it is has merit. And if scientist A makes a **hypothesis** that directly contradicts scientist B's theory, Dr. B will be even more keen to find a flaw in the hypothesis. The rest of the scientific community then sits in judgment while each side tries to prove its case.

Statistics are a necessary part of the scientific enterprise because we need objective means for evaluating competing claims. Otherwise, the scientist with the best reputation (like the Springfield parent with the loudest voice) would always win the debate.

But many people expect too much from statistics, believing that the "right" statistic should automatically and definitively answer any scientific question. These people are the ones who become disillusioned with statistics, because only people, not numbers, can answer questions. To appreciate the role of statistics in science, you must accept the fact that a statistic is nothing more than a rhetorical device—an aid in making an argument for or against some claim.

The logic of hypothesis testing outlined on the previous page reflects this point. Dr. Durley wants to claim that his data show deficits in elderly people's problem-solving abilities. So the first thing he does is anticipate the easiest argument a skeptic can make against his claim: there really is no deficit at all, and the difference found in this experiment is due to chance, a random fluke of sampling. (This boils

Statistics don't answer questions themselves; they help people debate their answers.

down to a counterclaim that Dr. Durley happens to have tested young people who were abnormally good or elderly people who were abnormally poor problem solvers.)

Using his **inferential statistic,** Dr. Durley can counter this counterclaim: if the two **populations** were identical, the probability of getting a difference in means as large as that observed would be less than .05. He's now on his way to making a compelling argument for his original claim.

The skeptics will not sit idly by. If they can't fall back on the random fluke explanation, they'll look for some other way to explain the difference observed in Dr. Durley's experiment (e.g., maybe elderly people are just as good at problem solving but don't deal with tests as well as younger people). This will lead them to perform their own experiment. Then Dr. Durley will perform a third experiment to counter the skeptics' experiment. And the scientific enterprise will go on.

1.3 Statistical Notation

You may not be happy about it (many behavioral science students aren't), but you are currently taking a math class. This means that you're going to have to deal with mathematical formulas and calculations. In this textbook, we will try to be as gentle as we can with formulas, expressing them in the simplest way possible. Nevertheless, you're going to have to get used to the idiosyncratic notation that we'll need to use.

In general, we'll introduce symbols as they come up, rather than overloading you up front with all the jargon. But here are a few basics to start out with:

- A set of data collected in an experiment is called a **variable.** Variables are designated by uppercase letters. Until Chapter 10, we will need to use only one variable in each experiment, which we'll always designate X. Thus if the scores in an experiment are 5, 3, 9, 10, and 4, we can say that for the first subject, $X = 5$, for the second subject $X = 3$, etc. (In Chapters 10 and 11, we'll discuss experiments in which there are two variables, with the second one receiving the label Y.)

Glossary Term: variable

- Almost all statistics take the number of scores in a dataset into account. We use the letter n to refer to the number of scores in a variable. When we only have scores for a sample (this will be the most common case), we use a lowercase n, while if we have scores for every member of a population, we use an uppercase N. For the dataset $\{5, 3, 9, 10, 4\}$, $n = 5$.

- One of the most common operations we need to perform in statistical computations is adding, or "summing," a set of numbers. The symbol Σ (this is the Greek capital letter sigma) is used in formulas to tell us that we need to add all the scores in a dataset. For example, if our dataset is {5, 3, 9, 10, 4}, ΣX means "add up all the X's":

$$\Sigma X = 5 + 3 + 9 + 10 + 4 = 31$$

Summing

- Formulas become tricky whenever they contain more than one operator, because the order of operations must be determined. For example, how should we evaluate the expression $\Sigma X + 4$? Should we sum all the X's first and then add 4 to the sum, or should we add 4 to each X before summing all the values? Mathemeticians have developed a long list of rules to specify the order of operations for any formula, but in this text we will need to make use of only a few of these rules:

Order of operations

1. If you see a set of parentheses, do the operations inside the parentheses first.
2. Then do the rest of the operations in the formula from left to right.
3. If you see two formula elements (numbers or symbols) or two sets of parentheses right next to each other (without a +, −, ×, or / in between), multiply the two elements together.
4. If the formula contains a horizontal line, do the operations above the line, then do the operations below the line, then divide the result from the top by the result from the bottom.

- Here are some examples of these rules in action, using the same set of X scores we've been working with on this page, {5, 3, 9, 10, 4}:

  For the first score in the dataset, $X + 3 = 8$
  For the first score in the dataset, $5 \times (X + 3) = 40$
  For the first score in the dataset, $5(X + 3) = 40$

  For the first score in the dataset, $\dfrac{5(X+3)}{4} = 10$

  $\Sigma X = 5 + 3 + 9 + 10 + 4 = 31$
  $\Sigma X + 3 = 31 + 3 = 34$
  $\Sigma(X + 3) = (5 + 3) + (3 + 3) + (9 + 3) + (10 + 3) + (4 + 3) = 46$

The Review Exercises for this chapter include a multipart problem (**Exercise 1.4**) that will allow you to practice the use of these rules. There are also problems that review several mathematical concepts that you should have covered in earlier math classes: exponents and roots (**Exercise 1.5**), rounding, fractions, decimals, and percentages (**Exercise 1.6**), and basic algebra (**Exercise 1.7** and **Exercise 1.8**).

Most of these problems are "regenerative," meaning that the computer will generate a new set of numbers so that you can do them as many times as you like. We suggest that you go through all of the exercises at least once, then spend some additional time practicing the ones you have trouble with until you're comfortable with your ability to do all the problems. A little extra practice now will save you lots of headaches down the road.

1.4 Chapter Summary/Review

This Chapter Summary/Review is interactive on the *Interactive Statistics* website and CD. Also be sure to go through all the **Review Exercises** for this chapter.

- Statistics can reduce a large amount of data into a more manageable and understandable form. For example, a mean reduces a group of numbers to a single number. However, a **statistic** is at best meaningless and at worst misleading without a proper standard to compare it to (**Section 1.1**).

- **Descriptive statistics**, such as the mean, function purely to describe a set of data (**Section 1.2**). **Inferential statistics** help us use data (and descriptive statistics calculated from the data) to draw conclusions (**Section 1.2**). More specifically, inferential statistics allow us to make inferences about a **population** on the basis of a sample (**Section 1.2.1**), through a procedure called hypothesis testing (**Section 1.2.2**), which will be covered in great detail in later chapters.

- A statistic itself can never provide an answer to a research question. Therefore, statistics are best considered rhetorical devices that aid researchers in making arguments on the basis of their data (**Section 1.2.3**).

Chapter 2
Looking at Data:
Frequency Distributions

2.1 Organizing Data

Consider the following experiment, designed to assess the effects of cell phone conversations on driving performance:

> The experiment uses a driving simulator, complete with a steering wheel and brake pedal. As subjects drive in the simulator for 30 minutes, they are confronted with 40 traffic "emergencies" (pedestrians darting out into the road, the car ahead of them suddenly stopping, etc.). Subjects are instructed to press the brake pedal as quickly as possible when an emergency occurs. If they brake within half a second of the emergency, they get one point. To

motivate the subjects to perform as well as possible, they are paid 50 cents per point at the end of the experiment.

Each subject is randomly placed in one of three conditions. In the Cell phone condition, subjects wear a "hands-free" cellular phone and carry on a conversation with an experimenter sitting in another room for the duration of the experiment. In the Radio condition, subjects listen to a radio station of their choosing throughout the experiment. In the Quiet condition, subjects drive without being distracted by either a cell phone conversation or the radio.

The point totals for 25 subjects in each condition are as follows:

Cell phone:	38 32 19 30 26 25 26 24 28 26 26 32 28 34 25 27 26 23 32 30 27 30 22 30 30
Radio:	37 40 30 35 32 34 40 38 39 34 29 33 31 37 37 35 29 34 28 36 34 38 34 39 38
Quiet:	37 34 37 35 37 35 35 32 37 30 31 38 25 34 35 37 32 31 31 36 36 36 39 31 40

Did cell phone use impair simulated driving performance in the study? Scanning the data for the three conditions, you may be able to find some evidence that it did. Try answering the following questions:

What was the lowest point total in each condition?

How many subjects in each condition had perfect scores (40 points)?

- Lowest point total in Cell phone condition: 19

- Lowest point total in Radio condition: 28

- Lowest point total in Quiet condition: 25

- Subjects with perfect scores in Cell phone condition: 0

- Subjects with perfect scores in Radio condition: 2

- Subjects with perfect scores in Quiet condition: 1

These figures hint that subjects in the Cell phone condition performed more poorly overall than those in the other two conditions:

The lowest point total in the Cell phone condition (19) is considerably lower than in the other two conditions (28 and 25), and no one in the Cell phone condition managed a perfect score.

But you probably found it fairly difficult to extract this information from the raw scores as given above, because the scores are completely unorganized. A simple way to organize a dataset is to put the scores in numerical order:

Cell phone:	19 22 23 24 25 25 26 26 26 26 26 27 27 28 28 30 30 30 30 30 32 32 32 34 38
Radio:	28 29 29 30 31 32 33 34 34 34 34 34 35 35 36 37 37 37 38 38 38 39 39 40 40
Quiet:	25 30 31 31 31 31 32 32 34 34 35 35 35 35 36 36 36 37 37 37 37 37 38 39 40

The **Data Transformations** Calculation Tool will order a set of scores for you. Just enter the scores in the "*X*" text box on the left side of the Tool (scores can be separated by spaces or commas), make sure "Sort X" is selected in the "Transformation" pull-down menu, click the "calculate" button (the small red button with the = sign), and the scores will appear in numerical order in the "Results" text box on the right side of the Tool. Try it out yourself on the data from the Cell phone condition before reading on.

From this representation of the data, it's much easier to pick out the minimum and maximum values asked for on the previous page. However, it is still difficult to see what's going on *in between* the minimums and maximums. To organize the data further, we might count how many subjects in each condition scored between 0–9 points, how many scored between 10–19 points, etc. For example, in the Cell phone condition, 0 subjects scored between 0–9, 1 scored between 10–19, 14 scored between 20–29, 10 scored between 30–39, and 0 subjects achieved a perfect score of 40. These values for all three conditions can be compiled into the following table:

	Condition		
Interval	**Cell phone**	**Radio**	**Quiet**
0–9	0	0	0
10–19	1	0	0
20–29	14	3	1
30–39	10	20	23
40–49	0	2	1

Interactive
Page 18

Now we can clearly see the adverse effects of cell phone use. For example, all but 4 of the 50 subjects in the Radio and Quiet conditions scored at least 30 points, but 15 out of the 25 subjects in the Cell phone condition failed to meet this criterion.

This type of table is called a **frequency distribution,** since it gives the frequency of occurrence for various intervals distributed along the scale of measurement for the variable. (Actually, the table above is a combination of three frequency distribution tables, one for each condition in the experiment.) The next section in this chapter describes how to construct frequency distribution tables.

Frequency distributions can also be represented with a type of graph called a **histogram.** A frequency distribution histogram includes one bar for each interval in a frequency distribution table, with the height of each bar proportional to the frequency for the interval. Histograms constructed from the table above are shown in Figure 2.1.

As demonstrated here, a histogram allows us to glean most of the same information we get from a table, and often allows us to see patterns more quickly and clearly. After describing frequency distribution tables in **Section 2.2,** we'll cover histograms and other types of data representations in **Section 2.3.** Then we'll finish the chapter in **Section 2.4** by detailing the information we can ascertain about a dataset from frequency distributions and previewing how we will use descriptive statistics to summarize this information.

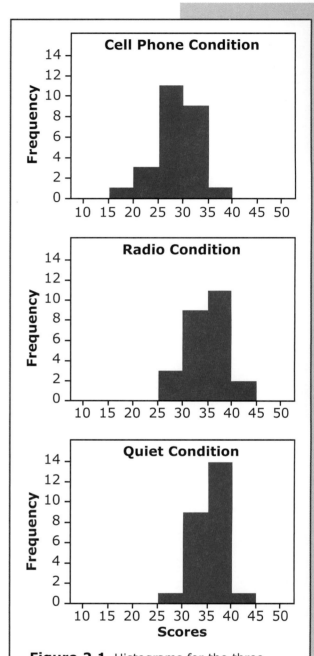

Figure 2.1 Histograms for the three conditions in the cell phone experiment.

Interactive
Page 19

2.2 Frequency Distribution Tables

The procedure for constructing a basic **frequency distribution** table from a set of data is quite straightforward:

1. First, break the measurement scale into intervals.

2. Then count and record how many values fall into each interval.

Basic procedure for constructing frequency distribution tables

The trickier of the two steps is usually deciding how to slice up the measurement scale. Sometimes we can simply use every possible score as its own interval. For example, say you keep track of the number of times you check your email each day for two weeks and come up with the following data:

7 6 6 3 7 6 5 5 2 6 9 3 5 4

Here, it makes sense to use intervals of one score each: 2, 3, 4, 5, 6, 7, 8, and 9. Counting up the scores in each of these "intervals," we get the following frequency distribution table:

Interval	f
2	1
3	2
4	1
5	3
6	4
7	2
8	0
9	1

Note that we leave a space for 8's in this table even though there were no scores with this value in the dataset. The reason for this is that it can be misleading not to do so. To see why, consider another set of scores. Say you ask your friend to collect the same data, and she records the following number of email checks per day for her fortnight:

2 8 1 9 4 2 1 2 1 8 8 3 3 2

Now say we construct the following frequency table:

Interval	f
1	3
2	4
3	2
4	1
8	3
9	1

If you weren't looking closely enough at this table, you might completely miss the fact that there were no days when your friend checked her email 5, 6, or 7 times. The proper frequency table for this set of data is:

Include a row for every interval even if there are no scores in that interval.

Interval	f
1	3
2	4
3	2
4	1
5	0
6	0
7	0
8	3
9	1

This table makes the absence of days in which email was checked 5, 6, or 7 times, which might be the most interesting aspect of these data, very clear.

Exercise 2.1 provides more practice with constructing simple frequency distribution tables like the one above.

The one-interval-per-score method of constructing **frequency distribution** tables only works when you have a very narrow range of scores. Say we expand our email study by asking 20 college-age subjects to count the number of emails they receive over a one-month period. We might get the following set of data (organized in numerical order):

> 17 18 21 30 31 32 36 39 40 41 43 47 48 49 52 57 59 60 63 75

If we listed every score as its own interval in the table for these data, we would need 59 (75 − 17 + 1) rows. Furthermore, the frequency for every row would be either 0 or 1. This combination of attributes would make for an essentially useless frequency table.

As we saw with the cell phone experiment data at the beginning of the chapter, the way to deal with sets of data like this one is to use intervals that span more than one score. But exactly how big should our intervals be?

The general answer to this question is that you should choose an interval size that produces a frequency table that is as informative as possible. That is, you want the table to provide an accurate picture of how the scores were distributed along the scale of measurement. Here are some more specific guidelines to keep in mind when deciding on interval sizes and other issues regarding frequency table construction:

- All the intervals in your table should be exactly the same width.

- Make sure that the intervals you choose cover the entire range of scores in the dataset. Also make sure that there isn't any overlap between intervals. You can address both of these concerns by

Guidelines for constructing frequency distribution tables

making sure that every possible score in the dataset fits in one and only one interval.

- Generally speaking, you want between 6–12 intervals in your table. If you have fewer than 6, scores tend to get lumped together too much, while if you have more than 12, the table tends to become too cumbersome.

- If possible, use a number for the interval width that readers of the table (including you!) will have an easy time working with. Widths of 1, 2, 5, 10, 20, 50, or 100 are good choices. Intervals of 13, on the other hand, will tend to make for a table that is difficult to parse.

- Try to construct your table such that the lower and upper bounds are also common numbers. For example, if your smallest data point is 17, you might want to start the table at 15 or 10. Preferably, the lower and upper bounds will also be multiples of the interval width.

The first two of these guidelines are the only ones that are absolutely necessary to follow. There may come a time, for example, when you find that a dataset is best described using 18 intervals of width 4, with the first interval starting at 7.

Now let's try to put these rules in practice. For the set of scores given at the top of this page, try to pick a good interval width, then decide what the lower boundary of the table should be and decide how many intervals should be included.

Most of the above guidelines are not set in stone.

Interactive Page 22

One good set of frequency table parameters for the data on the previous page would be:

- Interval width   10

- Lower boundary   0

- Number of intervals   8

An interval width of 5 would require at least 12 intervals, which is too many for a dataset this small (with this many intervals, there would be at most 3 scores in each interval). The next highest "round" number is 10, which is an especially easy-to-interpret interval.

Since the lowest score in the dataset is 17, you might think that it makes the most sense to start the table with the interval 10–19. However, recall that the data represents the number of emails received by each of a group of subjects over a one-month period. It certainly could have been possible that someone might have received fewer than 10 emails, and in fact it might be interesting to

It is often informative to include extra intervals at the top and bottom of the table.

know that no scores fall below 10. By including the interval 0–9, we make this fact explicit in the table.

With a lower boundary of 0 and an interval width of 10, we need at least 8 intervals to cover all the scores in the dataset. Here's what the table would look like with these parameters:

Interval	f
0–9	0
10–19	2
20–29	1
30–39	5
40–49	6
50–59	3
60–69	2
70–79	1

It's worth emphasizing again that these parameters are not the only ones we could have used for this dataset, nor are they necessarily the best ones. There would be nothing wrong, for example, with using an interval width of 7, a lower boundary of 12, and 9 intervals. **Exercise 2.3** and **Exercise 2.15** will give you some practice with and guidance in choosing good interval sizes for frequency tables.

This is also a good place to remind you of the purposes behind creating a **frequency distribution** table: to learn something about your data, and to communicate to others what you've learned. Following the parameter guidelines given on the previous page will usually maximize the utility of your frequency tables (and graphs, to be discussed in **Section 2.3**) for these purposes. But if a table with 7-unit intervals tells you something you didn't know before, then by all means construct your table with this interval width.

2.2.1 Relative Frequencies

Below, we see a combination of two **frequency distribution** tables. The left-hand column reproduces the table from the previous page, displaying monthly email volume for a group of 20 college students. The right-hand column shows a second set of frequencies, representing monthly email volume for a group of 80 senior citizens who live in a progressive retirement home in which every room is wired to the Internet.

College f	Interval	Senior f
0	0–9	36
2	10–19	26
1	20–29	8
5	30–39	5
6	40–49	3
3	50–59	2
2	60–69	0
1	70–79	0

Five college students sent between 30 and 39 emails during the month of the study. The same number of seniors sent this many emails. But this similarity in raw frequencies is misleading: as a proportion of the total number of subjects in each group, many more college students than seniors sent this many emails. We can fix this problem by including an additional column in our table that displays the percentage of scores falling in each interval:

College		Interval	Senior	
f	%		f	%
0	0%	0–9	36	45.0%
2	10%	10–19	26	32.5%
1	5%	20–29	8	10.0%
5	25%	30–39	5	6.3%
6	30%	40–49	3	3.8%
3	15%	50–59	2	2.5%
2	10%	60–69	0	0%
1	5%	70–79	0	0%

These percentages tell us the frequency of each interval relative to the total number of scores in the dataset, so we refer to them as **relative frequencies.** To find the relative frequency for each interval, take the frequency f for the interval, divide by the total number of scores n, and multiply by 100. That is:

$$\text{Percentage} = (f\,/\,n) \times 100$$

For the interval 30–39 in the tables above, we have:

$$\text{College \% } = (5\,/\,20) \times 100 = 25\%$$
$$\text{Senior \% } = (5\,/\,80) \times 100 = 6.25\% \text{ (rounded to}$$
$$6.3\% \text{ in the table)}$$

The relative frequencies tell us the true story for this interval: College students were four times more likely than seniors to send between 30 and 39 emails during the month of the study. In **Chapter 4,** we'll learn more about interpreting relative frequencies as **probability** values. Determining the probability of certain events will be a key step in the calculation and interpretation of **inferential statistics.**

Glossary Term:
relative frequencies

Relative frequencies are generally more useful than absolute frequencies when describing distributions.

2.2.2 Cumulative Frequencies

Imagine you're taking a course in which the professor gives weekly quizzes, each of which is graded on a 10-point scale (i.e., scores range from a low of 0 to a high of 10). On the first quiz, you receive a score of 7, and you wonder how well you did relative to the rest of the class. Anticipating that most students will be wondering this same thing, the professor provides a **frequency distribution** table of all the scores:

Interval	f	%	Cum. f	Cum. %
0	3	6%	3	6%
1	3	6%	6	12%
2	5	10%	11	22%
3	6	12%	17	34%
4	7	14%	24	48%
5	9	18%	33	66%
6	6	12%	39	78%
7	6	12%	45	90%
8	1	2%	46	92%
9	4	8%	50	100%
10	0	0%	50	100%

The third column of this table tells you that you were among the 12% of the class that scored 7/10 on the quiz. But the fourth and fifth columns are more informative. The fourth column, labeled "Cum. *f*," shows the **cumulative** frequency for each interval in the table. For example, the value 45 for the interval 7 tells us that 45 students in your class scored as well or worse than you. The cumulative frequency for any interval is determined by simply adding that interval's frequency to the frequencies of all intervals below it.

Even more useful is the column labeled "Cum. %", which gives the cumulative percentage (i.e., the **cumulative relative frequency**) for each interval. To calculate a cumulative percentage, divide the cumulative frequency for an interval by the total number of scores in the dataset. This column of the table tells us that 90% of students scored as well or worse than you on the quiz, meaning that only 100 − 90 = 10% of the students in the class did better than you. Not bad!

Interactive
Page 26

2.2.3 Dealing with Decimals

By now, you're hopefully beginning to get comfortable with the process of constructing **frequency distribution** tables. But let's throw one more monkey wrench into this process and see how we deal with it.

Imagine you're the coach of a high school track team. In a final practice race before an upcoming meet, 10 members of your team run the 200-meter dash in the following times (all expressed in seconds):

21.7 23.2 24.1 24.7 25.0 25.6 25.8 25.9 26.5 27.2

How can we construct a frequency table for these scores?

Since the 10 times span less than 6 seconds, it is reasonable to want to use an interval size of 1 second, which will make for an easily interpretable table. So we can start our table by setting out the following intervals:

Glossary Term:
cumulative frequency

Interval	*f*
20	?
21	?
22	?
23	?
24	?
25	?
26	?
27	?
28	?
29	?

The problem here that we haven't encountered before is that the scores we're dealing with have decimal places. For example, into which interval should we place the first time, 21.7 seconds?

Well, it seems pretty clear that this score belongs either in the interval labeled 21 or in the interval labeled 22. To decide which of these two intervals to choose, we need to be more explicit about what the intervals really mean:

a. The interval 21 could be taken to mean "all times greater than or equal to 21 seconds and less than 22 seconds." Under this definition, the interval could be more accurately expressed as 21.00000–21.99999 seconds.

b. Or, the interval 21 could be taken to mean "all times that, when rounded to the nearest second, come out to 21 seconds." Since we round numbers ending in .5 up and numbers ending in .4 down, this definition amounts to an interval of 20.50000–21.49999 seconds.

By definition (a), 21.7 seconds belongs in the interval 21, while by definition (b), this score belongs in the interval 22.

Unfortunately, there is no universally agreed-upon standard for how to define frequency table intervals. Some statisticians and some statistics packages use the first definition and others use the second. Luckily, though, if a dataset is large enough, it will make little difference which definition we choose: The distributions of scores will look about the same either way. For this book, **we will use definition (a),** for the practical reasons that it lets us avoid dealing with decimal places unless we really have to, and it makes tabulating a set of data easier.

Interactive Page 27

To recap our decision from the previous page, we're treating the interval 25 to mean all times from 25–26 seconds, including

In this book, the interval 20 is taken to mean 20.00000–20.99999.

25.000000000 seconds but excluding 26.000000000 (the latter time is included in the next interval).

Before leaving our discussion of decimals in **frequency distribution** tables, there are two more contingencies we have to deal with. First, consider the following set of data:

> 3.42 3.83 3.90 3.95 4.10 4.24 4.36 4.63 4.66 4.84
> 4.90 5.18 5.24 5.33 5.88

If we use an interval size of one, all the scores will be compressed into only 3 intervals, which is too few to get a good idea of the spread of the distribution. The solution is to use an interval size of less than one. For example, we could construct the following table:

Interval	f
2.5–2.9	0
3.0–3.4	1
3.5–3.9	3
4.0–4.4	3
4.5–4.9	4
5.0–5.4	3
5.5–5.9	1
6.0–6.4	0

Here, we treat the interval 3.0–3.4 to mean 3.000000–3.4999999 and 3.5–3.9 to mean 3.5000000–3.9999999. That's why 3.42 goes in the former interval and 3.95 goes in the latter.

> The interval 3.0–3.4 is taken to mean 3.000000–3.499999.

Finally, what do you think we should do with the following dataset?

> 40.1 44.9 45.0 48.0 49.1 52.4 53.0 53.3 59.9 63.2
> 64.8 66.7

Here, we need an interval size greater than one *and* we have to deal with the decimal places. No problem! Here's the frequency distribution table for these data:

Interval	f
35–39	0
40–44	2
45–49	3
50–54	3
55–59	1
60–64	2
65–69	1
70–74	0

As you can see, the interval 40–44 means all scores greater than or equal to 40.000000 and less than (but *not* equal to) 45.000000. Got it? If not, you can practice dealing with decimals (and practice calculating **relative** and **cumulative** frequencies) in **Exercise 2.4.**

Interactive Page 28

2.2.4 The Frequency Distributions Calculation Tool

It's important that you construct a number of **frequency distribution** tables by hand, so that you thoroughly understand the concept. But in the real research world, scientists almost always let computers do the work of tabulating data for them.

Besides saving time, there are two additional advantages to automating this process. First, it minimizes errors—computers are much less likely than humans to skip or misclassify a score. Second, it allows you to try several different representations of the distribution, helping to ensure that you get a good picture of how the scores are spread out. We've provided you with a Calculation Tool for tabulating (and graphing) frequency distributions.

To use the **Frequency Distributions** Calculation Tool, start by entering your data into the "Scores" text box on the left side of the Tool. To create a table, click the "Table" radio button near the top (by default, the Tool is set up to create histograms, which we'll learn about in the next section).

Next you have to choose the lower limit, interval width, and number of intervals you want to use, and enter the values in the appropriate text boxes on the right side of the Tool. You can click the "Suggest" link next to each box to have the Tool suggest an initial value, but it won't necessarily choose parameters that will make for the best possible frequency table.

Finally, click the "calculate" button, or hit the "enter" key if you've just typed in a value in one of the text boxes. A small window will be opened containing your table, which will include the following columns:

Int.	Interval
f	Frequency
%	Percentage
Cum. *f*	Cumulative Frequency
Cum. %	Cumulative Percentage

If you see a note below the table saying that "Not all scores are represented in the frequency table," the parameters you chose resulted in a set of intervals that didn't cover the entire range of your dataset. When you get this message, or when you want to see how the table looks with a new set of parameters, just go back to the Tool window, change the lower limit, interval width, and/or number of intervals, and hit the "calculate" button again.

Advantages of automated table generation

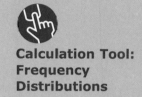

Calculation Tool: Frequency Distributions

When you generate a new table by clicking the "calculate" button or hitting the "enter" key, it is drawn in the same window as the last table you generated. This helps cut down on the number of windows you have open in your browser.

Sometimes, however, you may want to keep an old table open to compare it to the new one. If you click on the title of the Tool (that is, click on the words "Frequency Distributions" next to the "calculate" button at the top of the Tool), your new table will be opened in a new window, leaving the old window with your old table intact.

Click the title of the Frequency Distributions Calculation Tool to create a table in a new window.

To get some initial practice with the Frequency Distributions Calculation Tool, first construct a table for the following set of data by hand (hint: sort the data with the **Data Transformations** Calculation Tool first), then generate the same table using the Tool. Then try generating a new table with a different interval width using the Calculation Tool.

> 53 45 32 20 26 38 29 24 41 38 71 49 57 66 46 66
> 79 81 43 44

Interactive Page 29

2.2.5 Categorical Data

Recall the survey of senior citizens' email habits from **Section 2.2.1,** in which they tallied the number of emails they received over a month-long period. Suppose that at the end of the month we asked them to answer an additional question:

What brand of computer do you have in your room?

- Dell

- Gateway

- Compaq

- Apple

- Other

- I don't have a computer

The subjects' responses to this question will constitute a new set of data, but you should recognize that it is a qualitatively different type of data than the email volume scores we looked at previously. In statistical parlance, the email volume scores are called **interval data,** while the computer-brand responses are called **categorical data.**

Glossary Terms: interval data, categorical data

The distinction between interval and categorical data should be fairly intuitive. At the simplest level, an interval dataset is always a collection of numbers, while categorical data is a collection of, as the name suggests, categories.

Another distinction between the two measurement types is that there is no established way to order the "values" on a categorical scale. Is an Apple computer greater or less than a Compaq? Clearly, there is no one correct answer to this question. But two values from an interval scale can always be meaningfully compared: a person who received 10 emails in the month has a higher score than someone who received 5 emails.

A set of data must consist of numbers for us to be able to perform mathematical operations on them, so most of the statistics we will describe in this book are designed to analyze interval data. We bring categorical data up here because one of the few meaningful statistical analyses we can perform for this type of data is to construct a **frequency distribution** table. For example, the computer-brand data for our senior citizens might be described by the following table:

Brand	*f*	%
Dell	14	17.5%
Gateway	6	7.5%
Compaq	7	8.8%
Apple	4	5.0%
Other	11	13.8%
None	38	47.5%

Exercise 2.5 requires you to construct a frequency distribution table for a new set of categorical data.

Now that we've introduced you to the distinction between interval and categorical data, we'll largely ignore the latter until the very last chapter of the book, **Chapter 12**, where we'll introduce a host of procedures for dealing with categorical data.

2.3 Frequency Distribution Histograms

Glossary Term: histogram

Tables are useful for getting a sense of the **distribution** of a set of scores, but a picture is worth a thousand words (or numbers). To literally see the distribution, we can convert a frequency distribution table into a frequency distribution **histogram.** Here's how we do it (the letters below refer to parts a–g of Figure 2.2, which are shown on the *Interactive Statistics* website and CD; refer to the figure on the website and CD as you read these instructions):

a. Start by dividing your dataset into appropriate intervals (exactly as described in **Section 2.2**) and constructing a frequency distribution table. For a small dataset, we just need the raw frequency column of the table. We'll work with the following set of scores for this example:

Instructions for drawing a histogram

      6 12 14 14 15 19 20 20 22 22 23 24
      24 25 26 27 29 30 30 41

b. Draw a blank set of X and Y axes for the histogram. Your graph will be more aesthetically pleasing if the X axis is a bit wider than the Y axis is high.

c. On the X axis, mark off a set of $i + 1$ evenly spaced ticks, where i is the number of intervals in your table. For example, here we have 10 intervals, so we need 11 tick marks on the X axis.

d. Label the tick marks with the lower limit of each interval in the table. Label the rightmost tick with the lower limit of the highest interval plus the interval width. (This and other steps may sound more complicated than they really are; everything should be clear from Figure 2.2.)

e. Give the entire X axis as descriptive a label as possible. Since we haven't specified what our sample dataset represents, we'll just label the axis with the generic "Scores."

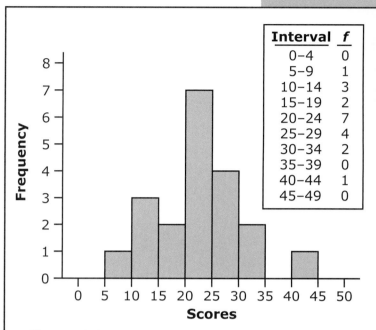

Interval	f
0–4	0
5–9	1
10–14	3
15–19	2
20–24	7
25–29	4
30–34	2
35–39	0
40–44	1
45–49	0

Figure 2.2 Steps in creating a histogram. This figure is interactive, with eight parts, on the *Interactive Statistics* website and CD.

f. On the Y axis, mark off and label at least as many ticks as the frequency count for the highest-populated interval in the table (also include a tick for 0 at the bottom of the axis). It's not a bad idea to add at least one extra tick, to leave a little room at the top of the graph. Label the whole Y axis "Frequency."

g. For the first interval in your table that has a frequency greater than 0 (in our example, this is the interval 5–9), draw a bar (rectangle) in your graph such that:

- The left side of the bar is even with the X-axis tick mark for the lower limit of the interval.
- The right side is even with the next tick mark on the right.
- The bottom of the bar is on the X axis.
- The top of the bar extends to the level of the Y-axis tick mark that corresponds to the frequency for that interval (in our example, we draw the first bar one unit high, since there is one score in the first interval).

h. Repeat step (g) for the rest of the intervals in your frequency table. Here, the second bar will be 3 units high, the third bar will be 2 units high, etc. The rectangles for adjacent intervals should touch each other, as shown in the figure. If an interval has a frequency of 0, leave blank the space for the interval in the graph.

And that's it! Grab a piece of paper and try the process yourself on the following dataset, using an interval width of 1.

3 3 4 4 4 4 5 5 6 6 7 8

Interactive Page 31

Figure 2.3 shows the **histogram** for the data given at the end of the previous page.

As you probably found out for yourself, drawing histograms by hand on paper isn't a lot of fun (it's pretty tedious even if you use graph paper). The **Frequency Distributions** Calculation Tool will automate this process. To use this Tool, enter the data, lower limit, interval width, and number of intervals exactly as described in **Section 2.2.4** for frequency tables, click the "Histogram" radio button, and then click the "calculate" button. (**Exercise 2.12** provides further guidance in using this Calculation Tool to calculate frequency distribution tables and histograms.)

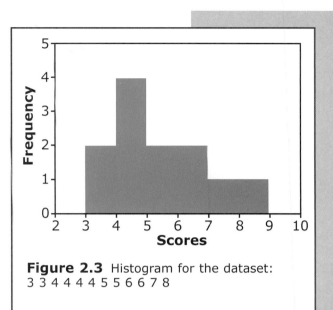

Figure 2.3 Histogram for the dataset: 3 3 4 4 4 4 5 5 6 6 7 8

You can try out the Calculation Tool on the data above if you wish. But resist the temptation to start using this Tool consistently right now. A bit of practice constructing histograms one bar at a time will help guarantee that you thoroughly understand how a histogram represents the underlying data.

This is where the **Histogram Helper** Activity comes in. Open it up and you'll see a series of text boxes in which you can fill in tick-mark labels for the X and Y axes. Decide how many ticks you need and what you want to label them and enter this information in the boxes. Then choose a bar color and an X-axis label (by default, the axis will just get labeled "Scores") and click the "Generate Shell" button.

Activity: Histogram Helper

An empty histogram shell will then be generated in the bottom part of the window. To construct your histogram, click inside the shell to build the bars needed to represent the data you're graphing.

For example, the sample data above requires the X-axis tick labels 2, 3, 4, 5, 6, 7, 8, 9, and 10 (technically, we don't need 2 and 10, but it's a good idea to leave a little padding on either side of your bars) and the Y-axis tick labels 0, 1, 2, 3, 4, and 5 (again, the 5 is not strictly necessary). Try entering these values into the Histogram Helper and clicking the "Generate Shell" button.

Now we fill in the first bar of our histogram by clicking just above the X axis between the 3 and 4 tick marks, then clicking again just above the square that was just produced to create a bar two units high. Next we click four squares between the X-axis tick marks 4 and 5, two squares between 5 and 6, and so on. (If you make a mistake and add a square you didn't want, just click on it again to erase it.)

The result should look just like Figure 2.3, which was in fact generated using the Histogram Helper. **Exercise 2.6, Exercise 2.7,** and **Exercise 2.8** ask you to construct histograms using this Activity. You can print the output of the Histogram Helper if you want a hard copy of a histogram you make.

Now that we have the basics of **distribution** graphing down, how do you think you would go about constructing a **histogram** from the following frequency distribution table?

Interval	f
0–4	0
5–9	3
10–14	3
15–19	9
20–24	12
25–29	24
30–34	15
35–39	18
40–44	12
45–49	6
50–54	3
55–59	0

Interactive Page 33

The tricky thing about graphing this **distribution** is that the most populous interval (25–29) includes 24 scores. There aren't enough spaces in the Histogram Helper for this many Y-axis tick marks, nor would you want to include this many if you were drawing the **histogram** by hand on a piece of paper.

The solution to this problem, as shown in Figure 2.4, is to have each Y-axis tick represent three scores instead of one (enter the values 3, 6, 9, 12, 15, 18, 21, and 24 as Y-axis tick labels in the **Histogram Helper**). So, for example, the interval 20–24, which includes 12 scores, gets a bar 12 / 3 = 4 units high.

Scaling the Y axis to graph large distributions

Luckily, the distribution we graphed here included frequencies that were all divisible by a common denominator (3), making it easy to draw the bars using the Histogram Helper. If you're not so fortunate with your next set of data, just round your frequencies up or down to the nearest multiple of your Y-axis tick scale. For example, if each tick represents 5 units, you would round a frequency of 8 up to 10 (and make the bar for that interval two squares high) and round a frequency of 7 down to 5 (this bar would be one square high). Or, if you're drawing your graph on paper, draw the bar for a frequency of 8 so that it extends above the "5" tick mark but not quite up to "10." (The **Frequency Distributions** Calculation Tool will also generate histograms with bars that are exactly the right height.)

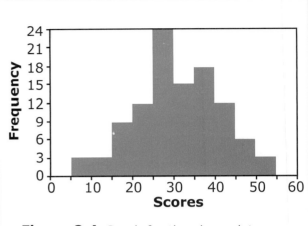

Figure 2.4 Graph for the above data, created using the Histogram Helper Activity.

2.3.1 Relative Frequency Histograms

This section and the next few sections describe additional ways to graph frequency distributions.

Figure 2.5b shows a **relative frequency distribution histogram**. Rather than representing the absolute number of scores in each interval, bars in this type of graph represent the **relative frequency** of each interval, compared to the total number of scores in the **distribution.**

Glossary Term: relative frequency distribution histogram

Note that the shape of the relative frequency **histogram** in Figure 2.5b is identical to the shape of the absolute frequency histogram in Figure 2.2, reproduced in part a of Figure 2.5. This similarity holds not just for this dataset, but for every distribution of scores.

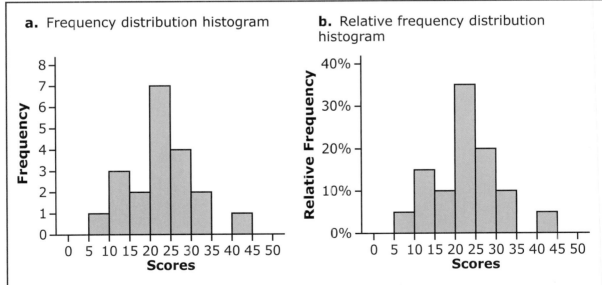

Figure 2.5 Frequency distribution histogram (a) and relative frequency distribution histogram (b) for the data first graphed in Figure 2.2.

As noted in **Section 2.2.1,** relative frequencies are generally more informative than absolute frequencies, since the relative frequencies from one dataset can be fairly compared to those from another dataset. For this reason, most distribution graphs you'll see in later chapters will plot relative frequencies. (Usually, we won't even show the scale on the Y axis; it will simply be labeled "Relative Frequency.")

Most histograms in this textbook will plot the relative frequencies of distributions.

Interactive
Page 35

2.3.2 Frequency Distribution Polygons

Figure 2.6 is a **frequency distribution polygon,** constructed by placing dots at the centers of the top sides of each of the bars in the **histogram,** then connecting the dots. A frequency distribution polygon can represent either absolute frequencies or **relative frequencies.**

Frequency polygons can also be smoothed out to form a **density curve,** as shown in Figure 2.7. Density curves usually represent idealized distributions of entire **populations,** rather than actual distributions of small **samples** of scores. For example, Figure 2.7 shows the density curve for the weights (in kilograms) of the population of 2-year-old girls in the USA.

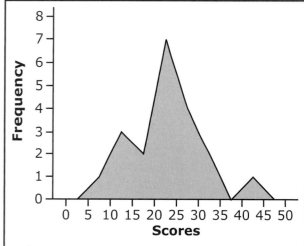

Figure 2.6 A frequency distribution polygon. This figure is animated on the *Interactive Statistics* website and CD.

Interactive
Page 36

2.3.3 Cumulative Frequency Distributions

Figure 2.8b illustrates a **cumulative relative frequency distribution polygon,** corresponding again to the data originally graphed in Figure 2.2 (reproduced below).

Glossary Term: cumulative frequency distribution polygon

Raw data:

6 12 14 14 15 19 20 20 22 22 23 24 24 25 26 27 29 30 30 41

Frequency Distribution Table:

Interval	*f*	%	Cum. *f*	Cum. %
0–4	0	0%	0	0%
5–9	1	5%	1	5%
10–14	3	15%	4	20%
15–19	2	10%	6	30%
20–24	7	35%	13	65%
25–29	4	20%	17	85%
30–34	2	10%	19	95%
35–39	0	0%	19	95%
40–44	1	5%	20	100%
45–49	0	0%	20	100%

The cumulative distribution plot in Figure 2.8b was made by placing a dot over the center of each interval on the X axis, at the Y-axis height that corresponds to the **cumulative frequency percentage** for the interval. This graph illustrates, for example, the fact that 65% of the scores in the dataset are in or below the interval 20–24.

Like **frequency distribution polygons,** cumulative distribution polygons can be smoothed out into precisely defined curves that represent large populations of scores. Figure 2.9 shows the cumulative distribution curve for the 2-year-old-girl weight data from Figure 2.7.

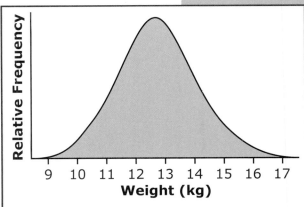

Figure 2.7 Population density curve for the weights of 2-year-old girls in the United States.

One thing that cumulative distribution curves are good for is estimating **percentiles** and **percentile ranks.** The percentile rank of a particular value is the percentage of individuals in the **population** that have scores less than or equal to that value. Turning this concept around, the *P*th percentile is the value for which *P*% of the population is less than or equal to the value.

Percentiles and percentile ranks

Given this definition, can you estimate (to the nearest kilogram) what the 50th percentile is for 2-year-old girls' weights from the cumulative distribution graph above?

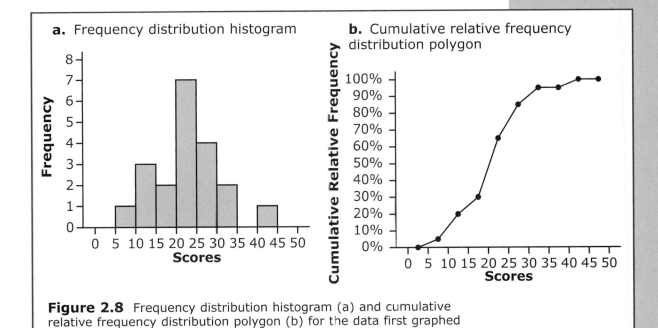

Figure 2.8 Frequency distribution histogram (a) and cumulative relative frequency distribution polygon (b) for the data first graphed in Figure 2.2 and reproduced above.

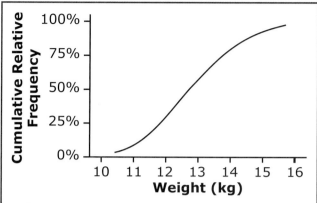

Figure 2.9 Cumulative relative frequency distribution curve for the weights of 2-year-old girls in the United States.

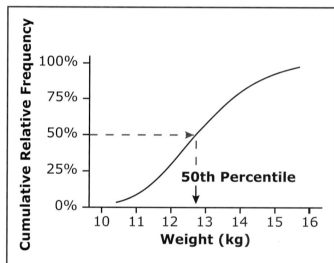

Figure 2.10 Determining a percentile value from a cumulative relative frequency curve. This figure is animated on the *Interactive Statistics* website and CD.

Interactive Page 37

We can estimate the 50th **percentile** by drawing a line from the 50% tick mark on the Y axis, over to the **cumulative distribution** curve, and down to the X axis. For 2-year-old girls, this value turns out to be about 12.7 (which you should have rounded up to 13) kilograms, as shown in Figure 2.10. We'll have more to say about percentiles in the next chapter.

Interactive Page 38

2.3.4 Stem and Leaf Diagrams

The final **frequency distribution** display we will consider is the **stem and leaf diagram**, invented by noted statistician John W. Tukey.

Glossary Term: stem and leaf diagram

The flowery name for this type of diagram comes from the process of separating each score in the dataset into the rightmost digit (the leaf) and the rest of the digits (the stem). If the score only has one digit, we use it as the leaf and 0 as the stem. So for the dataset we've been working with:

6 12 14 14 15 19 20 20 22 22 23 24 24 25 26 27 29
30 30 41

The stems and leaves for the first three scores are:

Stem	Leaf
0	6
1	2
1	4

To create the diagram, we list each of the stems on one line, then list out all the leaves to the right of each stem. Thus we get:

0 | 6

1 | 24459

2 | 00223445679

3 | 00

4 | 1

(Note that since there are two 14's in the dataset, the stem "1" gets two "4" leaves.)

A stem and leaf diagram looks very similar to a **histogram** turned on its side (Figure 2.11). It can be constructed quickly by hand, without the need for graph paper, and has the added advantage of showing exactly what the scores in the dataset are (rather than lumping all the scores in each interval together, as happens in a histogram).

However, stem and leaf diagrams only work well when the data are grouped into intervals of size 10, and they are cumbersome for large datasets. When invented in the 1960s, their ease of construction compensated for these limitations. Now, however, ubiquitous and easy-to-use computers can plot histograms faster than people can draw stems and leaves, so these diagrams are rarely used these days.

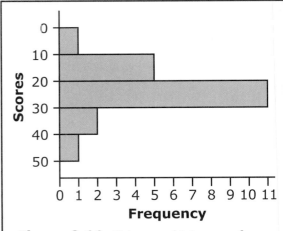

Figure 2.11 Sideways histogram for the data in the stem and leaf diagram above.

Interactive
Page 39

2.4 Describing Distributions

As we discussed in **Chapter 1,** the goal of **statistics** is to take large sets of data and describe them in a compact and meaningful way. In this chapter, we've learned how to construct tables and graphs that show how data are distributed across the scale of measurement. For example, Figure 2.12 reproduces the **histogram** first presented in Figure 2.2.

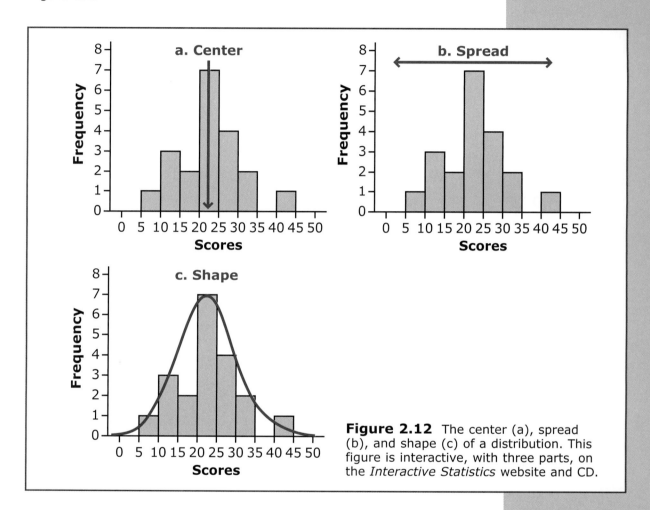

Figure 2.12 The center (a), spread (b), and shape (c) of a distribution. This figure is interactive, with three parts, on the *Interactive Statistics* website and CD.

The distribution pattern of a dataset is so central to the meaning of the dataset that we will often refer to a set of scores as a **"distribution."**

Once we've visualized a distribution with a histogram, we can go on to summarize what it looks like. Most distributions can be adequately described by answering three questions:

1. Where is the distribution centered on the X axis (that is, on the scale of measurement)? (Figure 2.12a)

2. How much do the scores spread out from this central location? (Figure 2.12b)

3. What is the general shape of the distribution? (Figure 2.12c)

Distributions can be compactly summarized by specifying their shape, center, and spread.

The center and spread of a distribution are conveniently summarized by statistics such as the **mean** and **standard deviation. Chapter 3** will be all about computing these and other related **descriptive statistics.**

Shape is more difficult to quantify with a statistic. Luckily, however, the shapes of most distributions can be described using a few well-defined terms.

Interactive
Page 40

- A **symmetrical** distribution is one in which the two halves of the distribution are mirror images of each other. The distributions in Figures 2.13a, d, and f are symmetrical.

- Asymmetrical distributions are usually **skewed** in one direction or another. In a positively skewed distribution (Figures 2.13b and e), scores are piled up on the left side and taper off to the right. In a negatively skewed distribution (Figure 2.13c), the big pileup is on the right and the **"tail"** is on the left. (The distribution in Figure

Terms used in describing distribution shapes

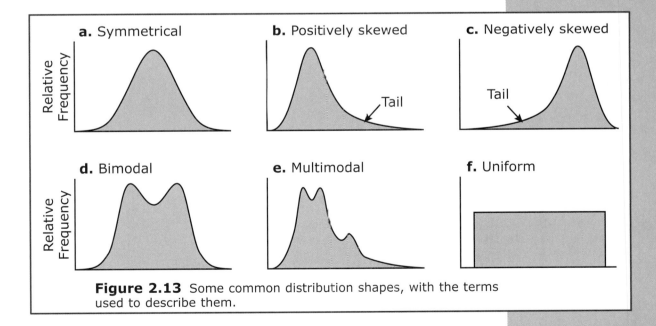

Figure 2.13 Some common distribution shapes, with the terms used to describe them.

2.13a is said to be two-tailed, since frequencies taper off on both the right and left sides).

- Most distributions, including those shown in Figures 2.13a, b, and c, have a single "peak" with frequencies sloping down on either side of this highest point. The score at which the peak occurs is called the **mode,** and these distributions that have one mode are called **unimodal**.

- Sometimes distributions are **bimodal** (Figure 2.13d), meaning that they have two peaks, often with one slightly higher than the other. Distributions with more than two modes are called **multi-modal** (Figure 2.13e).

- Some distributions don't have any mode at all. For example, if you roll a 6-sided die 600 times, you'll probably get about 100 1's, 100 2's, 100 3's, etc. This is called a **uniform** distribution, and is pictured in Figure 2.13f.

Exercise 2.13 and **Exercise 2.14** drill you on describing the shapes of distributions using various combinations of these terms.

2.5 Chapter Summary/Review

This Chapter Summary/Review is interactive on the *Interactive Statistics* website and CD. Also be sure to go through all the **Review Exercises** for this chapter.

- A frequency distribution table or graph records the number of scores in a dataset that fall into various intervals. This allows us to observe how scores are distributed along the scale of measurement for the variable (**Section 2.1**).

- To construct a basic **frequency distribution** table for a **variable**, we first slice up the measurement scale of the variable into equal intervals. Then we count and record how many scores in the variable fall in each interval. Three decisions must be made regarding the table: how wide the intervals will be, what the smallest interval in the table will be, and how many intervals will be included (**Section 2.2**). For example, given the following dataset:

      33 38 40 41 44 48 49 55 56 57 58 62

We could decide to use an interval width of 5, a starting point of 30, and 7 intervals. These parameters would lead to the following table:

Interval	f
30–34	1
35–39	1
40–44	3
45–49	2
50–54	0
55–59	4
60–64	1

Note that you should never skip intervals in a table, even if the frequency for an interval is 0. In this book, we consider an interval of 35–39 to include all scores between 35.000000 and 39.999999. Similarly, if we use an interval of 20.0–20.4, we would take it to mean 20.000000–20.499999 (**Section 2.2.3**).

- Frequency distribution tables can be made more useful by including columns for **relative frequency** (the percentage of scores falling into each interval) (**Section 2.2.1**), **cumulative frequency** (the number of scores falling in or below each interval), and **cumulative relative frequency** (**Section 2.2.2**). Adding these columns to the table above, we get:

Interval	f	%	Cum. f	Cum. %
30–34	1	8.3%	1	8.3%
35–39	1	8.3%	2	16.7%
40–44	3	25.0%	5	41.7%
45–49	2	16.7%	6	58.3%
50–54	0	0%	6	58.3%
55–59	4	33.3%	11	91.7%
60–64	1	8.3%	12	100.0%

- Frequency distribution tables can also be constructed for variables consisting of **categories** instead of numbers (**Section 2.2.5**).

- Frequency distributions can be graphically illustrated with histograms—bar graphs in which the height of each bar corresponds to the frequency (**Section 2.3**) or relative frequency (**Section 2.3.1**) of an interval from a frequency distribution table. In a frequency distribution polygon (**Section 2.3.2**), a line connects the top of each bar of a **histogram**. Cumulative frequency distribution histograms or **polygons** (**Section 2.3.3**) graph cumulative frequencies (these graphs can be used to estimate percentiles and percentile ranks). Figure 2.14 illustrates these graphs for the dataset above. Stem and leaf diagrams (**Section 2.3.4**) are a kind of hybrid between frequency distribution tables and histograms.

- The distribution pattern of a dataset can usually be described by specifying the center, spread, and shape of the distribution (**Section 2.4**). Center and spread are most commonly described by statistics called the mean and standard deviation, covered in the next chapter. Distribution shapes are described with terms such as **symmetrical**, positively or negatively skewed, **unimodal**, **bimodal**, and **uniform** (**Section 2.4**).

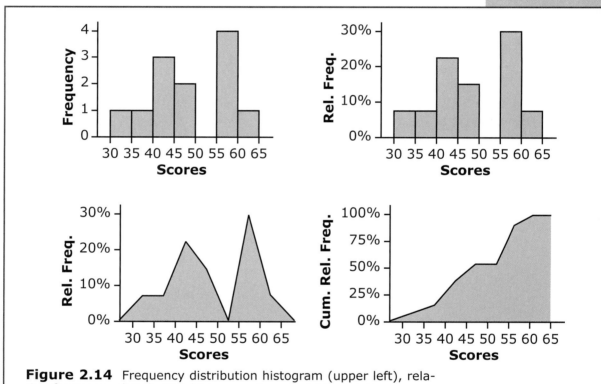

Figure 2.14 Frequency distribution histogram (upper left), relative frequency distribution histogram (upper right), relative frequency distribution polygon (lower left), and cumulative relative frequency distribution polygon (lower right) for the data shown in the table above.

Chapter 3
Describing Data: Measuring Center and Spread

3.1 The Need to Describe

Imagine that Ben, the Manager of Burger Town restaurant #456, is reviewing his weekly sales figures for the past year. The gross sales for each week, in thousands of dollars, were:

95 109 111 97 99 113 99 100 95 84 94 104 117
89 100 88 77 103 99 85 96 107 107 112 82 87
103 96 122 91 99 106 108 102 88 102 104 97 91
92 104 93 89 112 79 111 99 102 95 100 119 91

In **Chapter 2,** we learned that the first step in interpreting a set of **interval data** such as this is to visualize its distribution using a histogram. Use the **Histogram Helper Activity** or the **Frequency Distributions** Calculation Tool to graph Ben's data, then check your graph against the one on the next page.

Interactive Page 44

This graph tells Ben a great deal about his sales over the past year. The most common interval (the mode of the distribution) is 95–99 thousand dollars. Ben's restaurant made between 95 and 105 thousand dollars in close to half of the weeks of the year. The distribution is unimodal and roughly symmetrical (see **Section 2.4**). There don't seem to be any weeks in which the restaurant did terribly poorly, nor were there any incredibly good weeks.

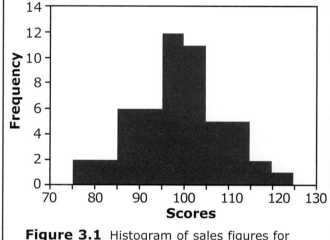

Figure 3.1 Histogram of sales figures for Ben's Burger Town restaurant.

Now consider the plight of Susan, the District Manager for Burger Town Corporation. Her job is to analyze the sales figures for all 22 of the Burger Towns she oversees. One frequency distribution histogram is often revealing, but 22 would be overwhelming. So Susan needs a shorthand way to summarize the sales figures of each restaurant.

As noted in **Section 2.4** and illustrated in Figure 2.12 (reproduced on page 43), the three most important distinguishing characteristics of a distribution are shape, center, and spread.

In terms of shape, Susan can probably assume that all 22 of the weekly sales distributions are pretty similar to Ben's, since this distribution is roughly "**normal**" (we'll have much to say about normal distributions in **Chapter 4**). She may rely on her individual restaurant managers to report any deviations from this shape.

To summarize the center and spread of the 22 distributions, Susan will use descriptive statistics. The most common measures of a dis-

Descriptive statistics summarize distributions.

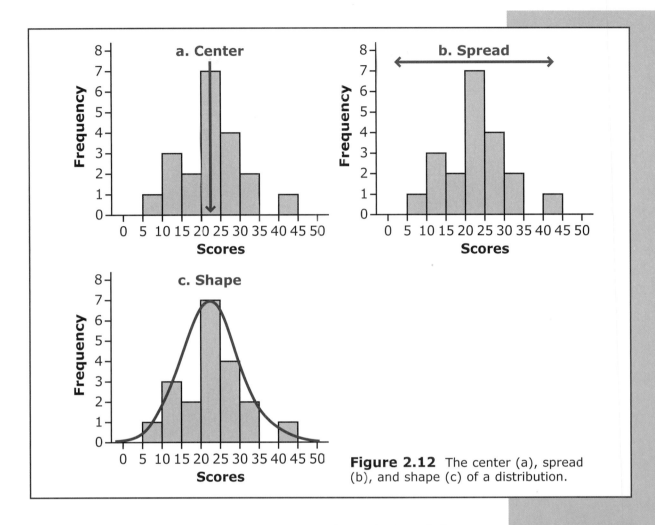

Figure 2.12 The center (a), spread (b), and shape (c) of a distribution.

tribution's center and spread are the mean and standard deviation, respectively, which we will cover in **Section 3.2** and **Section 3.3** of this chapter. In **Section 3.4,** we will cover resistant measures of center and spread, which are less widely used than the mean and standard deviation but more appropriate for certain datasets. **Section 3.5** will introduce the correlation, a descriptive statistic used to describe relationships *between* two distributions, as opposed to summarizing a single distribution. Finally, in **Section 3.6** we will learn how to combine measures of center and spread in standardized statistics called *z*-scores.

To preview what is to come, the mean and standard deviation of Ben's Burger Town will be 99.1 thousand dollars and 10.6 thousand dollars, respectively. Ben will report this two-number summary of his restaurant's performance to Susan. Now, let's imagine that the means and standard deviations of all 22 of Susan's restaurants are:

Burger Town #	Sales mean	Sales st. dev.
456	98.9	10.3
461	68.6	9.4
464	81.3	8.4
465	68.4	8.9
466	88.7	8.8
467	77.2	9.6
468	67.8	11.8
469	83.9	10.7
470	84.4	11.0
471	87.7	9.8
473	87.3	10.2
474	63.7	10.0
475	78.6	8.6
476	58.8	8.6
477	125.2	11.1
478	98.2	10.9
480	79.1	10.7
485	24.5	9.4
492	86.7	10.1
494	56.8	11.0
497	75.9	10.1
499	44.8	10.1

Susan can now do with these annual sales figures for 22 restaurants exactly what Ben did with his weekly sales figures for 52 weeks: graph them with a histogram, calculate their mean and standard deviation, and report this summary of her district's performance to *her* supervisor.

3.2 Measuring Center: The Mean

Say you ask 12 of your friends how many girlfriends or boyfriends they've had since high school. The results of your survey come out as follows:

9 5 12 0 4 3 3 6 1 4 5 8

Now suppose the friends you queried start asking how the survey came out. You could report all 12 scores, or you could show them a histogram of the data.

But most people aren't interested in expending the cognitive effort to interpret raw data or histograms. If they ask "how many girlfriends or boyfriends have your friends had," they probably really want to know about the *average* number of partners reported.

And when most people say "average," they have in mind the statistic formally known as the **mean,** defined as the sum of all the scores in the dataset divided by the total number of scores. If we're treating a set of scores (*X*'s) as a population, we symbolize the mean by the Greek letter μ (pronounced *mew*), whose formula is:

Glossary Term: mean

$$\mu = \frac{\Sigma X}{N}$$

(If you've forgotten what the Σ symbol means, refer back to **Section 1.3.**) For a sample of scores, the mean has a different symbol, $\overline{X}$ (say "X-bar"), but the same formula:

$$\overline{X} = \frac{\Sigma X}{N}$$

N represents the number of scores in a **population** and n the number of scores in a **sample**. For a refresher on the difference between populations and samples, see the **Box** on populations and samples (on p. 7). Remember that the same dataset can often be considered either a sample or a population. In this chapter, we'll usually assume that a set of scores is a sample, so we'll usually be using $\overline{X}$ to represent the mean.

Box: Populations and Samples

Activity: Formula Reference

Note that these formulas, as well as every other important formula we will discuss in this and later chapters, are always a couple of mouse clicks away in the **Formula Reference** Activity. A link to this Activity always appears in the upper part of the blue frame on the left side of this window.

The mean is easily calculated using a desk calculator such as the one on the left side of this window (click the "Desktop Calculator" link if it's not visible right now), so go ahead and find the mean for our sample dataset.

Interactive Page 46

The mean number of beaux and belles for our dataset is:

$$\overline{X} = (9 + 5 + 12 + 0 + 4 + 3 + 3 + 6 + 1 + 4 + 5 + 8) / 12$$
$$= 60 / 12 = 5.0$$

So to the question "how many girlfriends or boyfriends have your friends had" we can give the simple answer "5, on average." Easy, right? Now try answering the following question:

> Dr. Meene teaches two sections of introductory statistics, one with 20 students and the other with 28. On the first quiz, the average score for the former section is $\overline{X} = 6.8$, while the latter section averages $\overline{X} = 8.0$. What was the mean score for all of Dr. Meene's students combined?

Interactive Page 47

Your first instinct might be to find the mean score for the two sections by averaging the two averages:

$$\overline{X} = (6.8 + 8.0) / 2 = 7.4$$

The problem with this computation is that we were not asked to find the mean of the section means. Instead, we were asked to find the mean of all the students in the two sections. To do this, we need to figure out the numerator and the denominator for the mean formula. That is, we need to calculate the sum of all the scores and the total number of scores.

To calculate the numerator (ΣX), we first turn the mean formula around:

$$\Sigma X = \overline{X} \times n$$

Now we can calculate that ΣX for the first section is $6.8 \times 20 = 136$, and for the second section it is $8.0 \times 28 = 224$. Thus the total ΣX for the two sections combined is $136 + 224 = 360$.

The denominator for the mean formula is easier to compute: The total n for the two sections is 20 (n for the first section) + 28 (n for the second section) = 48.

Finally, we plug our values for ΣX and n into the mean formula:

$$\overline{X} = 360 / 48 = 7.5$$

Another way to perform the same calculation is to realize that 20 / 48 = .42 of Dr. Meene's total number of students are from the first section, while 28 / 48 = .58 of the students are from the second section. If we multiply the two section means by these values and add them together, we'll get the value we're looking for:

$$(.42 \times 6.8) + (.58 \times 8.0) = 7.5$$

Since the second mean is carrying more "weight" than the first in determining the overall mean, 7.5 is called the *weighted mean* of the two sections. (The values .42 and .58 are called weights.)

Exercise 3.1, Exercise 3.2, and **Exercise 3.3** let you practice the basics of calculating means; do these exercises a few times now if you have time. **Exercise 3.4** walks you through a slightly trickier problem: calculating the mean of a dataset when you have a frequency **distribution** table, rather than the raw scores, to work with.

3.3 Measuring Spread: The Standard Deviation

Let's return to the data from our boyfriend/girlfriend survey a couple of pages ago:
9 5 12 0 4 3 3 6 1 4 5 8

We've already established that the **mean** of this **sample** is 5.0. Now consider a second sample of scores:

5 5 5 5 5 5 5 5 5 5 5 5

The mean of this sample (which you should be able to determine without using a calculator!) is also 5.0. As these examples illustrate, the mean alone does not tell us everything we want to know about a dataset.

The difference between the two samples can be stated like this: In the first, there is considerable **variability** in the scores, while in the second, the scores do not vary at all.

To describe a distribution we need to specify how much the scores vary, as well as the average value.

So how do we summarize the amount of variability in a dataset? The simplest way is to calculate the **range,** which is defined as the largest score in the dataset minus the smallest score. For our original sample, the range is 12– 0 = 12. For the new sample, the range is 5 – 5 = 0.

Glossary Term: range

The problem with the range as a measure of variability is that it only takes into account two of the scores in the dataset. For example, the following two sets of scores have the same range:

a. 32 53 78 89 103 125 130

b. 32 78 79 80 82 85 130

But the middle scores in the two sets are clearly different. Scores in dataset (a) are scattered throughout the range, while in dataset (b) all but the most extreme scores are clustered around the middle of the range. In other words, the scores are generally more spread out in dataset (a) than in dataset (b), despite the fact that the smallest and largest scores in each are the same. (**Exercise 3.5** makes this same point.)

What we really want to do is to measure how far away the scores are, on average, from the center of the distribution. The standard way to do this is to calculate a statistic called the standard deviation. This statistic is quite a bit more complicated than the mean, so we'll take some time to develop its exact definition.

Interactive
Page **49**

The "deviation" in the term "standard deviation" refers to the difference between an individual score in a **distribution** and the **mean** of the dataset. For the following small dataset:

$$2\ 3\ 5\ 6\ 9$$

the mean is (2 + 3 + 5 + 6 + 9) / 5 = 25 / 5 = 5.0, so the deviations are:

X	μ	$X - \mu$ (deviation)
2	5	−3
3	5	−2
5	5	0
6	5	1
9	5	4

(For now, we'll treat this distribution as a **population;** later on we'll learn what we have to do to calculate standard deviations of **samples.**)

If we're looking for the average distance from each score to the mean, you might think that we would want to simply average these deviation scores. Go ahead and compute the mean of the deviations yourself.

Interactive Page 50

To compute this mean, we add the deviations together, then divide by the number of scores:

$$(-3 + -2 + 0 + 1 + 4) / 5 = 0 / 5 = 0$$

In fact, the mean of the deviations will be 0 for *any* set of scores, because the sum of the deviations will always be 0. Here's a quick algebraic demonstration of this for three scores, which can be extrapolated for a dataset of any size:

$$\begin{aligned}
\text{Sum of Deviations} &= (X_1 - \mu) + (X_2 - \mu) + (X_3 - \mu) \\
&= (X_1 + X_2 + X_3) - (\mu + \mu + \mu) \\
&= \Sigma X - 3(\mu) \\
&= \Sigma X - 3(\Sigma X / 3) \\
&= \Sigma X - \Sigma X \\
&= 0
\end{aligned}$$

Since the mean deviation will be 0 for any dataset, it is obviously not a very useful measure of **variability**! To solve this problem, we square the deviations before adding them up. The sum of the squared deviation scores is called the **sum of squares,** abbreviated *SS.* When we divide the *SS* for a population by the number of scores *N*, we get a **population parameter** called **variance.** Here's what the process looks like:

X	μ	$X - \mu$	$(X - \mu)^2$
2	5	−3	9
3	5	−2	4
5	5	0	0
6	5	1	1
9	5	4	16

	$(X - \mu)^2 = SS =$		30
	Variance = $SS / N =$		6.0

The variance itself is one measure of variability, but it is expressed in squared units (the units are squared because we squared the deviation scores). For example, if our small dataset above represented the number of children born to a set of 5 female dogs, the variance would be "6.0 puppies squared." Since there's no such thing as a squared puppy, it's difficult to interpret this number.

To create a more easily interpretable measure of variability, we transform the variance into another statistic: our long-foreshadowed standard deviation.

The sum (and mean) of the deviations for any dataset is always 0.

Glossary Terms: sum of squares (*SS*), variance

Variance is expressed in squared units.

Interactive
Page 51

Since the **variance** is expressed in squared units, we take the square root of the variance to translate back into standard units. This gives us the **standard deviation.** When treating a dataset as a **population,** the standard deviation is symbolized by the Greek letter σ. The population variance goes by the symbol σ^2. Here are the formulas:

$$\sigma^2 = \frac{SS}{N} \qquad \sigma = \sqrt{\frac{SS}{N}}$$

Glossary Term:
standard deviation

Figure 3.2 shows histograms of three **distributions** with the **mean** and population standard deviation superimposed as vertical and horizontal arrows, respectively. Note that in each case, the standard deviation stretches about halfway from the mean to the minimum or maximum score in the dataset.

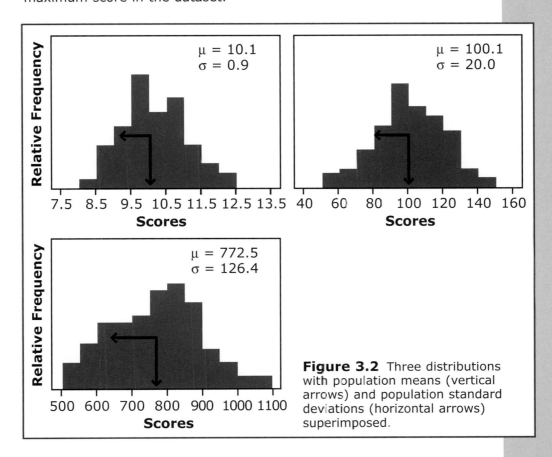

Figure 3.2 Three distributions with population means (vertical arrows) and population standard deviations (horizontal arrows) superimposed.

This observation is indicative of a general rule you can use to estimate standard deviations: Usually, the standard deviation is about one-fourth the size of the range. That is:

$$\sigma \approx range\ /\ 4$$

The same rule of thumb will apply to **sample** standard deviations, which we will discuss shortly.

The standard deviation of a distribution is generally about one-fourth of its range.

Interactive Page 52

3.3.1 The Standard Deviation Helper Activity

To accurately calculate a **standard deviation** by hand, you really need to draw up a table like the one we began two pages ago and finish off here:

X	μ	$X - \mu$	$(X - \mu)^2$
2	5	−3	9
3	5	−2	4
5	5	0	0
6	5	1	1
9	5	4	16
$\Sigma(X - \mu)^2 = SS =$			30
$\sigma^2 = SS\ /\ N =$			6.0
$\sigma = \sqrt{\sigma^2} =$			2.45

These calculations become quite tedious with anything more than a handful of scores. Luckily, spreadsheet programs and statistics packages will do the calculations much more quickly (and more accurately) than you can, and in **Section 3.3.3,** we'll introduce a Calculation Tool you can use to compute the standard deviation (along with the mean and other **descriptive statistics**) for datasets right in your web browser.

However, it's important for you to get some practice with the hand calculations, so that you're sure to understand what the standard deviation means and how it works. The **Standard Deviation Helper** Activity will help you organize, check, and practice these calculations.

To use the Helper, first enter the scores you want to analyze in the large text box in the upper left corner of the Helper Window. Or you can click the "generate" link to have the Activity create a random dataset. Choose whether to treat the dataset as a **population** or a **sample** (for now, leave the "population" radio button selected), then start filling in the blanks in the table:

- In the leftmost column, enter the scores in the same order in which they appear in the large text box. Then in the three boxes below the line in this column, sum the scores, enter the number of scores, and divide ΣX by N to calculate the mean.

- Enter μ in each of the boxes in the second column.

- In the third column, subtract the **mean** from each score to get the **deviation** scores. Remember that if a raw score X is less than the mean, the corresponding deviation score should be negative.

Activity: Standard Deviation Helper

- In the fourth column, square the deviations. Then add these up to compute the **sum of squares** and enter the sum in the first box below the line. Enter the number of scores below this, then divide *SS* by *N* to get the variance. Finally, take the square root of σ^2 to produce the standard deviation.

Note that you can press the "tab" key on your keyboard to move from field to field in the table.

To check your calculations, click the "Check Answers" button at any time, or simply hit the enter or return key after typing any value into a text box. Any incorrect answers will be colored red; correct calculations will remain in black.

Roll your mouse over the bullet (small circle) to the left of each text box to see a hint for what's supposed to go there. Click the bullet to show the correct value (it will appear in blue so you can keep track of which fields you needed help with).

Try this Activity now using the boyfriend/girlfriend data we've been working with. Enter the data into the Helper as instructed above, calculate *SS*, σ^2, and σ, transfer these values to the text boxes below, and click the "OK" button. Here are the data again:

9 5 12 0 4 3 3 6 1 4 5 8

The **variability** statistics for the girlfriend/boyfriend data (repeated once again above) are:

Sum of squares   126

Variance   11.45

Standard deviation   3.38

Practice calculating the **standard deviation** for a random dataset or two before moving on to the next section, which discusses the standard deviation of a **sample.**

3.3.2 The Standard Deviation of a Sample

Just as we had different formulas for the **mean** of a **population** and the mean of a **sample,** we also need to distinguish between the **standard deviation** of a population and a sample. But unlike the population and sample formulas for the mean, the population and sample formulas for the standard deviation differ substantially. Before we see the sample formula, let's discuss why it needs to be different from the population formula.

The formula for sample standard deviations differs substantially from the formula for population standard deviations.

Consider the distribution of 100 scores listed below and illustrated by the blue histogram in Figure 3.3, which we'll treat as a population. The population standard deviation for this dataset is 8.1.

1 4 4 5 5 6 6 7 8 8 10 10 10 10 11 11 12 12 12

12 12 13 14 14 14 15 15 15 16 16 16 16 17 17

17 18 18 18 19 19 19 19 20 20 20 20 20 20 20

20 21 21 21 21 21 21 21 22 22 22 23 23 23

23 23 23 23 24 24 24 24 25 25 25 26 26 26 26

26 27 27 27 28 28 28 28 29 30 30 31 31 32 33

33 33 34 36 37 41

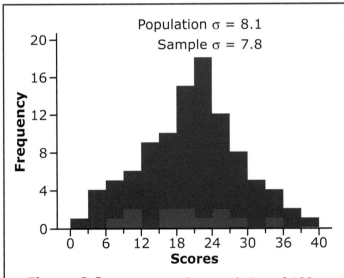

Figure 3.3 Histogram of a population of 100 scores (blue) and a sample of 12 scores drawn from this population (red). This figure is interactive on the *Interactive Statistics* website and CD.

A random sample of 12 scores from the population is also plotted, in red, in the figure. This sample is less variable than the population it was drawn from. This decrease in **variability** will be immediately apparent by comparing the ranges of the sample and the population (the only way the sample could have as large a range as the population would be if the sample happened to include both the smallest and largest scores in the population).

The limited variability of the sample is also reflected in the value of σ for the sample (7.8), which is smaller than σ for the whole population (8.1). On the *Interactive Statistics* website and CD, you can

Samples are usually less variable than underlying populations.

use this figure to generate as many different samples as you wish. If you collect enough **samples,** the average sample σ will always be smaller than the true **population parameter** (although you will occasionally observe samples that are more variable than the population). In statistical jargon, we say that the sample's **standard deviation** (when computed using the population formula) is a **biased** estimator of the **variability** in the population as a whole.

This explains why we have to alter the standard deviation formula for samples: We need to make a correction in order for the sample standard deviation to be an unbiased estimator of the population standard deviation.

Glossary Term: bias

Interactive Page 55

With only four scores, the **sample** will usually be much less variable than the **population.** But with 99 scores, the sample **distribution** will look very much like the population distribution, so the sample **variability** and population variability will be very similar. In general, the larger the **sample size,** the closer the sample's variability will tend to be to the population's variability.

The larger the sample, the closer the sample's variability will tend to be to the population's variability.

Therefore, to correct the bias in **sample standard deviations,** we want to change the computation such that small samples get corrected more than large samples. Here's how we do it. In the formula for population standard deviation, we divide the sum of squared **deviations** by the total number of scores in the population, N:

$$\sigma = \sqrt{\frac{SS}{N}} = \sqrt{\frac{\Sigma(X - \mu)^2}{N}}$$

But in the formula for sample standard deviation, we divide the sum of squared deviations by the number of scores in the sample *minus one:*

$$s = \sqrt{\frac{SS}{n-1}} = \sqrt{\frac{\Sigma(X - \overline{X})^2}{n-1}}$$

Divide SS by $n-1$ to calculate sample variance and standard deviation

As shown in the formula, the sample standard deviation is denoted by a lowercase s. **Sample variance** goes by the symbol s^2, and is equal to:

$$s^2 = \frac{SS}{n-1} = \frac{\Sigma(X - \overline{X})^2}{n-1}$$

Interactive Page 56

Here are the formulas for **population** and **sample standard deviation** again:

$$\sigma = \sqrt{\frac{SS}{N}} \qquad s = \sqrt{\frac{SS}{n-1}}$$

It can be mathematically proven that the underestimation of population σ^2's by sample σ^2's, illustrated in Figure 3.3, is perfectly compensated for by the $n - 1$ correction (although we won't subject you to the proof here). If $n = 4$, the correction will increase s^2 by 25% (compared to what it would be if we used the σ^2 formula to calculate sample variance), whereas if $n = 99$, the use of $n - 1$ will only change the computed variance by 1%.

In statistical jargon, we say that the sample variance (as well as the sample standard deviation, since it is derived from sample variance) has $n - 1$ **degrees of freedom,** for reasons outlined in the **Box** on degrees of freedom. The box also provides a second informal explanation for why $n - 1$ is the correct correction to apply.

Box: Degrees of Freedom

You can also empirically support the appropriateness of the correction using Figure 3.4 on the *Interactive Statistics* website or CD, which is just like Figure 3.3 but with text boxes added to show individual and cumulative sample standard deviations.

Generate a few samples, and you should see that s for each sample is usually closer to the standard deviation of the entire population (8.1) than σ for the sample. In other words, by using the sample standard deviation formula when we only have data for a sample, we usually get a better estimate of the standard deviation for the entire population. And as you add more and more samples, the **mean** sample standard deviation should get closer and closer to the true population standard deviation. (It is well worth the effort to prove this to yourself using the interactive figure on the *Interactive Statistics* website and CD.)

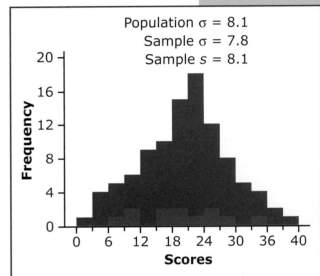

Population $\sigma = 8.1$
Sample $\sigma = 7.8$
Sample $s = 8.1$

Figure 3.4 Histograms of a population of 100 scores (blue) and a sample of 12 scores drawn from this population (red). This figure is interactive on the *Interactive Statistics* website and CD.

Exercise 3.8 quizzes you on your understanding of the differences between population and sample standard deviations.

Interactive
Page 57

It is important to be clear about what the **sample** correction does in the formulas for s and s^2: By dividing SS by $n - 1$, we get a sample **variance** that is an accurate, **unbiased** (that is, neither too large nor too small) estimate of the variance of the **population** from which the sample was drawn. If we calculated s^2 using the population standard deviation formula, we would consistently underestimate σ^2 for the whole population.

Box: Degrees of Freedom

The formula for sample variance (s^2), also called "mean squared deviation" is:

$$s^2 = \frac{SS}{n-1}$$

The denominator of this formula may seem a bit puzzling. If variance is the mean of the squared deviations, shouldn't we divide the sum of the squared deviations by the number of deviations, n? Why instead do we divide by $n - 1$?

This is a difficult question to answer simply, but we'll give it a try. To calculate the deviation for each score in the sample, we subtract the mean from the score (deviation = $X - \overline{X}$). To do this, we have to know the mean *before* we calculate the deviations. It is also the case that, by definition, the deviations (not the squared deviations, but the deviations themselves) must add up to exactly zero. The positive deviations always exactly cancel out the negative deviations.

When we put these two conditions together, we find that *the value of the last score in the sample is restricted by the values of the other scores in the sample*. This point is best understood through an example. Let's say that the mean of a sample of three scores is 10. If the first two scores are 7 (deviation = 7 − 10 = −3) and 12 (deviation = 12 − 10 = +2), the third score must be 11 (deviation = 11 − 10 = +1) in order for the sum of deviations to add to 0 (−3 + 2 + 1 = 0).

It doesn't matter which score is designated as the "last" one in the sample. If the first two scores are 7 and 12, the third score must be 11. Alternatively, we could say that if the first two scores are 7 and 11, the third score must be 12, or that if the first two are 11 and 12, the third must be 7. The point is that once you know two of the scores, the value of the third score is fixed.

When we calculate the variance of these scores, we're not averaging three unrelated numbers, because *one of the scores is tied to the values of the other two*. Therefore, to get the mean squared deviation, we divide the sum of the squared deviations by the number of deviations that are free to vary, $n - 1$. (If we had a sample of 4 scores, 3 would be free to vary and the fourth would be fixed; if n was 100, 99 would be free to vary and the hundredth would be fixed; etc.)

In mathematical jargon, we say that the variance statistic has $n - 1$ degrees of freedom (df). Furthermore, any other statistic that uses variance or standard deviation (the square root of variance) in its calculation also has a limited number of *df*. For example, the formula for the single-sample t statistic is:

$$t = \frac{\overline{X} - \mu}{s_{\overline{X}}} = \frac{\overline{X} - \mu}{s / \sqrt{n}}$$

Because s in this formula has $n - 1$ degrees of freedom, the t statistic also has this number of *df*.

The $n - 1$ correction does not completely compensate for the bias in sample **standard deviations,** but s as calculated using the $n - 1$ formula is still a much less biased estimator of σ than it would be if we divided by n.

Practically speaking, just remember that if you're treating a dataset as a sample, divide by $n - 1$. Divide by N only if your scores represent every individual in the population you're interested in.

When in doubt, use the sample formula—we rarely measure every single member of a population, so we're almost always dealing with samples.

You should now be ready to use the **Standard Deviation Helper** Activity to practice calculating s. Just click the "Sample" radio button below the data text box in the Activity, then try your hand on the following dataset:

$$11\ 8\ 12\ 8\ 5\ 3\ 7\ 6$$

(Don't forget to divide SS by $n - 1$ instead of n!) You should calculate s to be 2.98.

3.3.3 The Descriptive Statistics Calculation Tool

It is very important that you practice calculating **population** and **sample standard deviations** (as well as **means**) by hand, using pencil and paper, a desk calculator, and/or the **Standard Deviation Helper** Activity. **Exercise 3.6** covers the basic calculations, and **Exercise 3.7** asks you to apply your knowledge of the various formulas in a new way.

Once you thoroughly understand how these values are computed, you can use the **Descriptive Statistics** Calculation Tool to do the math for you. To use this Tool, simply enter the scores you wish to analyze in the large text box (scores can be separated by spaces, commas, or tabs), then click the "calculate" button. The other text boxes in the Tool will be filled in with the most commonly used **descriptive statistics:**

- The number of scores n in the dataset

- The minimum and maximum values in the dataset

- The **median** value (we'll get to this statistic in **Section 3.4.1**)

- The mean of the scores

- The sum of the squared deviations of the scores from the mean (SS)

- The standard deviation

Activity: Standard Deviation Helper

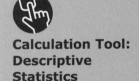

Calculation Tool: Descriptive Statistics

By default, the sample standard deviation will be calculated. To use the population formulas instead, click the "Population" radio button.

If you click the "calculate" button (or hit the "enter" or "return" key after entering the scores), all statistics will be rounded to two decimal places (e.g., 4.995 will be rounded up to 5.00, whereas 4.994 will be rounded down to 4.99). You will rarely need more than two decimal places of precision for the calculations you're expected to do in this book. But if you want to be more precise, click the title of the Calculation Tool (the words "Descriptive Statistics," just to the right of the calculate button) to show the statistics with up to 12 decimal places (e.g., the value 4.995 would show up as 4.995).

Try the tool out now on the following sample of scores:

> 11.8 11.3 10.0 8.1 9.7 11.2 9.4 10.8 8.5 10.7 8.5
> 12.4 7.4 7.4 7.4 7.8 9.9 8.9 9.0 11.7

Interactive
Page 59

You should get the following statistics:

- n 20

- Minimum   7.4

- Maximum   12.4

- $\bar{X}$ 9.60

- SS 48.77

- s 1.60

Again, we can't stress enough the importance of calculating these statistics by hand (on paper or using the Standard Deviation Helper Activity) over and over until you're completely comfortable with how they are computed. One good reason to practice the hand calculations is that you may be asked to do them on a test.

More important (from a pedagogical point of view, at least), we will be using the **mean, *SS*,** and **standard deviation** as intermediate calculations when we compute z-scores, t statistics, analyses of variance, Pearson correlations, and other **statistics** later in the course. In fact, we'll start in on z-scores later in this chapter. If you don't have a firm grasp of exactly how means, sums of squares, and standard deviations are computed, you're going to have trouble grasping the concepts behind these more complicated statistics.

Interactive Page **60**

3.4 Resistant Measures of Center and Spread

Consider the following scenario:

> Emily, a developmental psychologist, has recently gotten permission from all the parents of a preschool class at a local day care center to use their children as subjects in a series of cognitive development experiments. As a preliminary step in her research program, Emily administers to all the children a general cognitive abilities test. In the test, children do a number of simple tasks such as saying the alphabet, counting to 20, etc., and performance is graded according to a strict procedure. Past research has indicated that 3-year-olds (the average age of the kids Emily is testing) typically score about 10 on this test, with a standard deviation of about 1.5. Emily's subjects receive the following scores:

9 10 10 8 12 25 9 11 10 11

The **mean** and **standard deviation** of this dataset indicate that Emily's 10 children are smarter and more **variable** than typical 3-year-olds: The **sample** mean is 11.5, compared to the typical value of 10, and the sample standard deviation is 4.88, compared to the typical value of 1.5.

But a quick glance at a **histogram** of Emily's data (Figure 3.5, which you can produce yourself using the Frequency Distributions Calculation Tool) should lead you to question this conclusion. All but one of the children are clustered around the **population**-typical score of 10, but the one exception is *very* exceptional, achieving a score of 25.

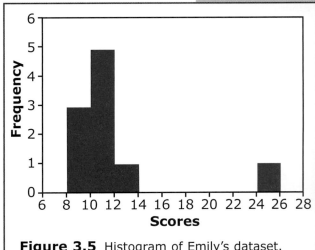

Figure 3.5 Histogram of Emily's dataset.

What would happen to the sample mean and standard deviation if this extreme score were not included? Recalculate the statistics for the other nine scores to find out.

Interactive Page **61**

Leaving out the single extreme score (25), the **mean** is reduced by 13% (from 11.5 to 10.0) and the **standard deviation** is slashed by 75% (from 4.88 to 1.22). The huge effect of the extreme score on the standard deviation is apparent when we look at the individual **deviations** and squared deviations (see table below): The squared deviation for the score of 25 (182.25) is over five times as large as all nine of the other squared deviations combined!

X	$\bar{X}$	$X - \bar{X}$	$(X - \bar{X})^2$
9	11.5	−2.5	6.25
10	11.5	−1.5	2.25
10	11.5	−1.5	2.25
8	11.5	−3.5	12.25
12	11.5	.5	.25
25	11.5	13.5	182.25
9	11.5	−2.5	6.25
11	11.5	−.5	.25
10	11.5	−1.5	2.25
11	11.5	−.5	.25

This example illustrates two important and related points about the mean and standard deviation. First of all, you should never blindly trust these two statistics to perfectly describe the center and spread of a distribution. Always use a **histogram** or other graphical display to inspect the raw data for extreme values like the one here, which are called **outliers** (because they lie outside the range of typical scores in the histogram).

Second, if you find an outlier, you probably shouldn't use the mean and standard deviation of all the scores in the dataset when describing the distribution. Doing so misrepresents the true center and spread of the scores.

So what should you do instead? One option is to calculate and report the mean and standard deviation leaving out the outlier(s), as we did above. A second option is to use alternative measures of center and spread that are **resistant** to the effects of extreme values. First we'll describe a pair of resistant measures, the median and the interquartile range, then we'll discuss when to use them.

Interactive Page 62

3.4.1 The Median

The **median** is the middle value in a distribution. When there is an odd number of scores in the distribution, this definition also describes the method for calculating the median: Order the scores and find the one in the middle (that is, the score that has an equal number of other scores to the left and to the right of it). Thus for the following dataset:

1 4 5 8 12

the median is 5. When a dataset contains an even number of scores, there is no one score in the middle, so the median is computed as the average (**mean**) of the two middle scores. So the median for these scores:

10 12 18 25 26 47

is (18 + 25) / 2 = 21.5.

Don't rely solely on statistics—always inspect a distribution's histogram.

Distributions with outliers need special attention when describing their center and spread.

Glossary Term: median

To make sure you understand these definitions, find the medians for the following datasets:

a. 100.9 99.6 95.5 102.7 104.4 102.5 97.8 84.9 115.9 106.9

b. 10 8 8 9 6 7 7

(Note that you can use the **Data Transformations** Calculation Tool to put the scores in order. *Don't* use the Descriptive Statistics Tool to find these medians; you can start using this Tool later, once you've had some practice finding medians by hand.)

Now that we know what the **median** is, let's look at why it's a **resistant** measure of the center of a **distribution.** Imagine you conducted a survey by standing on a street corner in New York City, stopping 10 random people one at a time, and ask them each how much money, in thousands of dollars, they made in the past year. Say you got the following responses from the first nine people:

$25,000 $30,000 $32,000 $38,000 $40,000

$41,000 $56,000 $60,000 $65,000

The **mean** of these nine incomes is $43,000, and the median is $40,000. So far, the two measures of center are pretty similar.

But suppose the tenth person you ask reports an income of $10,000,000. The mean for your sample will suddenly jump to $1,038,700, while the median will rise much more modestly, to $40,500. The median resists the extreme influence of this large value.

The median is less sensitive to the influence of outliers than the mean.

Similarly, the median for Emily's cognitive abilities scores, arranged in order and repeated below, is 10 regardless of whether or not the extreme value (25) is included.

Emily's data: 8 9 9 10 10 10 11 11 12 25

3.4.2 The Interquartile Range

In **Section 2.3.3,** we introduced the concept of **percentiles** and **percentile ranks:**

- The Pth percentile is the score in a distribution that is greater than P% of the other scores in the distribution and less than $(1 - P)$% of the other scores.

- The percentile rank of a score is the percentage of other scores falling below it in the distribution.

Review: percentiles and percentile ranks

Given the definitions of percentiles and the **median,** you should realize that the median is always equivalent to a particular percentile. Which one?

Interactive Page 65

The **median** is always the 50th **percentile** of a distribution: One half of the other values in the **distribution** fall below it, while the other half of the values fall above it.

Median = 50th percentile

While the median of a distribution is almost always quite easy to determine, most other percentiles are trickier to calculate for small datasets. (When a **population**'s distribution is defined by a smooth curve, as described in **Section 2.3.2,** any percentile can be calculated quickly and precisely. We'll see how to do this in **Section 4.2.**)

In addition to the 50th percentile, four other percentiles are also easy to determine:

• The 0th percentile is the lowest score in the dataset.

• The 100th percentile is the highest score in the dataset.

• The 25th percentile, which is called the **first quartile,** is the median of the scores below the 50th percentile.

• Similarly, the 75th percentile, or **third quartile,** is the median of the scores above the median of the entire distribution.

Consider, for example, the following dataset:

8 10 12 14 16 17 18 20 20

• The **median** of these nine scores is the 5th highest value, 16.

• The 0th **percentile** is the lowest score, 8, and the 100th percentile is the highest score, 20.

• The **first quartile** is the median of the four scores below 16 (8, 10, 12, and 14), which turns out to be (10 + 12) / 2 = 11.

• The **third quartile** is the median of the four scores above 16 (17, 18, 20, and 20): (18 + 20) / 2 = 19.

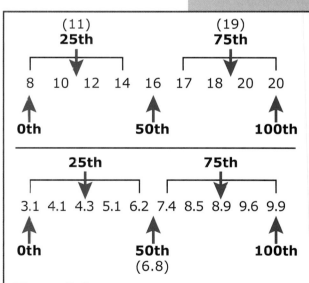

Figure 3.6 Locations of the 0th, 25th, 50th, 75th, and 100th percentiles in two datasets.

Figure 3.6 illustrates these five percentiles for the dataset above and an additional dataset that has an even number of scores. Note that when there are an even number of scores, you should include the score just below the median (6.2 in the dataset shown in the figure) when determining the 25th percentile, since this score is below the 50th percentile. Likewise, include the score just above the median when determining the 75th percentile.

Now try determining these percentiles for our income-level data from a few pages ago, translated below into thousands of dollars (i.e., 30 = $30,000):

$$25 \ 30 \ 32 \ 38 \ 40 \ 41 \ 56 \ 60 \ 65 \ 10000$$

You should find that the 0th, 25th, 50th, 75th, and 100th percentiles are 25, 32, 40.5, 60, and 10000, respectively.

The **range,** you should recall, is the difference between the highest and lowest scores in a **distribution.** We can now redefine the range as the difference between the 0th and 100th **percentiles.**

Like the **standard deviation** and **mean,** the range can be overwhelmed by **outliers.** For the first nine scores in our income data, the standard deviation is 14.12 and the range is 65 − 25 = 40. But when we add in the multimillionaire, s jumps to 3148.71 and the range leaps to 10000 − 25 = 9975.

To avoid being unduly influenced by outliers, we can use the **interquartile range** (IQR) as our measure of spread. The IQR is defined as the difference between the first quartile (25th percentile) and third quartile (75th percentile). So for the income data, the IQR is 60 − 32 = 28 with the outlier and 58 − 31 = 27 without the outlier. Pretty resistant!

Try **Exercise 3.9** a few times now to make sure you're comfortable calculating the median, quartiles, and IQR.

3.4.3 When to Use Resistant Measures

We've now covered two sets of descriptive statistics for center and spread: the **mean** and **standard deviation,** and the **median** and **interquartile range**. It's time now to address the issue of when to use which statistic.

The first rule of thumb is that **resistant** and **nonresistant** measures should not be mixed and matched. The mean and standard deviation are both based on the premise that all the scores in a **distribution** should be incorporated into the **descriptive statistic.** On the other hand, the mathematical concept of **percentiles** forms the foundation for both the median and IQR. Combining the mean with

Glossary Term: interquartile range

Do not mix resistant and nonresistant statistics.

the IQR or the median with the standard deviation would mix up these principles and result in a confusing description of the data.

The second point to make is that by default, it is best to use the mean and standard deviation. One reason for this preference is that the premise behind these measures, noted in the previous paragraph, is a worthy one: If possible, all the scores in a distribution should contribute something to the measures of center and spread.

Another reason we prefer to use the mean and standard deviation is that the **inferential statistics** based on these values are more powerful than the inferential statistics associated with the median and IQR. That is, it's easier to find evidence for a significant experimental effect using mean-based inferential statistics than using median-based inferential statistics.

Interactive
Page 68

Though the **mean** and **standard deviation** are used in the majority of research situations, there are some situations in which it is impossible to calculate these statistics. Usually, this occurs when a dataset includes open-ended or infinite values.

For example, suppose we're interested in studying how long it takes people to learn to use a new computer program. The first six subjects might take 10, 34, 17, 45, 51, and 22 minutes to learn the program. So far, so good. But the seventh subject might struggle with the program for a whole hour (60 minutes) and then give up. What score should we assign to this final subject? Would he have figured out the program in 70 minutes? 120 minutes? 240 minutes? We don't know, and without a definite score, we can't compute a mean.

However, we can compute the **median.** The scores, in order, are:

10 17 22 34 45 51 60+

where "60+" stands for the person who never finished. Although we don't know how long he would have taken to learn the program, we know he would have taken longer than everyone else. So the median of this distribution is the fourth score in the distribution, 34. The 25th **percentile** is the median of the scores below this one, (17 + 22) / 2 = 19.5, and the 75th percentile is the median of the scores above this one, (34 + 45) / 2 = 48. The IQR s thus 48 – 19.5 = 28.5.

If we can't compute the mean and standard deviation, the median and IQR still provide a good description of the center and spread of the **distribution.** As we saw in **Section 3.4,** there are also some situations in which the mean and standard deviation are computable, but don't appear to accurately reflect the distribution's center and spread. These are the most common situations in which the median and IQR are used.

Use the mean and standard deviation rather than the median and IQR if possible.

Open-ended or infinite values preclude calculating the mean and standard deviation.

Interactive Page 69

Here's the scenario from **Section 3.4** again:

> Emily administers a general cognitive abilities test to 10 3-year-old children. Past research has indicated that 3-year-olds typically score about 10 on this test, with a standard deviation of about 1.5. Emily's subjects receive the following scores:
>
> 9 10 10 8 12 25 9 11 10 11

The mean and standard deviation of these scores is 11.5 and 4.88, respectively. But the histogram of the scores (shown in Figure 3.5, repeated below) indicates that one of the scores is an outlier.

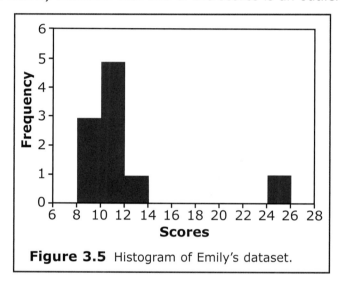

Figure 3.5 Histogram of Emily's dataset.

Clearly, the **mean** and **standard deviation** of these scores don't accurately describe the center and spread of most of Emily's subjects. There are three possible remedies to such situations:

1. We can exclude the extreme score(s) and compute the mean and standard deviation for the remaining scores. The term **trimmed mean** is used to describe a mean calculated from a set of scores with outliers excluded.

2. We can keep all the scores and use a **transformation** to try to reel in the outliers. A detailed discussion of the use of data transformations for this purpose is beyond the scope of this textbook, but see the **Box** on transformations for a brief overview of how this technique works.

3. We can keep all the scores and use the **median** and **IQR,** which are resistant to extreme effects of outliers, as our measures of center and spread.

Unfortunately, there are no hard and fast rules on which of these three policies to adopt in which situations. When distributions are severely **skewed** (see **Section 2.4** and Figure 3.7a), the mean of

Glossary Terms:
trimmed mean,
transformation

**Box:
Transformations**

Box: Transformations

Consider the distribution of 50 response times (RT's) graphed in the histogram at right. As you can see, the distribution is positively skewed, and a few RT's fall particularly far to the right of most of the scores in the distribution. However, there isn't enough of a separation between the scores at far right and the bulk of the RT's to really call those at far right outliers.

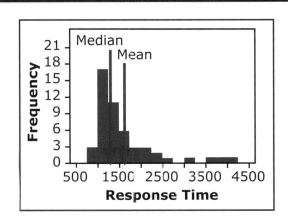

How should we summarize the central tendency of this distribution? The mean RT is 1611. As shown by the right arrow in the figure, this value does not appear to adequately describe where the center of the distribution is. The median (1330) does a better job of locating the center.

But as noted in the text, the most commonly used inferential statistics, such as t tests and analysis of variance (ANOVA), are based on the mean, not the median. It would not be appropriate to report medians if we use these inferential statistics. Furthermore, these inferential statistics don't work as well (that is, they are less likely to detect experimental effects) with skewed distributions such as this one.

One solution to these dilemmas is to transform the data before analyzing it. One of the most common transformations, especially for RT's, is to take the base-10 logarithm (abbreviated "log") of each score.* Here are 10 of the RT's in the dataset graphed above, along with their log transforms:

RT	log(RT)
1530	3.18
1928	3.29
1213	3.08
1147	3.06
1862	3.27
1181	3.07
4097	3.61
1413	3.15
3195	3.50
947	2.98

The following figure shows the histogram of all 50 of the log(RT)'s. As you can see, it is much less severely skewed, and the scores on the far right of

*The base-10 logarithm of a number is the power to which 10 must be raised to equal that number. For example, log(100) = 2, since 10 to the second power (i.e., 10 squared) equals 100.

the original distribution have been pulled in toward the center. As a result, the mean (3.17) is much more similar to the median (3.12), and now both statistics do an adequate job of indicating the center of the distribution. We can also calculate our inferential statistics using the log-transformed data instead of the raw RT's.

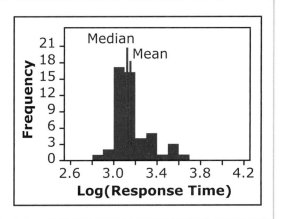

all the datapoints overestimates the typical score in the distribution, and there is no obvious value to use as the cutoff for extreme values that should be excluded. Medians are therefore often used to describe the centers of such distributions.

On the other hand, if most of the values in a distribution appear **symmetrical,** as in Figure 3.7b, most statisticians deem it acceptable to use a trimmed mean. For example, if Emily scours the previous research and finds that no child has ever scored higher than 25 on her test, she is justified in excluding this extreme score from her dataset and reporting the trimmed mean.

Dealing with extreme datapoints is one of the most troublesome aspects of data analysis for many researchers because it requires judgment calls, and critics can always complain that the wrong decisions were made.

The best way to avoid such criticism is to be consistent. For example, many researchers adopt the rule that any value greater than or less than 2.5 standard deviations (or 2, or 3 standard deviations, depending on the researcher's preference) away from the mean

> When dealing with extreme datapoints, the most important rule is to be consistent.

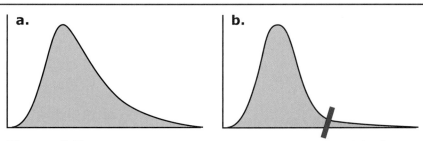

Figure 3.7 Two skewed distributions. For distribution (a), there is no obvious cutoff point to decide which values are outliers. But in distribution (b), trimming values beyond somewhere around the red line will leave a roughly symmetrical distribution.

should be excluded. A researcher gets herself in trouble when she excludes values greater than 2 standard deviations in her first experiment, uses the median in a second experiment, then uses a trimmed mean again in the third experiment.

A final important point about outliers is this: While extreme scores are often difficult to deal with, sometimes the outliers are the most interesting things about a dataset. If no 3-year-old has ever scored 25 on Emily's test, maybe she should concentrate on figuring out how that kid got so smart!

Exercise 3.10 and **Exercise 3.11** ask you to think more about situations in which one might want to use resistant measures.

3.5 Correlation

Remember Susan, the District Manager for Burger Town? When we last left her, she was looking at the annual sales figures for her 22 restaurants. Suppose Susan suspects that the employee turnover rate at the restaurants is affecting sales. To investigate this issue, she asks all her restaurant managers to report how many employees have quit their jobs at each restaurant. The employee turnover data (the proportion of employees at each store who quit during the past year), along with the sales figures, are shown in the following table:

Burger Town #	Sales (thousands of dollars)	Employee turnover rate
456	98.9	.11
461	68.6	.19
464	81.3	.15
465	68.4	.35
466	88.7	.13
467	77.2	.16
468	67.8	.25
469	83.9	.16
470	84.4	.17
471	87.7	.19
473	87.3	.25
474	63.7	.53
475	78.6	.32
476	58.8	.39
477	125.2	.10
478	98.2	.12
480	79.1	.29
485	24.5	.47
492	86.7	.15
494	56.8	.22
497	75.9	.20
499	44.8	.34

You should recognize that measures of center and spread will not help Susan answer the question she wants to address with these data. The **mean** sales figure is 76.7 thousand dollars, and the mean employee turnover rate is .24, but these statistics tell us nothing about the **relationship** between sales and employee turnover.

Relationships between pairs of **variables** are assessed using a completely different type of statistic, called a **correlation.** A high correlation indicates a strong relationship between the two variables, whereas a low correlation suggests that one variable has little to do with the other. Relationships can be visualized using a type of graph called a **scatterplot,** shown in Figure 3.8. The relationship shown there between sales and employee turnover is quite strong (i.e., there is a high correlation).

Correlations are considerably more complicated to compute and understand than means and standard deviations. Also, as we'll see in the next chapter, correlational research is qualitatively different than research that uses the mean (or median) as the primary **descriptive statistic.** Therefore, we'll wait to describe correlations in detail until after we've described the **inferential statistics** that are associated with means.

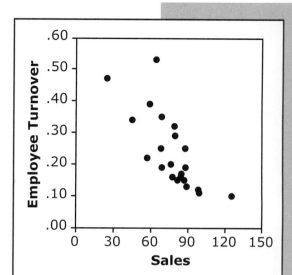

Figure 3.8 Scatterplot of Susan's sales and employee turnover data. Each dot in the graph represents an individual restaurant: The vertical position of the dot indicates the restaurant's employee turnover rate, while the dot's horizontal position indicates its sales volume.

3.6 *z*-Scores

Consider the following description of "Billy":

> Billy walks with a limp because he has rheumatoid arthritis. His teeth are not what they used to be, his hair is falling out, and his breathing appears strained. And most recently, he contracted leukemia. His doctor says that Billy is dying. But he's led a long, full life. After all, he is almost 15 years old.

That last sentence may seem incongruous with the rest of the description. But if we tell you now that Billy is a Labrador retriever, the fact that he's dying of old age shouldn't seem so surprising.

This example highlights a fundamental difficulty with interpreting **statistics** that was first discussed in **Section 1.1:** When contextual information is lacking, numbers can be meaningless and even deceiving.

Statistics taken out of context are meaningless at best, deceiving at worst.

In an attempt to put dogs' ages into context, many people use the rule of thumb that one "dog year" is equal to seven "human years." Using this rule, we would multiply 15 by 7 and say that Billy is 95 in human years, a statistic that makes it clear that he has lived to a ripe old age.

The statistical term for putting numbers into context is **standardization.** The dog-year rule standardizes dogs' ages to match humans' ages, which we are more familiar with. In formal statistics, the most common way to standardize a number is to **transform** it into something called a *z*-score.

Interactive
Page **72**

Like dog years, *z*-scores are intended to take a number that might be uninformative or confusing on its own and transform it into another number that can be readily interpreted. Say you're talking to your friend Ron Lemon, who owns a used-car lot. Ron mentions that his dealership sold 65 cars last week. How should you interpret this statistic? Did Ron have a good week, a bad week, or one that's just about average?

The last word of the previous paragraph provides a hint about the first factor we consider in constructing a *z*-score: It's always useful to compare an individual score to the **mean** (average) score for the **population** in question. Here, the population is the set of all the weeks since Ron's dealership opened. If Ron has averaged 100 cars sold per week over this period, he had a pretty rough time of it last week, while if he moves an average of 30 cars per week, he's probably pretty pleased with himself.

The second factor we take into consideration when constructing *z*-scores is **variability.** Let's assume that Ron averages 80 cars per week. This tells us that last week was worse than normal, but how much worse was it? If his sales regularly jump around by up to 50 cars or more every week, a week where he sold 15 fewer than average is not that big a deal. On the other hand, if his sales figures rarely go up or down by more than 5 cars from the mean, this past week was disastrous.

As we learned earlier in the chapter, the measure of variability that goes along with the mean is the **standard deviation.** Suppose the standard deviation of Ron's weekly sales is 10. Now we can say that Ron's sales last week were 1.5 standard deviations below the mean. And there you have it—the formula for *z*-scores:

$$z = \frac{X - \mu}{\sigma}$$

For Ron's sales last week, *z* is calculated like so:

$$z = \frac{X - \mu}{\sigma} = \frac{65 - 80}{10} = \frac{-15}{10} = -1.5$$

Glossary Terms: standardization, *z*-score

The first step in constructing a *z*-score is to compare the raw score to the population mean.

The second step in constructing a *z*-score is to compare the raw score to the population standard deviation.

Formula: *z*-scores

(Note that we use σ, not *s*, in the formula for *z*. When working with *z*-scores, we are always treating the dataset in question as if it were an entire population.)

Now, you may be thinking that –1.5 is no more meaningful a number than 65. In fact, in a sense it's less meaningful, because at least the raw statistic (65) is expressed in meaningful units (number of cars). The **z-score** is a unit-less number, and it has a fractional part to boot.

However, the unit-free nature of *z*-scores is actually an asset, once you're used to working with them. To see why, consider Ron's sister-in-law Lucy Sugar. Lucy owns a cookie factory, and she sold 65000 cookies last week. How did her sales compare to Ron's? Well, Lucy sold 1000 times as many cookies as Ron sold cars, but this comparison is meaningless because the units are completely different. Comparing cars to cookies is worse than comparing apples to oranges!

But if we know that Lucy averages 95000 cookie sales per week with a standard deviation of 20000, we can calculate her *z*-score and compare it to Ron's, because *z*-scores don't have units. Can you figure out Lucy's *z*-score on your own?

z-Scores from different populations can be compared becuase *z*-scores are unit-free statistics.

Interactive
Page 73

Lucy averages 95000 cookie sales per week, with a standard deviation of 20000. Last week she sold 65000 cookies. The **z-score** for last week's sales is calculated as follows:

$$z = \frac{X - \mu}{\sigma} = \frac{65 - 80}{10} = \frac{-15}{10} = -1.5$$

Since the *z*-score for Lucy's sales is identical to that for Ron's sales, we can conclude that they had comparably poor weeks. Without some kind of **standardization** procedure, we would have no way of comparing Lucy's 65000 cookies to Ron's 65 cars.

There are other ways to standardize besides the *z*-score formula. For example, the administrators of the Scholastic Aptitude Test standardize SAT scores so that the mean is 500 and the standard deviation is 100. But *z*-scores are the most common way to standardize, and they are particularly useful because you can learn two important things about an individual right away by looking at his/her/its *z*-score:

1. The sign (+ or –) tells you whether the individual's score is above (+) or below (–) the average score of the population it comes from.

2. The absolute value tells you how many standard deviations the individual is away from the mean. In other words, a small value, whether negative or positive, indicates an individual that is fairly typical of the population; the larger the value, the more atypical the individual.

Quick facts you can learn about an individual by inspecting a *z*-score.

Interactive
Page **74**

3.6.1 The z-Score Transformation

Consider the small population of scores listed and graphed in a histogram in Figure 3.9. If we were to **transform** all of these raw scores into **z-scores,** what effect would this transformation have on the distribution?

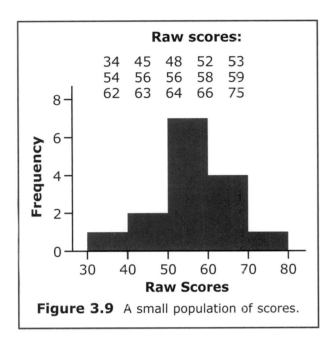

Figure 3.9 A small population of scores.

Interactive
Page **75**

The histogram on the right side of Figure 3.10 shows the transformed distribution, which has three notable properties:

1. The **mean** of the population of z-scores is 0.0.

2. The **standard deviation** of the z-score population is 1.0.

3. The shape of the z-score **distribution** is identical to the shape of the raw score distribution.

In fact, these properties will hold for *any* distribution of z-scores, as long as the entire population of raw scores is transformed and included in the distribution. Here's why:

1. Subtracting a constant from every score in a distribution shifts the distribution up or down by the constant amount. The **z-score** formula shifts every score down by an amount equal to the mean of the raw score distribution. Thus the mean itself becomes 0 in the z-score distribution.

2. Dividing every score in a distribution by a constant divides the standard deviation by the constant amount. The z-score formula divides every score by the standard deviation of the raw scores, σ. Thus the standard deviation of the z-scores will be $\sigma / \sigma = 1.0$.

Characteristics of
z-transformed
distributions

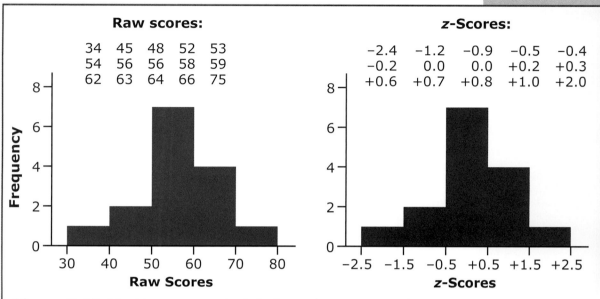

Figure 3.10 The histogram on the left shows the raw scores from Figure 3.9. The histogram on the right shows the same dataset transformed into z-scores.

3. Subtracting and dividing every score in a distribution by constant amounts will do nothing to the relative locations of each of the scores in the distribution. For example, every score that is above the mean in the original distribution will remain above the mean in the z-score distribution. Similarly, every score that is exactly two standard deviations below the mean in the original distribution will remain exactly two standard deviations below the mean in the z-score distribution. Thus the shape of the distribution will be preserved across the transformation.

It should be noted that these properties do not hold for all **transformations.** For example, taking the square root of every score in a population yields a new distribution with a reduced mean, a reduced standard deviation, and an altered shape. The z-score formula is specially constructed to convert the mean and standard deviation of a distribution to 0 and 1, respectively, while preserving the distribution's shape.

3.6.2 From z-Scores to Raw Scores

Suppose your friend Ron later tells you that sales have picked up since his dismal 65-car week. In fact, in the past week the **z-score** for his sales was a hefty +1.3. Remembering that his **mean** sales per week is 80 cars, with a **standard deviation** of 10, can you calculate how many cars he sold in the last week?

Interactive
Page **77**

Transforming from **z-scores** back to raw scores is a matter of simple algebra:

$$z = \frac{X - \mu}{\sigma} \quad (z)(\sigma) = X - \mu \quad X = \mu + (z)(\sigma)$$

For Ron's sales this past week (a z-score of 1.3, with $\mu = 80$ and $\sigma = 10$), we have:

$$X = \mu + (z)(\sigma) = 80 + (1.3)(10) = 93$$

The z-score formula and its inverse are pretty easy to use by hand, but calculating lots of z-scores can get pretty tedious. This is where the **z-Scores** Calculation Tool comes in handy. To compute a z-score, enter the raw score, the **population mean,** and the **population standard deviation** in the appropriate boxes of the Tool and click the "calculate" button.

To go the other way, enter the mean, standard deviation, and z-score, leaving the box for the raw score blank. Click the calculate button and the Tool will compute the raw score that would produce the given z-score. (Actually, the Tool will fill in any one of the four terms in the z-score formula if its text box is left blank and the other three terms are filled in.)

The figure on the right side of this Calculation Tool shows the location of the z-score in the normal distribution. We'll discuss this distribution in **Chapter 4.**

As with means and standard deviations, make sure you're comfortable computing z-scores by hand before you start relying exclusively on the Calculation Tool. **Exercise 3.12, Exercise 3.13,** and **Exercise 3.14** will give you copious amounts of practice.

Formula:
raw scores from
z-scores.

Calculation Tool:
z-Scores

Interactive
Page **78**

3.7 Chapter Summary/Review

This Chapter Summary/Review is interactive on the
Interactive Statistics **website and CD.** Also be sure to go through all the **Review Exercises** for this chapter.

- The most common measure of the **central tendency** of a **distribution** of scores is the **mean,** calculated by summing the scores and dividing by the number of scores. The mean of a population is symbolized by μ and the mean of a sample by $\bar{X}$. The formulas are (**Section 3.2**):

$$\mu = \frac{\sum X}{N} \quad \bar{X} = \frac{\sum X}{n}$$

- The most common measure of the variability of a distribution of scores is the **standard deviation,** symbolized by σ when the distribution is considered to be an entire **population** and s when the distribution is considered to be a **sample.** To compute either standard deviation, we subtract the mean of the distribution from each score in the distribution, then square each of these deviation scores. The sum of the squared **deviations** is called the **sum of squares** (*SS*). For populations, we divide *SS* by the number of scores *N* to get a statistic called the **variance** (σ^2); σ is the square root of σ^2. The formulas are (**Section 3.3**):

$$SS = \Sigma(X - \mu)^2 \qquad \sigma^2 = SS \,/\, N \qquad \sigma = \sqrt{\frac{SS}{N}}$$

- When dealing with a sample of a larger population, the variance and standard deviation formulas include a correction for the fact that samples are generally less variable than the populations from which they are drawn. The correction is to divide by *n* − 1 instead of *N*. In part because of this correction, we say that the sample standard deviation has *n* − 1 degrees of freedom (**df**). The formulas are (**Section 3.3.2**):

$$s^2 = SS \,/\, (n-1) = \frac{\Sigma(X - \bar{X})^2}{n-1} \qquad s = \sqrt{\frac{SS}{n-1}} = \sqrt{\frac{\Sigma(X - \bar{X})^2}{n-1}}$$

- Two alternative statistics for measuring central tendency and **variability** are the **median** and **interquartile range** (IQR). The median (also called the 50th **percentile**) is the middle score of a dataset when there is an odd number of scores, or the mean of the two middle scores when there is an even number of scores (**Section 3.4.1**). The IQR is the 75th percentile (the median of the scores above the 50th percentile) minus the 25th percentile (the median of the scores below the 50th percentile) (**Section 3.4.2**).

- Compared to the mean and standard deviation, the median and IQR are influenced much less by **outliers**—datapoints on the extreme right or extreme left of a distribution. For this reason, the median and IQR are called resistant measures (**Section 3.4**) and are used as alternatives to the mean and standard deviation to describe distributions with outliers. Another alternative for dealing with outliers is to exclude the extreme scores and report **trimmed means** and standard deviations (**Section 3.4.3**).

- The mean, median, standard deviation, and IQR are all used to describe a single variable. When we wish to describe the relationship between two variables, we use a statistic called the **correlation.** This complex and useful statistic will be described in **Chapter 11** (**Section 3.5**).

- To **standardize** a score or statistic, we **transform** it in a way that puts the number into a more meaningful context (**Section 3.6**). The **z-score** is a standardized value produced by taking a raw score, subtracting from it the mean of the population the score came from, then dividing by the population's standard deviation. The formula is (**Section 3.6**):

$$z = \frac{X - \mu}{\sigma}$$

- A cursory examination of an individual's z-score reveals two important pieces of information about the individual: the sign of the z-score tells us whether the score is above (+) or below (−) the mean of the population, and the absolute value of the z-score tells us how far away the score is from the mean, in standard deviation units (**Section 3.6**).

- Transforming an entire population of raw scores into z-scores preserves the shape of the **distribution** of scores, but shifts the center of the distribution to 0 and transforms the standard deviation of the distribution to 1.0 (**Section 3.6.1**). z-Scores can be transformed into raw scores by rearranging the terms in the z-score formula (**Section 3.6.2**):

$$z = (X - \mu) / \sigma$$
$$(z)(\sigma) = X - \mu$$
$$X = \mu + (z)(\sigma)$$

Chapter 4
Preparing To Test Hypotheses

4.1 Probability

Take a coin, flip it up in the air, catch it, and hold it on the back of one hand, covered by the other hand. Is the head side of the coin facing up or down?

The answer, of course, is that until you see the coin, you have no idea which side is up. It could be heads, but equally likely, it could be tails. A flip of a coin is a prototypical example of a **random**

The outcome of a single random event is unpredictable.

event: an event whose outcome is uncertain. Other random events include the gender of an unborn child, tomorrow's weather, and whether or not a batter will reach base safely in a baseball game.

Now consider a different but related question. Suppose you flipped your coin 100 times. How many of these 100 flips would come up heads? Well, we still can't know for sure unless we actually perform the flips. But we can make a pretty good guess that we would get about 50 heads and 50 tails. In fact, an English statistician named Pearson, who apparently had far too much time on his hands, once flipped a coin 24000 times. Of these flips, 12,012 came up heads and 11,988 tails, very close to the 50-50 ratio we would predict.

These observations lead us to an important point about random events. Although the outcome of a single such event is uncertain, a regular pattern of outcomes will usually emerge if we observe the event occurring over and over again. Thus in the long run, random events aren't really so random after all.

> Over many repetitions, the outcomes of random events are regular and predictable.

In the case of coin flips, this regularity doesn't buy us much. Since the likelihoods of heads and tails are exactly the same, Pearson's effort doesn't help us predict the outcome of the next coin flip.

But for other random phenomena, an assessment of the **probability** of a particular outcome—the proportion of times that outcome will occur over many repetitions of an event—can be quite valuable. Should you go on a picnic this weekend? Consult your local meteorologist first. If the chances of rain are 90%, you're probably better off staying home. Should a manager send up a particular pinch hitter in a crucial situation in a baseball game? Consider his batting average, which is an estimate of the probability that he will get a base hit.

> Glossary Term: probability

In this course, we will use probability theory to help us evaluate **statistics** that represent the outcomes of experiments. More specifically (but still pretty vaguely), we will formulate a **hypothesis** (a formal guess) about an experimental phenomenon, then evaluate the probability that a given statistic is consistent with this hypothesis.

The logic of this hypothesis testing procedure is complicated, and we'll wait a couple more chapters before spelling it out. This chapter will provide some of the tools we need to test hypotheses, including the rules we use to assess the probabilities of random events.

4.1.1 Misunderstandings of Probability

Interactive Page 81

Many aspects of **probability** theory are fairly intuitive. For example, we have little difficulty interpreting probability estimates given by weather forecasters, and if you read that the chances of winning your state lottery are 1 in 10,000,000, you will probably recognize that these are extremely long odds.

But some aspects of random phenomena are regularly misunderstood by most people. For example, suppose you were to flip a coin 10 times, recording each flip as a head (H) or tail (T). Which of the following series of results would you be more surprised to see?

H T H H T T H T H T

H T T T H T H H T T

Most people will judge the second series to be less likely, because it includes a run of three tails in a row. Since we know the probabilities of heads and tails are equal, we expect these two outcomes to alternate back and forth—that is, we expect that in every two coin flips, one will be heads and one will be tails.

However, this expectation is wrong, as you can easily prove to yourself by actually performing the experiment a few times instead of just thinking about it. If you flip a coin 10 times, there is better than an 8 in 10 chance of getting a run of three or more heads or tails in a row! So more often than not, you will in fact see such a run.

The confusion about chance processes demonstrated in this example also applies to many other random phenomena. For example, a gambler who wins on three consecutive spins of the roulette wheel may feel he's on a "hot streak" and bet more than he can afford on the next spin. Or, if he's lost several times in a row, he may feel he is "due" for a win and again make an overly large wager. Neither of these beliefs is correct: The likelihood of any number on the wheel coming up is exactly the same regardless of any previous outcomes. In other words, the roulette wheel has no memory, even though gamblers often act as if it does.

As these examples demonstrate, probability is a fascinating subject, even if you're not mathematically inclined. Another example of a probability problem with a nonintuitive solution is provided in **The Monty Hall Problem** Activity.

But for the most part, in this course we'll stick to just the simpler aspects of probability theory necessary to pursue our goal of testing statistical hypotheses. The next section will start us off with some basic probability terminology and rules.

Interactive
Page 82

4.1.2 Probability Terminology and Rules

The portion of **probability** theory we will need to employ is pretty intuitive. In fact, you probably already know the rules we'll need, even though you may not know that you know them. Mostly, we just need to introduce some formal terminology and then show how probability rules apply in certain situations.

First, let's formally define an **event** as a random phenomenon with a set of possible outcomes. For example, a coin flip is an event with two possible outcomes (heads or tails), a single card dealt from a

Outcomes of random events are "streakier" than people expect them to be.

Activity: The Monty Hall Problem

Glossary Term: event

standard deck is an event with 52 possible outcomes (ace of spades, two of hearts, etc.), and a roll of a six-sided die is an event with six possible outcomes.

We will call the set of all possible outcomes of an event the **sample space** of the event, and denote the sample space with an uppercase *S*. For the dice example, *S* = {1, 2, 3, 4, 5, 6} (Figure 4.1). For a coin flip, *S* = {heads, tails}.

Figure 4.1 The six possible outcomes of a roll of a standard die.

The probability of an outcome is an estimate of the proportion of times we would get that outcome if the event were repeated an infinite number of times. That's a long-winded way of saying that the probability is the likelihood of the event occurring. We denote a probability estimate with a lowercase *p*. If we denote an outcome with a letter such as *A*, we can shorten "the probability that outcome *A* will occur in a single instance of an event" to *p*(A). (A probability estimate is usually called a probability value, or *p*-value for short.)

There are a number of ways to estimate probability values. One is to observe the phenomenon many times and record the outcomes, as in the case of baseball batting averages. Another is to just make a subjective guess, based on whatever information you have available (if it's winter in Seattle and there are clouds in the sky, you might estimate the probability of rain at 90%). Most elegantly, we can sometimes use established probability rules, along with assumptions about the event in question, to make mathematical estimates of probabilities, as illustrated below.

Methods for estimating probability values

Interactive Page 83

Rule number one in our abbreviated version of **probability** theory is that *something* has to happen. Formally, we state this rule as $p(S) = 1$: There is a 100% chance that we'll get one of the outcomes in the sample space *S* of an event.

Probability rule 1: $p(S) = 1$

Although it might seem that this rule is so simple that it must be useless, it sometimes comes in quite handy. For example, all six outcomes of a roll of a die are equally likely. Therefore, the probability of any one particular number coming up is 1, the probability that *some* number will come up, divided by 6, the number of possible outcomes.

Now, here's a slightly more difficult question. We've just established that the probability of rolling a 2 with a fair die, *p*(2), is 1/6. Likewise, *p*(3) is also 1/6. Given these two probability estimates, what's the probability that a roll of the die will come up with either a 2 *or* a 3?

Interactive
Page 84

Given that $p(2) = 1/6$ and $p(3) = 1/6$, the **probability** of rolling either a 2 or a 3 with a die is $1/6 + 1/6 = 2/6$ (or 1/3). This example illustrates probability rule number two, which we will call the addition rule:

Probability rule 2: the addition rule

If outcomes A and B are disjoint, $p(A \text{ or } B) = p(A) + p(B)$

The qualifier for this rule needs to be explained. **Disjoint** outcomes are outcomes that can never occur at the same time. In our example the outcomes are disjoint because it's not possible for both the 2 and the 3 side of the die to be up at the same time. Therefore, the probability of one outcome or the other happening is the sum of the individual probabilities of each outcome:

Glossary Term: disjoint outcomes

$p(\text{roll 2 or roll 3}) = p(\text{roll 2}) + p(\text{roll 3}) = 1/6 + 1/6 = 1/3$

However, it's not difficult to think of outcomes of events that are not disjoint. For example, the probability of being dealt a spade from a deck of cards is 1/4, and the probability of being dealt a face card (jack, queen, or king) is 3/13. But the probability of being dealt a spade or a face card is not $1/4 + 3/13$. The card you're dealt could be consistent with both outcomes (for example, you might get the queen of spades), so the outcomes are not disjoint and the addition rule cannot be used.

Although we can't use the addition rule directly here, it is certainly possible to determine the exact probability of being dealt a spade or a face card. In fact, you can probably do it on your own if you take a little time.

Interactive
Page 85

To determine the **probability** that a card dealt from a standard deck is a spade or a face card, we can consider each of the 52 cards in the deck as a separate outcome, each with an equal probability of 1/52 of occurring. These outcomes are all **disjoint** (the dealt card cannot be both the ace of spades and the two of hearts), so we can add 1/52 for each of the cards that meets our criteria (see Figure 4.2):

$$p(\text{spade or face card}) = p(\text{ace of spades}) + p(2 \text{ of spades}) + p(3 \text{ of spades}) + \ldots + p(\text{jack of diamonds}) + p(\text{queen of diamonds}) + \ldots + p(\text{king of hearts}) = 22/52$$

Note that this probability calculation did not involve any new rules, just a somewhat creative application of the addition rule. The example illustrates the fact that if we can express the outcome we're interested in as a collection of simpler, disjoint outcomes, we can calculate the probability of the outcome of interest by adding up the probabilities of the "sub-outcomes."

There are other probability rules that we won't go into here, because they're not necessary for calculating and interpreting the inferential statistics we'll encounter in this course. But see this **Box**

Box: The Multiplication Rule for Probabilities

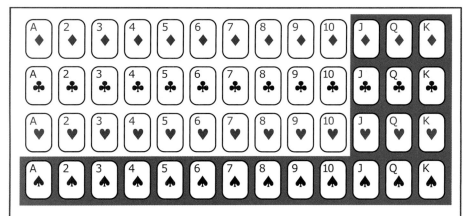

Figure 4.2 The 52 cards in a standard deck, with the cards satisfying the outcome (spade or face card) highlighted in blue. To assess the probability of this outcome, we simply count these cards (there are 22 of them) and divide by the total number of possibilities (52).

for a brief discussion of the multiplication rule, which is used to calculate probabilities of more complex (and sometimes moneymaking) events such as rolling a combined total of seven with two dice.

Box: The Multiplication Rule for Probabilities

We've seen in the text that the chances of rolling a three on a six-sided die are 1/6, or about .17. We've also seen that the chances of rolling either a three or a four are 1/6 + 1/6 = 2/6, or .33. Now let's ask one more question: If you roll two dice, what are the chances that you'll roll a three with die A and a four with die B?

Your first instinct might be to use the addition rule again and postulate that this probability is also 1/6 + 1/6 = 2/6. But you should suppress this instinct, because we're not asking an "or" question here; we're asking about the probability of one outcome *and* another outcome occurring at the same time. For such probability questions, we use the multiplication rule:

If outcomes A and B are independent, $p(A \text{ and } B) = p(A) \times p(B)$

As with the addition rule, there is an important condition for the multiplication rule: the two outcomes in question must be independent. We will have more to say about independence in later chapters, but briefly, two outcomes are independent if the probability of one does not depend on the probability of the other. Two dice rolls are independent: the outcome of the first roll will have no effect on the outcome of the second. Thus the chances of a three on die A and a four on die B are 1/6 × 1/6 = 1/36, or about .028.

Now, let's see how the multiplication and addition rules can be combined to calculate complex probabilities. Say you're playing a game of chance and will win if the combined total of your next roll of die A and die B is exactly seven. How can the probability of this outcome be computed?

Box: The Multiplication Rule for Probabilities *(continued)*

First, we must enumerate all the outcomes that we're interested in—all the combinations of die A and die B that add to seven. Here they are:

A	B	Total
1	6	7
2	5	7
3	4	7
4	3	7
5	2	7
6	1	7

Now, we can use the multiplication rule to find the probability of any one of these combinations: they are all $1/6 \times 1/6 = 1/36$. To find the probability of any one of the six outcomes occurring, we can use the addition rule, because they are all disjoint outcomes (the two dice cannot come up 1–6 and 2–5 at the same time). Thus the final probability is $1/36 + 1/36 + 1/36 + 1/36 + 1/36 + 1/36 = 6/36 = 1/6$. You have about a .17 chance of winning if you need to roll a seven.

It is important to avoid overextending the multiplication rule. The rule can only be used when the outcomes in question are independent. For an example of nonindependence, we can consider card deals. If you deal two cards from a shuffled deck (call the cards A and B), what's the probability that both will be from the suit of spades? Since 13/52 of the cards are spades, the probability of A being a spade is .25.

But the chances of B being a spade depend on the suit of A. If A was a spade, then only 12 of the 51 cards left are spades, so p(B is a spade) is .235. On the other hand, if A was a heart, diamond, or club, then 13/51 of the remaining cards are spades, so p(B is a spade) = .255. The outcome of B is affected by the outcome of A, so A and B are not independent, and the multiplication rule cannot be used.

We will not need to make use of the multiplication rule very much in this course, which is why our discussion of it is limited to this Box.

Here are a few more things you should know and keep in mind about probabilities:

- Probabilities can be expressed as fractions, decimals, or percentages. In cases such as dice rolls and card deals, it is convenient to use fractions, since the denominator specifies the total number of possible outcomes for the event. But when the number of possible outcomes is more difficult to define, it's better to use decimals or percentages. In this course, we will usually use decimals—starting in a few chapters, you'll see the expression "$p <$ .05," meaning "the probability of this outcome is less than .05," many, many times. If you're uncomfortable working with decimal places, you can always multiply a decimal by 100 to get a percent, and/or convert the decimal to a fraction (.05 translates to 5% or 1/20).

- The probability of any one outcome for an event must fall within the range .0 to 1.0. This is a corollary to our first rule: If the probability of the set of all outcomes is 1.0, the probability of a single outcome has to be between 0 (there's no chance that it will happen) and 1 (it's guaranteed to happen).

Probabilities must be between .0 and 1.0.

- Similarly, if you add up all the individual probablities of disjoint outcomes for an event, the sum had better come out to 100% (or something less than 100% if it's admitted that not all possible outcomes are specified). Beware the soccer commentator who bravely predicts a 50% chance that Brazil will win the World Cup, a 40% chance that Germany will win, and a 30% chance that France will win.

The probabilities of all disjoint outcomes for an event must add to 1.0.

Exercises 4.1–4.5 will test your understanding of the basic probability rules presented in this section. If possible, you should run through these brief exercises now before going on, since these rules are essential for understanding much of the rest of the chapter.

Interactive Page 86

4.1.3 Probability and Distributions

Up to now, we have been considering **probabilities** of events such as card deals and dice rolls, in which the possible outcomes are easy to enumerate and predict. Next we'll look at how probability theory applies to the considerably messier world of behavioral science experiment results. Consider the following scenario:

> Students at Professor Edwards' university are asked to rate their courses at the conclusion of each semester. Instructors see their course ratings soon after the semester ends, but the ratings are kept anonymous. One term, Professor Edwards' 20 students gave the following ratings of his course (the scale for the ratings runs from 1 to 9):
>
> 4 9 2 7 5 3 7 3 5 4 9 5 1 5 2 7 6 9 3 5
>
> The following semester, one of his former students comes to visit him during his office hours. He wonders whether or not this person liked his course. Say we define a positive rating as a 6 or higher. What is the probability that a randomly chosen individual from the population of Professor Edwards' 20 students gave such a rating?

Although we could work the answer out using the raw scores given above, probability questions such as this turn out to be easier to answer when we plot the **distribution** of scores in a **histogram**. Figure 4.3 shows such a histogram.

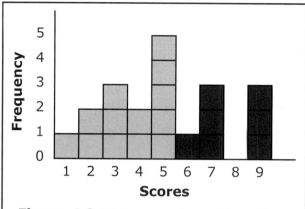

Figure 4.3 Histogram of a small population of scores, with the individuals whose scores are ≥ 6 highlighted. This figure is interactive, with four parts, on the *Interactive Statistics* website and CD.

We can evaluate p(6 or higher) in the same way we evaluated p(spade or face card) for the card deal on the previous page. All we have to do is count the number of scores that meet the **criterion** (i.e., the outcome) we're interested in, a simple task given the histogram above. In this case, the relevant criterion is $\{X \geq 6\}$.

Interactive
Page 87

When we represent a **population** of scores with a histogram, **probability** can be defined graphically as the proportion of the area of the histogram corresponding to the condition in question. This allows us to answer probability questions by measuring the appropriate area. For example, 15% of the area of the histogram falls to the left of 3, so the probability of randomly drawing an $X < 3$ is .15. (Do **Exercise 4.6** a few times to get more practice answering these basic probability questions.)

This graphical definition of probability is useful in the case of finite populations like Professor Edwards' class, but it becomes essential when we consider theoretically infinite populations, as we'll see in the next section.

Interactive
Page 88

4.2 Normal Distributions

What do you think the following sets of numbers might have in common?

- Heights of male recruits entering the U.S. Army

- IQ scores for fifth graders in San Francisco public schools

Probability can be defined graphically as a proportion of the area of a histogram.

- Driving times for road trips from Boston, MA to Daytona Beach, FL

The answer is that if one were to construct frequency distributions of each of these sets of numbers, the shapes of all three distributions would be almost identical. The means and standard deviations would vary, but each would take on the "bell-shaped" form shown in Figure 4.4. This shape is called a **normal** curve. Open **The Normal Distribution** Activity to learn more about it.

One prominent feature of the normal curve is that it is perfectly **symmetrical.** This means that exactly half of the curve (and half of the distribution) falls to the left of the **mean** of any distribution, with the other half falling to the right of the mean.

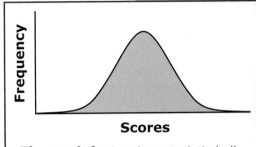

Figure 4.4 The characteristic bell shape of the normal curve.

Now, consider the following question. Say you have a friend at Harvard University who is planning to spend his spring break in Daytona Beach. He knows that the average driving time for this trip is 10 hours, and wants to know the probability that it will take him more than 10 hours to make it.

Solving this probability question should be a snap if you put together the fact that normal distributions are symmetrical with the principle introduced in the previous section, that probabilities can be equated with proportions of the area of a histogram. What is the likelihood that your friend will require more than 10 hours to get to the beach?

Interactive Page 89

On the previous page, we established the following facts. First, driving times for road trips from Boston to Daytona Beach are normally distributed, with a mean of 10 hours. And second, the distribution of driving times, like all normal distributions, is perfectly symmetrical.

Putting these two facts together (see Figure 4.5), we can determine that 50% of the area of the histogram plotting the distribution of Boston to Daytona Beach driving times falls to the right of 10 hours. Therefore, the probability that any one randomly selected driver will take more than 10 hours is $p = .50$.

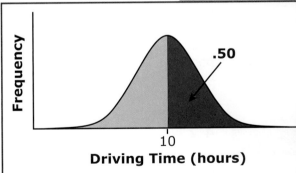

Figure 4.5 The probability of randomly selecting a score greater than the mean is exactly $p = .50$ for any normal distribution (here, the mean is 10 hours).

The normal curve is mathematically defined* by the following formula:

$$\text{Frequency} = \frac{1}{\sigma\sqrt{2\pi}}\, e^{-(X-\mu)^2/2\sigma^2}$$

You'll be happy to know that you'll probably never have to see this formula again.

But the fact that the normal curve is so precisely defined means that we can precisely determine the area under this curve (that is, the proportion of the area of a normal distribution's histogram) for any range of values. To prevent our discussion from getting too abstract, let's use a concrete example to illustrate how and why we would do this:

> Barney, deputy to the chief of a small town's police department, has a strong belief in the value of tall police officers for intimidating criminals. He approaches his boss, Andy, about instituting a new policy to only hire men over 6 feet 2 inches tall as officers. But Andy, who has heard too many of Barney's hare-brained schemes, responds that, in addition to being sexist and probably illegal, he's afraid this policy would limit the potential officer pool to a very small proportion of the population. Can Andy determine exactly how likely it is that potential male applicants would meet the 6-foot-2 requirement?

Suppose that the heights of all adult men in the world are normally distributed, with **mean** 69 inches (that's 5 feet 9 inches), and **standard deviation** 5 inches. If you inspect the formula for the normal curve, you will see that the only variables in the formula are μ and σ.

So, given the mean, the standard deviation, and the normal curve formula, we can construct a histogram of the heights of all adult men. We can then go on to highlight the portion of the histogram that corresponds to Barney's height requirement ($X > 74$ inches), as shown in Figure 4.6.

All we need to do now to address Andy's concern is figure out exactly what proportion of the area of this histogram is highlighted. One

---

*In theory, the shape of a normal distributed population's density curve is precisely defined by the ugly mathematical formula given above. However, frequency distributions for sets of real numbers almost never exactly match this ideal—most naturally occurring populations are only *approximately* normally distributed. In upcoming sections and chapters, the remarkable properties of normal distributions will often play crucial roles in statistical procedures. Unless otherwise noted, you can assume that these properties hold up even for approximately normal distributions.

The normal curve is precisely defined by a mathematical formula.

The only variables in the normal curve formula are the mean and standard deviation of the distribution.

way to do this would be to superimpose a grid on the histogram and count the number of boxes covering the highlighted portion. Roll your mouse over the words SHOW GRID in the interactive Figure 4.6 on the *Interactive Statistics* website and CD to reveal such a grid, and you should find that 14 of the 76 grid squares are highlighted. Thus we could estimate the probability of randomly selecting an adult man 6 feet 2 or taller at 14 / 76 = .18.

But because of the precisely defined normal curve, we can do better than just estimate; we can calculate the value exactly.

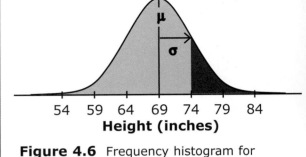

Figure 4.6 Frequency histogram for heights of adult men. Members of the population with heights greater than 74 inches are highlighted. This figure is interactive on the *Interactive Statistics* website and CD.

Interactive Page 90

Figure 4.7a repeats the **histogram** for heights of adult men, with the portion of men who fit the **criterion** {*X* > 74 inches} highlighted. We know that the **probability** that a prospective police officer would meet this criterion is equal to the proportion of the area under the curve that is highlighted.

If you had the wherewithal to do so, you could calculate this area using the normal curve formula directly. Fortunately for the less mathematically inclined, we have a Calculation Tool to do this for us. There's a small catch however: To use the Tool, you first have to convert the raw scores in the population of interest into **z-scores,** as shown in Figure 4.7b.

You should recall that the formula for z-scores is:

$$z = \frac{X - \mu}{\sigma}$$

For a height of *X* = 74 inches,
z = (74 − 69) / 5 = 1.0.

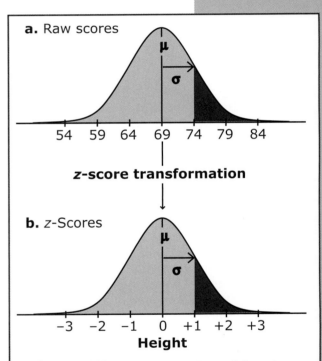

a. Raw scores

z-score transformation

b. z-Scores

Height

Figure 4.7 Histograms of raw (a) and z-transformed (b) heights of adult men. Members of the population with heights greater than 74 inches are highlighted in both histograms.

Once you have this z-score, you can use the **Normal Distribution Area** Calculation Tool to look up the area you're interested in. Enter z (1.0) in the left-hand text box of the Tool and click the radio button for the portion of the distribution you're interested in. Here, we

Calculation Tool: Normal Distribution Area

want to know the proportion of the area beyond z in one "**tail**" of the distribution, so click the center radio button (see **Section 2.4** if you've forgotten what a distribution's tail is). The area we're looking for, .1587, should then appear in the text box labeled "Area under Curve." The area will also be illustrated in the figure on the right side of the Tool. Figure 4.8 shows what the Calculation Tool should look like once the area has been calculated. See **Exercise 4.7, Exercise 4.8,** and **Exercise 4.9** for practice with the basic process of determining probabilities using z-scores and the normal distribution.

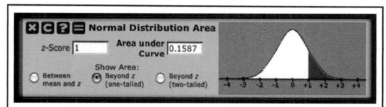

Figure 4.8 Screenshot of the Normal Distribution Area Calculation Tool, illustrating the calculation of the area beyond $z = 1.0$ in a normal distribution.

If you don't have access to this Calculation Tool (for example, when you're sitting for a test), you may be asked to look in a table to find the p-value associated with a z-score. The **Box** on statistical tables explains how to do this.

Box: Statistical Tables

Box: Statistical Tables

In the days before computers, the only realistic way to find the p-value associated with a given z-score was to use a table. Figure 1 below shows part of the Normal Distribution Table used for this purpose (this full table and others are found in the Appendix to this book and on the *Interactive Statistics* website and CD). Your professor may also ask you to use this table to look up p's when you are sitting for an exam and can't access the *Interactive Statistics* Calculation Tool or another statistics program.

z	.00	.01	.02	.03	.04	.05	.06	.07	.08	.09
1.6	.0548	.0537	.0526	.0516	.0505	.0495	.0485	.0475	.0465	.0455
1.7	.0446	.0436	.0427	.0418	.0409	.0401	.0392	.0384	.0375	.0367
1.8	.0359	.0351	.0344	.0336	.0329	.0322	.0314	.0307	.0301	.0294
1.9	.0287	.0281	.0274	.0268	.0262	.0256	.0250	.0244	.0239	.0233
2.0	.0228	.0222	.0217	.0212	.0207	.0202	.0197	.0192	.0188	.0183

Figure 1 Excerpt from the Standard Normal Table. This figure is interactive on the *Interactive Statistics* website and CD.

Box: Statistical Tables *(continued)*

To use this table, you must first find the whole number and first decimal place of your z-score in the column of numbers on the far left. If, for example, we were interested in a z-score of 1.76, we would find the row for 1.7, shaded blue in the figure.

Next, we read across the table to find the column for the second decimal place of our z-score (.06 in our example, the column for which is shaded red in Figure 1).

The number at the intersection of the row and column you've just located (.0392, shaded green in the figure) provides the information you need to calculate the p-value you're looking for. More specifically, this number shows the area under the normal distribution and to the right of this z-score, as illustrated in Figure 2. This area corresponds to the probability of randomly drawing a z-score this large or larger from a normal distribution.

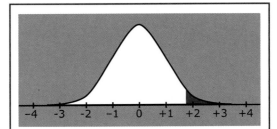

Figure 2 Area under the normal distribution for z-scores greater than or equal to 1.76.

The probability of randomly drawing a z-score as small as −1.76 or smaller is exactly the same: .0392. If you need the "two-tailed" probability, multiply the p-value from the table by two, thus giving you the likelihood of randomly drawing a z-score ≤ −1.76 OR ≥ +1.76.

Our Normal Distribution Table only lists z-scores up to z-scores of 3.5 or so. The reason it doesn't go beyond this is that the likelihood of randomly drawing a z-score of 3.7 or above is about .0001. That's already pretty unlikely, so there's rarely any need to be more precise than this. If you need to state the probability value for a z-score greater than 3.69, just say "p < .001."

Probability values associated with t, F, and χ^2 statistics (covered in Chapters 6, 10, and 12, respectively) can also be looked up in a table. (If you haven't made it to these chapters yet, you can stop reading now and come back to this box when you encounter these statistics later.) For these statistics, however, the situation is complicated by the fact that the exact shape of the t, F, and χ^2 distributions depends on the degrees of freedom for the test statistic in which we are interested.

This complication forces us to abridge and reformat the tables for these statistics. Figure 3 shows part of the t Distribution Table. Each row in the table represents a df value and each column represents a p-value. Suppose we calculate a t statistic of 2.19 with 15 df. To find the one-tailed p-value, go to the row of the table for 15 df (shaded blue in the figure), then read across the table until you find the two t values that straddle your calculated t. These values are 2.131 and 2.249, as highlighted in green in the figure.

Now move up to the column headings of the table and find the p-values for the columns you just identified (for our example, .025 and .02, circled in Figure 3). The one-tailed p-value for your test statistic is somewhere between these two values. Thus for $t(15) = 2.19$, you would report that $.02 < p < .025$. Or, more succinctly, you might just say that $p < .025$. If you need a two-tailed p-value for a t test, multiply the tabled p-value by two (in our example, the two-tailed p is between .04 and .05).

Box: Statistical Tables *(continued)*

df	.25	.20	.15	.10	.05	.025	.02	.01	.005
14	0.692	0.868	1.076	1.345	1.761	2.145	2.264	2.625	2.977
15	0.691	0.866	1.074	1.341	1.753	2.131	2.249	2.602	2.947
16	0.690	0.865	1.071	1.337	1.746	2.120	2.235	2.583	2.921
df	.25	.20	.15	.10	.05	.025	.02	.01	.005
17	0.689	0.863	1.069	1.333	1.740	2.110	2.224	2.567	2.898
18	0.688	0.862	1.067	1.330	1.734	2.101	2.214	2.552	2.878

df	.25	.20	.15	.10	.05	.025	.02	.01	.005
14	0.692	0.868	1.076	1.345	1.761	2.145	2.264	2.625	2.977
15	0.691	0.866	1.074	1.341	1.753	2.131	2.249	2.602	2.947
16	0.690	0.865	1.071	1.337	1.746	2.120	2.235	2.583	2.921
df	.25	.20	.15	.10	.05	.025	.02	.01	.005
17	0.689	0.863	1.069	1.333	1.740	2.110	2.224	2.567	2.898
18	0.688	0.862	1.067	1.330	1.734	2.101	2.214	2.552	2.878

Figure 3 Excerpt from the *t* Distribution Table. This figure is interactive on the *Interactive Statistics* website and CD.

The Chi-Square Distribution Table is used in exactly the same way as the *t* Table, since each χ^2 statistic is also associated with one *df* value. An *F* statistic, however, has two *df* values, one for the numerator and one for the denominator of the *F* ratio. Figure 4 shows part of the *F* table and illustrates how to use it to find the *p* value for $F = 2.53$ with 3 *df* in the numerator and 20 *df* in the denominator.

df_E	p	1	2	3	4	5	6	7
19	.10	2.99	2.61	2.40	2.27	2.18	2.11	2.06
	.05	4.38	3.52	3.13	2.90	2.74	2.63	2.54
	.01	8.18	5.93	5.01	4.50	4.17	3.94	3.77
	.001	15.08	10.16	8.28	7.27	6.62	6.18	5.85
20	.10	2.97	2.59	2.38	2.25	2.16	2.09	2.04
	.05	4.35	3.49	3.10	2.87	2.71	2.60	2.51
	.01	8.10	5.85	4.94	4.43	4.10	3.87	3.70
	.001	14.82	9.95	8.10	7.10	6.46	6.02	5.69

(column header: df_B)

Figure 4 Excerpt from the *F* Distribution Table. This figure is interactive on the *Interactive Statistics* website and CD.

First find the intersection of the set of rows corresponding to the denominator *df* (20) and the column corresponding to the numerator *df* (3). Then scan the set of *F* values in this block to find the values that straddle your test statistic. In this case, the bracketing values are 2.38 and 3.10, which according to their row headings correspond to *p*-values of .10 and .05. Therefore, the *p*-value for $F(3, 20) = 2.53$ is less than .10 but greater than .05.

4.2.1 Area Calculations Using the Addition Rule

Now let's consider a more complex **probability** problem:

> The mean annual rainfall amount for the Amherst, Massachusetts, area is 36 inches, with a standard

deviation of 4 inches. Annual rainfall amounts are normally distributed. Farm crops suffer if there is either too much or too little rainfall in a year, and local farmers would like to know the likelihood of such years. It is determined that annual rainfall amounts greater than 45 inches or less than 30 inches lead to significant damage to local crops. What is the probability that any one year will produce an amount of rainfall outside this range?

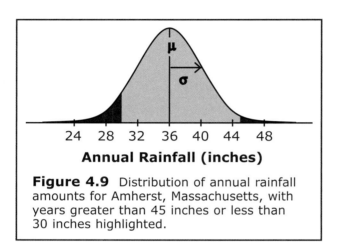

Annual Rainfall (inches)

Figure 4.9 Distribution of annual rainfall amounts for Amherst, Massachusetts, with years greater than 45 inches or less than 30 inches highlighted.

Figure 4.9 shows the distribution of annual rainfall amounts, with the amounts that are of concern to Amherst farmers highlighted. In the previous section we learned how to calculate the area of a **normal** distribution in one **tail**, but here we need to know the area of a more complex section of the **distribution**, including bits in both tails.

To calculate this area, we first use the **Normal Distribution Area** Calculation Tool just as before to calculate the areas in each of the two tails separately. Then, invoking the "or" rule for combining probability estimates of **disjoint** outcomes, we add these two areas together to find the probability of a year with greater than 45 inches *or* less than 30 inches of rainfall. (These two outcomes are disjoint because it's impossible to have both less than 30 inches *and* greater than 45 inches of rainfall in the same year.)

For the area of interest in the right-hand tail of the distribution, the area (and *p*-value) is calculated as follows:

1. Determine the *z*-score for 45 inches of rainfall: $z = 45 - 36 / 4 = 2.25$. (You can either calculate this value by hand or use the **z-Scores** Calculation Tool.)

2. Enter this *z*-score into the **Normal Distribution Area** Calculation Tool and click the "Beyond *z* (one-tailed)" radio button (or click

Calculating the probability of an outcome with multiple disjoint criteria

the "calculate" button if this radio button is already chosen). The figure on the right side of the Tool will show the appropriate area graphically, and the "Area under Curve" text box will show the value .0122, telling us that 1.22% of the area of the normal curve falls to the right of this *z*-score.

Interactive Page 92

The **probability** of Amherst, Massachusetts getting greater than 45 inches or less than 30 inches of rainfall in a year (Figure 4.9) is calculated like this:

	Criterion	*z*-Score	Area
Right tail	$X > 45$	2.25	.0122
Left tail	$X < 30$	-1.50	+ .0668
Total area = total probability			= .0790

Here, we combine a small area in the right tail of the **normal** distribution with a larger area in the left tail to get the total probability we're interested in. For the probability calculations we'll be doing in this course, however, we will more often need to combine equal areas from the two **tails.** That is, we will often want to evaluate the probability of getting a score as extreme or more extreme than some **z** value.

We often need to calculate the probability of a *z*-score being as extreme or more extreme than some criterion value.

For example, what is the probability of an adult man being greater than 2 standard deviations away from (that is, either above or below) the average height? Figure 4.10 shows the area we need to calculate to answer this probability question.

We could solve this problem by finding the one-tailed area beyond *z* separately for *z*-scores of +2.0 and –2.0, then adding these two areas together. Or, since the normal distribution is **symmetrical,** we could calculate the area for either tail alone and multiply by two to get the area in the two tails combined.

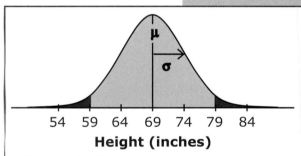

Figure 4.10 Distribution of adult men's heights, with individuals more than 2 standard deviations away from the mean highlighted.

But because this is such a common thing to have to do, the Normal Distribution Area Tool has an option to calculate the "two-tailed" area/probability directly. Just enter 2.0 in the *z*-score text box, click the "Beyond *z* (two-tailed)" radio button, and the appropriate probability value (.0455) will appear in the area text box. (What do you think would happen if you entered –2.0 for *z* and calculated the two-tailed area? Try it and find out!)

Do **Exercise 4.10** and **Exercise 4.11** for more practice using the addition rule along with the normal distribution and z-scores to calculate probabilities.

Interactive Page 93

4.2.2 The 68-95-99.7 Rule

A shortcut for estimating areas and probabilities for normal distributions is provided by the "68-95-99.7 rule":

- About 68% of the area of any normal distribution (the area in green in Figure 4.11a) falls within one standard deviation of μ.

- About 95% of the area (the area in green plus the area in purple) falls within two standard deviations of μ.

- About 99.7% of the area (the areas in green, purple, and orange combined) falls within three standard deviations of μ.

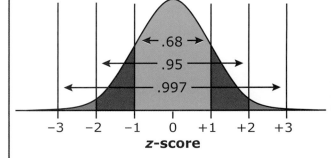

Figure 4.11 The 68-95-99.7 rule. This figure is interactive, with two parts, on the *Interactive Statistics* website and CD.

If the scores in the distribution in question are converted to z-scores, the 68-95-99.7 rule provides the following facts about the z-score of a randomly selected member of the population (Figure 4.11b):

- There is a 68% chance that the z-score will fall between −1.0 and +1.0.

- The probability is .95 that the z-score falls between −2.0 and +2.0.

- The z-score will fall between −3.0 and +3.0 with probability .997.

You should commit Figure 4.11 to memory. Whenever you're considering a probability question involving a normal distribution, you can sketch the figure to help visualize the area (and probability) you're being asked to assess. Sometimes you can even calculate the needed area without using the **Normal Distribution Area** Tool.

For example, Figure 4.12 shows the distribution of adult men's heights again, with the 68-95-99.7 rule superimposed and the critical value we considered in **Section 4.2**, 74 inches, marked with an arrow. Earlier, we used the Calculation Tool to determine the area under the curve falling to the right of this value (which in turn gives us the probability that a randomly selected police applicant will be 74 inches or taller). To determine this area using Figure 4.12 (without using the Calculation Tool), we reason as follows:

1. We know that 68% of the area under the curve falls between μ − σ = 69 − 5 = 64 inches and μ + σ = 69 + 5 = 74 inches.

2. Because normal distributions are symmetrical, one half of this area, 34%, falls between the mean (69 inches) and 74 inches.

3. We also know that 50% of the area falls to the right of the mean.

4. Therefore, the area falling to the right of 74 inches is 50% − 34% = 16%, and the probability that police applicants will meet the height requirement is .16.

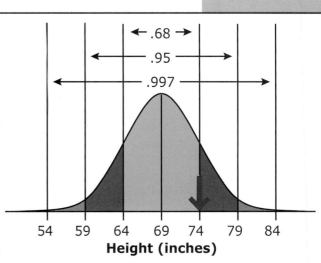

Figure 4.12 The 68-95-99.7 rule applied to the adult men's heights problem from **Section 4.2.**

This estimate is very close to the exact probability (.1587) we calculated using the **Normal Distribution Area** Calculation Tool. (See **Exercise 4.12** for another example of a probability problem you can solve using just the 68-95-99.7 rule.)

4.3 The Central Limit Theorem

Consider the following research situation:

> Judy, a marine biology graduate student, is studying the long-term effects of oil spills on marine animals. She suspects that a spill in one particular cove is affecting the health of the red rock crabs (*Cancer productus*) in the cove. One sign of ill health in crabs is small size, so Judy plans to determine the average claw-to-claw length of the cove's red rock crab population. How should she go about this task?

Well, Judy will presumably head to the tide pools of the cove in question and start trying to catch and measure her crabs. But no matter how good a crab catcher she is, she will never be able to measure every single red rock crab in the cove—even if she measures 1000 of them, how can she be sure that there weren't another 500 that were hidden away under some rock that she couldn't move?

The best Judy will be able to do is collect a **sample** of individuals from this **population**. (Now would be a good time to review the Box on **Populations and Samples,** p. 7.) Assuming the sample is **representative** of the population as a whole, she will then be able to

Researchers usually use sample means to estimate population means.

use the average length of the crabs in her sample as an estimate of the average length for the whole population.

In the second half of this chapter, we'll examine two crucial questions faced by researchers like Judy who want to draw conclusions about populations on the basis of samples:

1. How many crabs does she need to sample to get a good estimate of the mean of the whole population?

2. How close should we expect her **sample mean** to be to the actual **population mean**?

The answers to these two questions are closely intertwined. Briefly, the answer to the first question is "The larger the **sample size,** the better," and the answer to the second is "It depends on the sample size." As we will see shortly, a remarkable mathematical proposition called the **Central Limit Theorem** gives us the ability to make firm predictions about the accuracy of population estimates based on samples.

> The larger the sample, the closer we can expect its mean to be to the population mean.

Once you finish this chapter and understand the Central Limit Theorem and its implications, you'll finally be ready to tackle the eagerly awaited hypothesis testing procedure in the next chapter.

Our study of the relationship between samples and populations begins with a new concept: the **distribution of sample means.**

4.3.1 A Distribution of Sample Means

Consider the following thought experiment:

> Imagine we were to obtain a large vat and 5000 Ping-Pong balls. On 1000 of the Ping-Pong balls, we write the number "6" with a black magic marker. Similarly, we write the numbers "8," "10," "12," and "14" on another 1000 balls each. Then we throw all 5000 balls into the vat and stir them up with a very large spoon.

Since we assembled this **population** of Ping-Pong balls ourselves, we know exactly what the mean of all the numbers in the vat is. Employing the **Descriptive Statistics** Calculation Tool if you need to, see if you can determine this population mean yourself.

The **population mean,** μ, of all 5000 numbers in the vat is $(6 + 8 + 10 + 12 + 14) / 5 = 10$.

Now, suppose we reach into the vat and pull out a sample of $n = 2$ Ping-Pong balls. What would you expect the mean of these two numbers to be? Since the population mean is 10, we might hope that the sample mean would also be 10. And in fact, there's a pretty good chance that this is what we would get. Five different combinations of balls will produce this **sample mean:**

Ball 1	Ball 2	Sample Mean $\overline{X}$
6	14	10
8	12	10
10	10	10
12	8	10
14	6	10

However, while 10 is the most likely sample mean to get, there are a number of other possibilities: if we draw a 7 and a 9, the sample mean will be 8; if we draw a 10 and a 14, the sample mean will be 12, etc. In fact, for this well-defined population, we can precisely specify all the possible sample means. This **distribution of sample means** is shown in histogram form in Figure 4.13.

Glossary Term: distribution of sample means

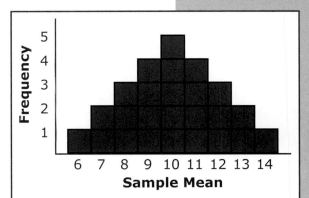

Figure 4.13 The distribution of sample means for samples of two balls in the thought experiment described in the text. This figure is interactive on the *Interactive Statistics* website and CD.

Technically, the distribution of sample means will only look exactly as shown in Figure 4.13 if we sample our Ping-Pong balls with replacement—that is, if we put the first ball back in the vat before we draw out the second one. However, given the large number of balls we're using, our distribution of sample means would look pretty much the same even if we drew each pair of balls out at the same time.

Interactive
Page 97

Now, having established what the **distribution of sample means** looks like for our thought experiment (Figure 4.14b), let's compare it to the original **distribution** of numbers in the vat (Figure 4.14a). (We will often use the shorter phrase "**sampling distribution**" in lieu of "distribution of sample means." We could construct sampling distributions of other statistics, such as sample standard deviations, but in this course, we'll only make use of sampling distributions of means.)

The first thing to notice in Figure 4.14 is that the mean of the sampling distribution is identical to the mean of the original "scores" (the numbers written on the Ping-Pong balls). Both means are 10.0, as you can confirm for yourself with the **Descriptive Statistics** Calculation Tool if you wish. This very useful relationship holds for any sampling distribution of any population.

The mean of a sampling distribution equals the population mean.

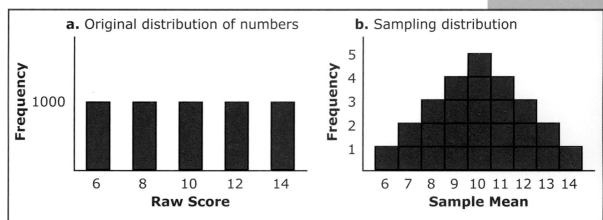

Figure 4.14 The original distribution of "raw scores" (numbers written on Ping-Pong balls) in our thought experiment (a), and the distribution of means of two-ball samples taken from this population (b).

The **Central Limit Theorem** formally states this and two additional relationships between sampling distributions and their underlying populations, one regarding the shape of the distributions and the other regarding **variability.** These other relationships will be easier to see if we consider a few more sampling distributions first. So let's think about what would happen if we drew larger or smaller samples from our imaginary vat of Ping-Pong balls.

Interactive Page 98

4.3.2 More Sampling Distributions

In the previous section, we constructed a **distribution of sample means** for a **sample size** of 2. (Recall that sample size is denoted by a lowercase n.) What do you think the sampling distribution for our thought experiment would look like with $n = 5$?

Since the population is described so precisely, we can once again specify this sampling distribution precisely. But it will be trickier than before, because with $n = 5$, there are 3125 (5 to the fifth power) possible samples, instead of the 25 possible samples with $n = 2$. Here are a few of the samples we could get:

Ball 1	Ball 2	Ball 3	Ball 4	Ball 5	Sample Mean $\overline{X}$
6	6	6	6	6	6
6	6	6	6	8	6.4
6	6	6	6	10	6.8
6	8	10	12	14	10
10	12	8	12	8	10
12	12	12	10	8	10.8

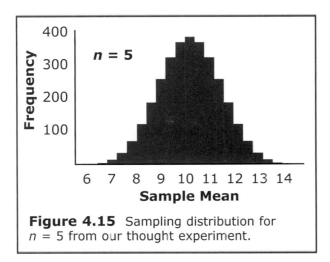

Figure 4.15 Sampling distribution for $n = 5$ from our thought experiment.

With all 3125 possible samples included, the histogram for the sampling distribution with $n = 5$ is as shown in Figure 4.15. The shape of this sampling distribution should remind you of the **normal** distribution even more than the one for $n = 2$.

This trend will continue as n gets larger and larger, and is one of the main tenets of the Central Limit Theorem. But what happens at the other end of the sample size spectrum? The smallest possible **sample** is a single ball ($n = 1$). What do you think the shape of the sampling distribution will look like for a sample size of 1?

There are five possible "samples" of one ball each:

Ball 1	Sample Mean $\bar{X}$
6	6
8	8
10	10
12	12
14	14

The shape of this sampling distribution is exactly the same as the shape of the original distribution of raw scores.

Figure 4.16 plots the three **sampling distributions** we've considered already plus the distribution for $n = 20$ (for the latter sample size, there are 95,367,431,640,625 possible samples!). The trend in shape toward normality is evident here.

As the sample size grows, sampling distributions become more normal in shape.

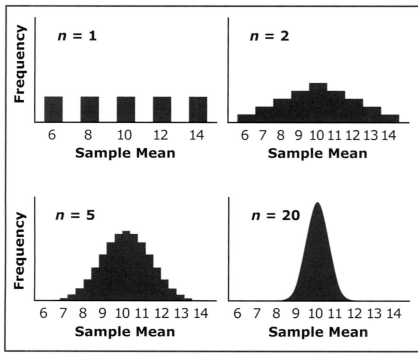

Figure 4.16 Sampling distributions for our thought experiment with n = 1, 2, 5, and 20. Two trends accompanying increases in sample size are clearly visible: the shape of the distribution becomes closer to the normal distribution and the standard deviation is reduced.

A second trend should also be evident in Figure 4.16. As **sample size** increases, the sampling distribution becomes more and more "squished" toward the center. That is, **sample means** tend to become more and more concentrated around the **population mean** of the raw scores (10 in this case), while sample means toward the tails of the distribution become less and less frequent.

To get an idea of why this trend occurs, consider that the number of possible samples that have a mean of 6 remains constant at 1, no matter how high the sample size. (With n = 1, the sample that produces a mean of 6 is {6}; with n = 2, the sample {6, 6} produces this mean, and so on.) On the other hand, the number of samples that have a mean of 10 grows exponentially as sample size increases: with n = 1, there is only one such sample, but with n = 2 there are 5, and with n = 5 there are 381.

This trend can be described as a decrease in the **standard deviation** of the sampling distribution with increases in sample size. This trend will also be formalized by the **Central Limit Theorem,** which we will formally state on the next page.

The variability in a sampling distribution decreases as sample size increases.

Interactive
Page **100**

Finally, may we present . . . (cue the trumpet fanfare) . . . the **Central Limit Theorem** (CLT):

For any **population** of raw scores with mean μ and standard deviation σ, the **distribution of sample means** for samples of size *n* will have a mean of μ, a standard deviation of σ / √*n* , and will approach the shape of the **normal** curve as *n* approaches infinity.

The CLT is extremely versatile: it provides the **central tendency, variability,** and shape of any sampling distribution from any population, no matter what the central tendency, variability, and shape of the original population is. We've already seen the tenets of the theorem illustrated for a population with a uniform distribution (our imaginary vat of Ping-Pong balls). Explore the **Central Limit Theorem** Activity to see the theorem work on populations that are normally distributed and positively **skewed.**

Since the CLT forms the bedrock of most of the **inferential statistics** we'll discuss in this course, we'll go through each of its three parts separately in the following three sections.

4.3.3 The Mean of a Sampling Distribution

The first and simplest part of the **CLT** states that the mean of a **sampling distribution** equals the mean of the underlying population. Although this statement is more intuitive than the other parts of the CLT, it is no less important. To see why, let's back up a bit and consider why sampling distributions are important to researchers in the first place.

Recall the story of Judy the marine biologist, who is studying red rock crabs. Say she captures and measures 25 crabs, and determines that the average length for these individuals is 125 mm. It seems reasonable to conclude that the mean length for the entire population is probably pretty close to the mean length for Judy's sample.

But this conclusion is actually based on two underlying assumptions. First, we are assuming that Judy's sample is near the mean of the sampling distribution. We can't know for sure whether or not this assumption is true; it could be that, for whatever reason, Judy happened to find lots of puny red rock crabs, and that her sample mean is actually much smaller than the typical sample mean. In the absence of any counterevidence, though, we will give Judy the benefit of the doubt and assume that her sample is in fact a typical one.

The second assumption is that the mean of the sampling distribution is the same as the mean of the underlying population, as stated by the CLT. If we could not make this assumption, then the mean of a sample would tell us nothing about the mean of the population.

For example, what if the mean length for the population of red rock crabs was 100 mm, but the mean of the sampling distribution was

125 mm (see Figure 4.17a)? If this were the situation, Judy's sample mean would be perfectly typical, but it would be way off from the population mean.

Luckily, we never have to face this situation, because the CLT guarantees that the **sampling distribution** will be centered on the same point as the underlying **population,** as shown in Figure 4.17b. If the mean of a particular sample is close to the mean of the sampling distribution, it will also be close to the mean of the raw scores.

The mean of a sampling distribution is technically called the "expected value of $\bar{X}$" and abbreviated $\mu_{\bar{x}}$, since we expect sample means to be close to this value. However, since $\mu_{\bar{x}}$ is always identical to the population mean (Figure 4.17b), we will use the symbol μ to refer to the center of the sampling distribution as well as to the center of the underlying population.

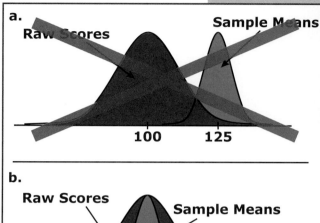

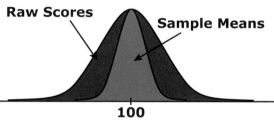

Figure 4.17 a) The situation that Judy would face if the mean of the sampling distribution was not equal to the mean of the underlying population. Fortunately, the CLT guarantees that this will never happen, which is why there is a big red "X" over this part of the figure. b) In fact, the CLT tells us that the sampling distribution will always be centered on the mean of the underlying raw scores, as shown in this part of the figure.

4.3.4 The Standard Deviation of a Sampling Distribution

The second postulate of the **CLT** deals with the **variability** of a **sampling distribution.** We already saw informally in our Ping-Pong ball thought experiment that the standard deviation of the distribution of sample means decreases as the sample size *n* increases. The CLT quantifies this observation, telling us that the standard deviation of a sampling distribution is equal to the **standard deviation** of the underlying population, σ, divided by the square root of the **sample size,** $\sqrt{n}$.

The standard deviation of the sampling distribution is such a crucial part of so many **inferential statistics** that it gets its own name, the **standard error of** $\bar{X}$ (usually shortened to "standard error"), and its own symbol, $\sigma_{\bar{x}}$. Here's the formal formula for the standard error:

$$\text{standard error} = \sigma_{\bar{x}} = \sigma / \sqrt{n}$$

The best way to understand mathematical relationships such as the one between standard error and sample size is often to consider what happens at the extremes. As we've already noted, the smallest possible sample from any population includes only a single individ-

Glossary Term: standard error

ual, $n = 1$. At the other end of the spectrum, many populations are, at least theoretically, infinite. For example, if the population is all human beings, it includes everyone who's ever been born, is still living, and/or will be born in the future.

Say we're researching an infinitely large population of scores with a population standard deviation of 10. What would be the standard error of this population for sample sizes of 1 and infinity? (If you want to use the standard error formula to calculate your answer, substitute some very large number, like 100,000,000, for infinity, then round the answer to the nearest .01.)

Interactive Page 103

Applying our formula for **standard error,** when $n = 1$, the standard error is $10 / \sqrt{1} = 10$. We've already seen that with a **sample size** of 1, the sampling distribution is identical to the original population distribution. So it makes sense that the standard deviation of the **sampling distribution** would be identical to the standard deviation of the population in this case.

For $n = 1$, standard error = population standard deviation

When n is infinitely large, the standard error is $10 / \sqrt{\infty}$. The square root of infinity is infinity, and any number divided by infinity is 0, so the standard deviation of the sampling distribution for an infinite sample size is 0. This holds for any population with any σ: As n approaches ∞, the standard error will approach 0.

For $n = $ infinity, standard error = 0

How should you interpret a standard deviation of 0? It just means that there is absolutely no **variability** for the **distribution,** and if you think about it, this makes sense as well. To see why, let's return to our Ping-Pong ball experiment again. We put 5000 balls in the vat, so the largest possible sample size for this population is 5000. There's only one sample this large, which will include all the balls in the vat. Therefore, there's only one possible mean for this sample size. And since there's only one possible mean, there's no opportunity for variation in the sampling distribution, so its standard deviation is 0.

Interactive Page 104

Looking at the relationship between sample size and **standard error** practically, instead of mathematically, we can say that the larger the sample, the more accurate we can expect the sample mean to be at predicting the population mean:

Larger samples generally produce more accurate estimates of the population mean.

- With very small sample sizes, the standard deviation of the **sampling distribution** will be similar to the standard deviation of the underlying population, so a sample mean may be far from the population mean.

- With medium-sized samples, the standard error will be smaller than the population standard deviation, but we must still expect some amount of variability—any given sample mean should not be expected to reflect the population mean precisely.

- With very large samples, the standard error becomes very small, and we can be confident that the mean of a sample is extremely close to the population mean.

Of course, the standard deviation of the underlying population also helps determine the standard error. We focus on the influence of sample size on standard error, though, because experimenters have much more control over this factor. Indeed, it is often impossible to do anything about the variability in the population, but we can usually design an experiment to include as many subjects as we desire. (We will take advantage of this fact when we discuss **power** analysis in **Chapter 7.**)

4.3.5 The Shape of a Sampling Distribution

The final proposition of the **CLT** is that the shape of a **sampling distribution** gets closer to a **normal** curve as *n* gets larger. We saw this trend illustrated in Figure 4.16, repeated below, and you can see it even more vividly in the **Central Limit Theorem** Activity, where you can construct a sampling distribution for any sample size you like.

Activity: The Central Limit Theorem

The population standard deviation has a large effect on standard error, but is usually immutable.

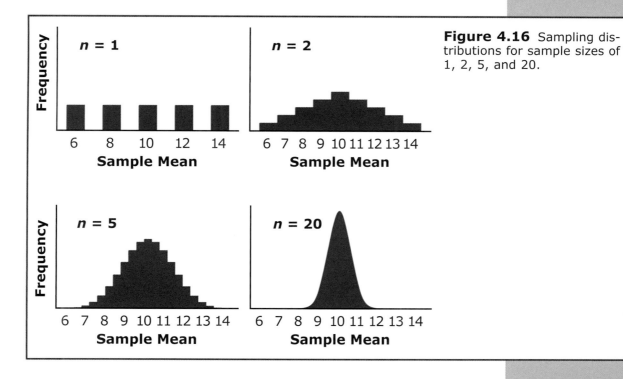

Figure 4.16 Sampling distributions for sample sizes of 1, 2, 5, and 20.

By the time sample size reaches about 20–30, the sampling distribution is virtually indistinguishable from a perfect normal curve, regardless of the shape of the underlying population distribution. And if the original population was itself fairly normal, the sampling distribution will be normal at even smaller sample sizes.

For sample sizes of 20 or more, sampling distributions are almost always perfectly normal.

Interactive
Page 106

It is this final part of the **CLT** that really sets the stage for the **hypothesis** testing procedure in the next chapter. Since we know that sampling distributions will be **normal** (as long as sample size is large enough), we can use the tools we've developed for working with normal distributions to help us draw inferences about **populations** from **samples.**

For an example of how this will work, let's return to Judy's research again. Here's the situation:

A brief preview of the logic of hypothesis testing

> Judy, a marine biology graduate student, is studying the long-term effects of oil spills on marine animals. She suspects that a spill in one particular cove is affecting the health of the red rock crabs (*Cancer productus*) in the cove. One sign of ill health in crabs is small size, so Judy plans to determine the average claw-to-claw length of the cove's red rock crab population. How should she go about this task?

Let's suppose that Judy knows from prior research that for the general population of red rock crabs (i.e., crabs from coves that haven't had oil spills wash up on their shores), the average length is 155 mm, with a standard deviation of 50 mm.

Furthermore, let's assume, for argument's sake, that Judy is wrong about the adverse effects of the oil spill on crabs in her cove. That is, assume that the population from the oil-damaged cove has the same characteristics as the general population: an average length of 155 mm, with $\sigma = 50$.

Now suppose Judy collects 25 crabs from her research cove. Given the assumed μ and σ and this sample size, the CLT tells us that the **sampling distribution** for Judy's research project should look like the one pictured in Figure 4.18:

- The mean of the sampling distribution should equal the mean of the population (155 mm).

- The standard deviation of the sampling distribution (also known as the **standard error**) should equal $\sigma / \sqrt{n} = 50 / \sqrt{25} = 10$.

- The shape of the sampling distribution should be normal.

For Judy's sample, say the average crab length turns out to be 125 mm. This value is indicated in Figure 4.18 by the red arrow. As you can see, this sampling distribution contains very few sample means this small. Put another way,

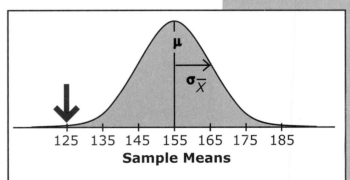

Sample Means

Figure 4.18 Distribution of sample means for $n = 25$ from a population with $\mu = 155$ and $\sigma = 50$. The location of Judy's observed sample mean in this distribution is shown by the red arrow.

the **probability** of getting a sample mean this small from a population with the given parameters ($\mu = 155$, $\sigma = 50$) appears to be quite low.

In fact, *since the sampling distribution is normal, we can calculate this probability exactly.* You can try to figure out how to do this on your own now if you wish, but we'll spend most of the next chapter walking you through the procedure. The end result will be a probability value of .003, which certainly qualifies as "quite low."

Now we reason backward as follows. If the population of red rock crabs in Judy's research cove had a mean of 155 mm, there would be very little chance of getting a sample mean of 125 mm. And yet we did get a sample mean this small. Therefore, the assumption is probably wrong, the population mean is probably *not* 155 mm, and Judy's initial suspicions are probably justified: The crabs in this cove are smaller than normal.

Easy, right? Well, don't be alarmed if this page confused you a bit. The logic of hypothesis testing is pretty tricky, which is why much of this book is dedicated to it. Just make sure you firmly grasp the implications of the CLT discussed in this chapter (for starters, go through **Exercise 4.13, Exercise 4.14,** and **Exercise 4.15**) before moving on to the next one.

4.4 Chapter Summary/Review

This Chapter Summary/Review is interactive on the *Interactive Statistics* website and CD. Also be sure to go through all the **Review Exercises** for this chapter.

- The outcome of a **random event** (e.g., the flip of a coin) is uncertain. However, over a long series of such events, a regular and predictable pattern of outcomes will emerge. The **probability** (or likelihood) of an outcome of a random event is defined as the proportion of times the outcome will occur over many trials of the event. For example, the probability of a coin flip coming up heads is .50, or 50%, or 1/2 (**Section 4.1**).

- The first rule of probability theory is that there is a 100% probability of some outcome when a random event occurs (that is, something has to happen) (**Section 4.1.2**). The addition rule states that for two **disjoint** outcomes A and B (outcomes that cannot occur simultaneously), the probability of one or the other outcome occurring is equal to $p(A) + p(B)$ (**Section 4.1.2**).

- From here on in this course, we will often have need to consider the probability of randomly drawing a particular score or range of scores from a population. For example, we might want to know the probability of drawing a score greater than or equal to six from the population in Figure 4.3, reproduced in part below. When a population is depicted graphically like this, probability is equivalent to the proportion of area of the **histogram** that corresponds to the specified condition. Here, 7/20 of the area of the distribution lies above 6, so the probability value is $p = .35$ (**Section 4.1.3**).

- For **normally** distributed populations (**Section 4.2**), we can use the mathematical regularity of the normal curve to determine the probability of randomly drawing a score from a particular range. Such calculations are made easier if raw scores are first transformed into z-scores, so that the **Normal Distribution Area** Calculation Tool can be used. For example, the probability of drawing an individual from a normally distributed population whose **z-score** is greater than +1 is known to be exactly .1587 (**Section 4.2**). More complex probability judgments, such as the likelihood of drawing a z-score greater than 1.0 or less than −1.0 are calculable using the addition rule (**Section 4.2.1**).

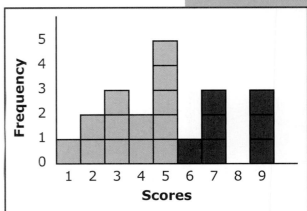

Figure 4.3 Histogram of a small population of scores, with the individuals whose scores are ≥ 6 highlighted.

- To make quick probability estimates for normal distributions, remember the 68-95-99.7 rule: The probability of randomly drawing an individual with a z-score between −1 and +1 is about .68, the probability of drawing a z-score between −2 and +2 is .95, and the probability of drawing a z-score between −3 and +3 is .997 (see Figure 4.11b, reproduced at right) (**Section 4.2.2**).

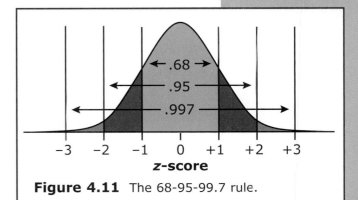

Figure 4.11 The 68-95-99.7 rule.

- The **distribution of sample means** for a population, often called the sampling distribution, specifies the relative frequency of **means** of **samples** of a certain **size** that are drawn from the **population.** For example, the means of all possible samples of two individuals from the population {6, 8, 10, 12, 14} are graphed in the histogram of Figure 4.13, repeated on the next page (**Section 4.3.1**).

- All sampling distributions are centered on the mean of the underlying population (that is, the mean of a sampling distribution is always equal to the mean of the population from which the samples are drawn). However, larger samples are likely to be more **representative** of the population than smaller samples, so sampling distributions get "squeezed" (i.e., the standard deviation of the sampling distribution is reduced) as sample size gets larger (**Section 4.3.2**).

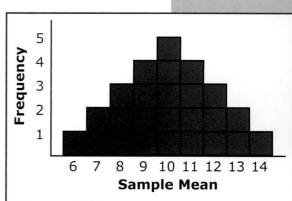

Figure 4.13 The distribution of sample means for samples of 2 individuals drawn from the population {6, 8, 10, 12, 14}. This figure is interactive on the *Interactive Statistics* website and CD.

- The **Central Limit Theorem** codifies these two facts about sampling distributions, and adds a third postulate regarding the shape of sampling distributions:

  ○ The mean of a sampling distribution will equal μ, the mean of the underlying population. Technically, the mean of the sampling distribution is called the expected value of $\overline{X}$. However, we use the symbol μ to refer to both the population mean and the sampling distribution mean, since their values are always identical (**Section 4.3.3**).

  ○ The standard deviation of a sampling distribution goes by the name standard error and the symbol $\sigma_{\overline{x}}$. The CLT states that $\sigma_{\overline{x}} = \sigma / \sqrt{n}$, where n is the sample size (**Section 4.3.4**).

  ○ The shape of a sampling distribution will approach the normal curve as n approaches infinity. In fact, distributions of sample means are usually almost perfectly normal whenever n is 20 or more (**Section 4.3.5**).

Chapter 5
From Samples to Populations: Hypothesis Testing with *z*

Interactive Page 109

5.1 What Counts as a "Significant" Increase?

Consider the following newspaper article:

> SPRINGFIELD (AP)—A group of 625 high school seniors today filed a class action suit against GetIn.com, a startup company that offers an online training course designed to boost SAT scores. The students claim that completing GetIn.com's course did not substantially raise their test scores, as advertised. The average verbal SAT score is 500, with a standard deviation of 100. The 625 plaintiffs averaged a verbal score of 512, an improvement that they argue is not consistent with GetIn.com's claim that graduates of their course score "significantly higher" than the overall average.

The plaintiffs in this case argue that an increase of 12 points over the average SAT score is not "**significant**." The defendant (GetIn.com) will presumably argue the opposite—that this magnitude of increase *is* significant. How can we evaluate these competing claims?

As you will learn in this chapter, research claims made by scientists follow the same structure as the claim made by the plaintiffs in this case, and a long-established procedure exists to determine whether or not there is sufficient evidence for such claims.

Basically, the procedure is as follows:

The four steps of hypothesis testing

- First, we establish the two competing claims, or **hypotheses,** in formal terms.

- Second, we collect and describe the **sample** data that is meant to serve as evidence for or against the hypotheses.

- Third, we compute a **test statistic** that compares the average score of our sample with the averages we would expect from each of the hypotheses we're evaluating. Test statistics also take into account the amount of **variability** inherent in the scores and the size of our sample.

- Fourth and finally, we use the test statistic to evaluate the two hypotheses.

In this chapter, we will work our way through the logic (some would say perverse logic) of hypothesis testing, then flesh out the four-step procedure outlined above using one of the simplest test statistics, the **z-score.** However, z-scores are actually not all that useful for analyzing real-world experiments. In upcoming chapters, we will introduce the **t test,** a more practical **inferential statistic,** and will go over some important considerations we must keep in mind when interpreting the results of hypothesis tests.

Most of the rest of this course will be concerned with teaching you how to use other test statistics to evaluate the results of different research situations. But the four-step procedure and the interpretational considerations introduced here will remain the same for all of these test statistics. If you don't get a solid grounding of the principles of hypothesis testing from this chapter, you will probably find yourself (and your quiz grades!) sinking quickly in chapters to come.

This is not meant to scare you, just to gently warn you: Get your thinking cap on and concentrate on the pages ahead, because putting in a little extra effort on this chapter will pay dividends down the road.

Interactive Page 110

5.2 The Logic of Hypothesis Testing

Let's review the facts of the fictional case introduced in the previous section:

- **Mean** score for all SAT-takers: 500

- **Standard deviation** for all SAT-takers: 100

- Number of plaintiffs who took GetIn.com's online training course: 625

- Average SAT score for plaintiffs: 512

You should recognize that this situation is analogous to an experiment in which a **sample** is taken from a **population** and given an **experimental** treatment. The population is all SAT-takers, the sample is the plaintiffs in the case, and the treatment is GetIn.com's training course. The variable in the "experiment" is SAT score.

The central question we want to answer in any experiment of this sort is "what is the effect of the treatment?" In most situations, we make the simplifying assumption that if the treatment has any effect at all, it will show up as a change in the mean score for subjects that undergo the treatment, compared to the general population of subjects that don't undergo the treatment. We also assume that the shape and standard deviation of the **distribution** of scores will be the same for the general and experimental populations. Figure 5.1 reflects these assumptions and illustrates two possible effects of GetIn.com's training course.

The arrow in Figure 5.1 indicates the mean score for the 625 plaintiffs in the court case—in statistical terms, this is the sample mean $\overline{X} = 512$. It should be clear that the treatment effect most consistent with this observed sample mean is the one illustrated in the bottom half of the figure.

However, *we cannot know for sure that the mean of the entire experimental population is the same as the mean for the sample.*

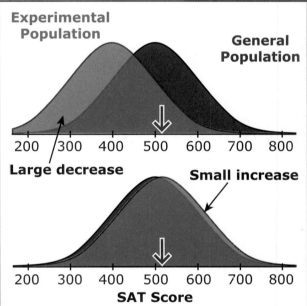

Figure 5.1 Two possible treatment effects of GetIn.com's training course on SAT scores (this figure is interactive, with five parts, on the *Interactive Statistics* website and CD). The blue distribution represents the general population of SAT-takers. The effect of the training course is assumed to be a shift in this distribution, either to the left if people tend to score worse after taking the course, or to the right if course-takers tend to score better. Possible distributions of course-takers' scores are shown in green.

It is possible, for example, that the actual mean of the experimental population is identical to the mean of the general population, $\mu = 500$. This is exactly the contention of the plaintiffs in our fictional case: They claim that the training course had no meaningful effect on their SAT scores.

GetIn.com makes the counterclaim that the course does have a meaningful effect. Note that this claim does not say exactly what the mean of the experimental population is. Rather, it simply says that the experimental population mean is not the same as the general (nonexperimental) population mean.

Interactive Page 111

Although you might not have noticed, we just completed the first step in the **hypothesis** testing procedure: we established two competing hypotheses to evaluate. In the language of hypothesis testing, a claim of no effect is known as the **null hypothesis,** abbreviated H_0 (the subscript signifies "zero effect"). Here,

$$H_0: \mu \text{ for course-takers} = \mu \text{ for all SAT-takers} = 500$$

A claim made against the null hypothesis is called the **alternative hypothesis** and is abbreviated H_a. For our case,

$$H_a: \mu \text{ for course-takers} \neq 500$$

You may be thinking that GetIn.com is really arguing not just that their course changes scores, but that it improves them. However, the more general alternative hypothesis stated here, known as a **bidirectional** hypothesis, is easier to reconcile with the logic of hypothesis testing, for reasons that should become clear soon. We'll stick with bidirectional H_a's for now, then cover **directional** hypotheses, where the direction of the effect is specified, in **Section 5.4.**

It's easier to prove a hypothesis false than to prove one true. Now, here's where the logic of hypothesis testing starts to get tricky. The rules of logic say that it is virtually impossible to conclusively prove a hypothesis true. On the other hand, it is relatively easy to prove some hypotheses false. Without going too deeply into the philosophy of logic, these principles are well illustrated by considering the hypothesis "all dogs have four legs." To prove this statement true, we would have to examine every single dog that ever lived, is living, or will live—an impossible task. But to prove the hypothesis false, we just need to find one dog with three legs.

Because of this dichotomy, we don't even try to prove hypotheses true with **inferential statistics,** at least not directly. Instead, we set up a null hypothesis and an alternative hypothesis and work toward proving the null hypothesis false. If we succeed in discrediting the null hypothesis, we indirectly support the alternative hypothesis.

Glossary Term: null hypothesis

Glossary Term: alternative hypothesis

Interactive Page 112

On the previous page, we established that according to the logic of **hypothesis** testing, we try to disprove the claim that a treatment has no effect in order to indirectly support the claim that the treatment does have an effect.

Now for one more wrinkle. We established earlier that we can completely discredit the hypothesis that "all dogs have four legs" by finding just one dog with three legs. Unfortunately, in experiments of the type we deal with in this book, we can never be 100% confident that the **null hypothesis** is incorrect. So not only can we not prove that a treatment has an effect—we also can't conclusively prove that the treatment has no effect!

The explanation for why we cannot ever completely disprove the null hypothesis has two parts. First, experimenters make hypotheses about averages, not absolutes. For example, GetIn.com is not claiming that *everyone* who takes their course scores above 500, just that *on average* course-takers score higher than 500.

Now, if we could assess every single member of the **population** in question, we could come to a firm conclusion about averages. For example, when the U.S. government takes its decennial census, it attempts to interview every single person in the population. To the extent that this is possible, we can make absolute claims about population means, such as "the average American family has 1.864 children."

But an experiment only tests a **sample** of the population. By assessing a sample we can make an educated guess about the mean of the whole population, but we cannot say anything with complete certainty.

So what exactly can we say about the null and alternative hypotheses on the basis of an experimental sample? Well, we can't say for *sure* whether or not the null hypothesis is true or false, but **test statistics** are designed to give us the exact **probability** that the null hypothesis is true, given the evidence from our experiment. (The phrase "exact probability" may seem like an oxymoron, but it makes sense to mathematicians.)

For our fictional court case, we can determine that given the performance of the 625 plaintiffs, there is only a .0027 chance (that's less than 3/1000) that people completing the GetIn.com training course receive the same mean SAT score as other SAT-takers. Remember that evidence that the null hypothesis is false (or, more accurately, evidence that the null hypothesis is unlikely to be true) counts as indirect evidence that the **alternative hypothesis** is true. Odds of 3 in 1000 are pretty long, so GetIn.com will be able to make a strong case for their alternative hypothesis that people who take their course score better than average on the SAT.

We can't completely disprove the null hypothesis.

Test statistics help us evaluate the probability that the null hypothesis is true.

Are you confused yet? If so, don't worry: The first step in understanding an intellectually challenging concept such as hypothesis testing is to appreciate just how complex the concept is. At some point later in the chapter, you will hopefully have an "aha experience," where you realize that you now actually know how hypothesis testing works. For now, try to get your mind around the idea that to establish evidence *supporting* the idea that a treatment *has* an effect (H_a), we attempt to show that it's very *unlikely* that the treatment *has no effect* (H_0).

We've now outlined all four steps in the **hypothesis** testing procedure for our court case:

We start by establishing the two competing hypotheses, which we now know are called the **null** (H_0) and **alternative** (H_a) hypotheses. For our court case, H_0 says that the mean SAT score for the population of all students who take GetIn.com's training course is 500 (i.e., the mean of all course-takers = the mean of all SAT-takers); H_a counters that the mean for this population is not equal to 500.

Next, we collect and describe the sample data. The plaintiffs did this by establishing that the mean of their 625 SAT scores was 512.

Then we calculate a **test statistic.** We haven't seen exactly how to do this yet, but we've already established all the components that we need: the sample size and mean, the population mean (500), and the population standard deviation (100).

Finally, we evaluate the hypotheses in light of the evidence provided by the test statistic. Again, we haven't discussed this in detail yet, but it was foreshadowed that the results will indicate a probability of about .003 that the null hypothesis is correct. Given the extreme improbability of the null hypothesis, we will conclude that the alternative hypothesis is almost certainly true.

These four steps are graphically summarized in the **Hypothesis Testing Steps** Activity. Run through this Activity briefly now and leave it open to refer back to as you go through the next section, which will explain how to calculate and interpret our first test statistic, the **z-score.**

5.3 Hypothesis Testing with *z*-Scores

Here's a research example we can use as we go through the procedure for **hypothesis** testing with the **z** statistic.

> Amy, a doctoral student in psychology, is studying the effects of divorce on children. As part of her dissertation, she gives a questionnaire designed to measure depression to a sample of 10-year-old children whose parents

Summary of hypothesis testing steps

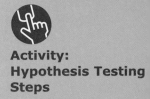

Activity: Hypothesis Testing Steps

were recently divorced. For the general population of 10-year-olds, scores on the questionnaire form a normal distribution with $\mu = 50$ and $\sigma = 12$. Higher scores indicate greater levels of depression. Amy's hypothesis is that children of divorced parents will be more depressed than the general population of kids. She assesses a small sample of children of divorced parents and obtains depression scores of 55, 79, 60, 66, 44, 62, 66, 51, and 39. Was Amy's hypothesis confirmed?

Let's work through the four steps of hypothesis testing. First, we must establish the two competing hypotheses in Amy's experiment, the **null hypothesis** (H_0) and the **alternative hypothesis** (H_a).

5.3.1 Step 1: Establish the Hypotheses

$$H_0: \mu \text{ for children of divorce} = 50$$
$$\text{and}$$
$$H_a: \mu \text{ for children of divorce} \neq 50$$

For H_0, we always specify that the mean of the experimental population (in this case, children of divorced parents) is equal to the mean of the general population (all 10-year-old children). In other words, the null hypothesis predicts no effect (a "null effect") as a result of the experimental treatment. (In the present case, as well as in others we will encounter, the treatment is not technically experimental, since the researcher did not randomly assign children's parents to be divorced. However, the logic of hypothesis testing works just the same.)

The null hypothesis always predicts no treatment effect.

The alternative hypothesis always predicts that the treatment did have an effect—that the mean of the experimental population is different from the mean of the general population. Amy's specific hypothesis is that children of divorced parents are more depressed than other kids—that μ for children of divorce is greater than 50. As noted earlier, though, the logic of hypothesis testing is most clear when the **alternative hypothesis** is the exact opposite of the null hypothesis; hence we use $\mu \neq 50$ for H_a.

The alternative hypothesis always predicts an effect.

Figure 5.2 illustrates the null and alternative hypotheses graphically:

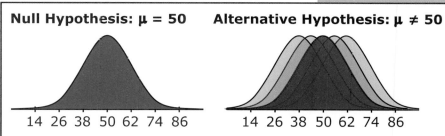

Null Hypothesis: $\mu = 50$ **Alternative Hypothesis: $\mu \neq 50$**

14 26 38 50 62 74 86   14 26 38 50 62 74 86

Figure 5.2 The two competing hypotheses for Amy's experiment (this figure is animated on the *Interactive Statistics* website and CD). The null hypothesis H_0 states that the experimental distribution overlaps the general population distribution precisely, and hence that the experimental population has exactly the same mean (50) as the general population. In contrast, the alternative hypothesis H_a states that the experimental distribution (green) is shifted by some unspecified amount compared to the general population distribution (blue), so the experimental population mean is not equal to the general population mean.

Now that we've specified the hypotheses, we move on to the second step of hypothesis testing, collecting and describing the sample data.

5.3.2 Step 2: Collect and Describe the Sample Data

The second step in the **hypothesis** testing procedure is to collect and describe the **sample** data. The data for our current example are given in the next-to-last sentence of the paragraph:

> As part of her doctoral dissertation, Amy gives a questionnaire designed to measure depression to a sample of 10-year-old children whose parents were recently divorced. For the general population of 10-year-olds, scores on the questionnaire form a normal distribution with $\mu = 50$ and $\sigma = 12$. Higher scores indicate greater levels of depression. Amy's hypothesis is that children of divorced parents will be more depressed than the general population of kids. She assesses a small sample of children of divorced parents and obtains depression scores of 55, 79, 60, 66, 44, 62, 66, 51, and 39. Was Amy's hypothesis confirmed?

The most important **descriptive statistic** for the hypothesis testing procedure with z is the sample mean, $\overline{X}$. We also need to know the number of subjects in the sample (n). Many **test statistics** also require the sample standard deviation, but for hypothesis tests with **z-scores** this is not necessary.

The sample size and mean are $n = 9$ and $\overline{X} = 58.0$ for Amy's data. These **descriptive statistics** seem quite straightforward when compared to the roundabout logic involved in the rest of the **hypothesis** testing procedure. Thus many students (and, unfortunately, quite a few researchers) tend to lose sight of the importance of descriptive statistics when considering research results. In fact, collecting and describing the data in an experiment is usually the most time-consuming and always the most important part of the research enterprise.

It is the descriptive statistics that tell us what happened in an experiment—most fundamentally, how large an effect the experimental treatment had on the subject population. **Inferential statistics** were created to help us interpret these descriptives. Sometimes an inferential statistic will lead us to discount an effect that otherwise looks quite meaningful. Other times an inferential statistic will help boost the importance of an effect that looks somewhat meager. But we should never lose sight of the fact that the effects themselves are reflected in the descriptive statistics, not in the inferential statistics.

The purpose of inferential statistics should be to help us interpret descriptive statistics.

Now let's move on to step three in the hypothesis testing procedure, calculating the **test statistic.**

5.3.3 Step 3: Calculate the Test Statistic

Now that we've established our hypotheses and collected and described our data, we're ready for the step that we've approached but never actually tackled, computing the inferential **test statistic.** Before we start, a word of warning: What follows is the most important and tricky part of hypothesis testing. In this and the following section, we're going to tie together concepts you've learned in previous chapters regarding **z-scores, probability,** and sampling. Get yourself a cup of coffee and prepare to read the next few pages over a few times.

We have our two competing **hypotheses:**

$$H_0: \mu = 50 \qquad \text{and} \qquad H_a: \mu \neq 50$$

and we've collected a **sample** of $n = 9$ subjects who produced a mean score $\overline{X} = 58.0$. Remember that the logic of hypothesis testing dictates that we cannot hope to prove either hypothesis true. But we can attempt to prove the **null hypothesis** false.

Clearly, $58.0 \neq 50$, so the **descriptive statistics** appear to indicate that the null hypothesis is incorrect. But we have to consider the possibility that our sample mean (which only includes nine children, after all) does not perfectly reflect the population mean of all children whose parents have divorced. Indeed, if you think about it, it's extremely unlikely that any **sample mean** (especially one from a sample this small) would be exactly the same as the overall **population mean.**

In other words, the 8-point difference between our sample mean and the population mean that is specified by the null hypothesis may not be due to the fact that the children in the experiment have divorced parents, but instead might be due to random chance.

> A deviation of the sample mean from the null hypothesis may be due to chance.

We can never completely rule out the random-chance explanation, but we can assess how likely it is. The following pages describe how we do this.

Our goal is to calculate a statistic that will allow us to assess the **probability** that the sample mean of 58 differs from the hypothesized mean of 50 solely on the basis of random chance. Here's what we do:

> Inferential statistics test the random-chance explanation of effects.

- Assume that the **null hypothesis** is true. That is, assume that μ for children of divorce = 50.

- Also assume that the distribution of scores in the experimental population has the same standard deviation (12) as the general population (we'll say more about this and other **assumptions** in **Section 5.5**).

- In **Chapter 4,** we learned several important facts that the **Central Limit Theorem** tells us about the **distribution of sample means** drawn from any population. First, such a **sampling distribution** is (at least approximately) **normal,** as long as the **sample size** is greater than about 10. Second, the mean of a sampling distribution is equal to the mean of the underlying population. Third, the standard deviation of a sampling distribution, called the **standard error** and denoted by the symbol $\sigma_{\bar{x}}$, is equal to the standard deviation of the underlying population divided by the square root of the sample size: $\sigma_{\bar{x}} = \sigma / \sqrt{n}$. (Review The **Central Limit Theorem** Activity if you need a brush-up on the CLT.)

Activity: The Central Limit Theorem

Putting all this information together, we can surmise that the distribution of means for all possible samples with $n = 9$ from an experimental population with a mean of 50 and a standard deviation of 12 will be a new distribution that is normally shaped, has a mean of 50, and has a standard deviation of $12 / \sqrt{9} = 4$. The situation is graphically summarized in Figure 5.3.

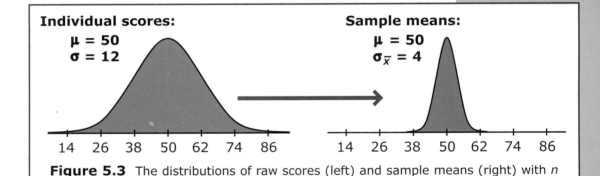

Individual scores:
μ = 50
σ = 12

Sample means:
μ = 50
$\sigma_{\bar{x}} = 4$

14 26 38 50 62 74 86   14 26 38 50 62 74 86

Figure 5.3 The distributions of raw scores (left) and sample means (right) with $n = 9$ for Amy's research study, assuming that the null hypothesis is correct (μ = 50).

Now take a look at the graph on the right of Figure 5.3. Drawing on your knowledge of extracting probability estimates from normal curves (remember the 68-95-99.7 rule from **Section 4.2.2**?), how likely does it appear to be that Amy's sample mean of 58.0 was drawn from the sampling distribution shown in this histogram?

Interactive Page 120

On the previous page, we asked how likely it is that Amy's sample mean of 58.0 was drawn from the **distribution of sample means** shown on the right side of Figure 5.3. The general answer is "very unlikely"—means this high for samples of nine scores appear to be quite uncommon for populations in which μ is 50. To evaluate exactly *how* unlikely it is to draw a value this extreme from a normal distribution, we can use **z-scores,** as first described in **Section 4.2.** Recall the original definitional formula for the *z*-score of a single individual given a population mean and population standard deviation:

$$z = \frac{X - \mu}{\sigma}$$

To compute a *z*-score for a sample mean, we just have to substitute the sample mean for the single score, the mean of the sampling distribution for the population mean (remember that these are equivalent, so there's actually no substitution here), and the **standard error** for the standard deviation. The formula thus becomes:

$$z = \frac{\overline{X} - \mu}{\sigma_{\overline{X}}}$$

Try using this formula now to calculate the *z*-score test statistic for Amy's experiment. Recall that the sample mean is 58.0, the hypothesized population mean is 50, the sample size is 9, and the population standard deviation is 12. (Also remember that the standard error $\sigma_{\overline{X}} = \sigma / \sqrt{n}$.)

Here are the essential facts we've established for Amy's experiment:

- **Null hypothesis** H_0: $\mu = 50$

- **Alternative hypothesis** H_a: $\mu \neq 50$

- **Sample mean** $\overline{X} = 58.0$

- **Population standard deviation** $\sigma = 12$

- **Sample size** $n = 9$

- **Standard error:** $\sigma_{\overline{X}} = 4$

z is computed as follows:

$$z = \frac{\overline{X} - \mu}{\sigma_{\overline{X}}} = \frac{58 - 50}{4} = 2.0$$

This **z-score** (2.0) is our long-awaited **test statistic**! Now we're ready to go on to the fourth and final step in hypothesis testing: evaluating the null and alternative hypotheses in light of this test statistic.

5.3.4 Step 4: Evaluate the Hypotheses

Instead of calculating Amy's *z*-score test statistic by hand, we could use the **z-Scores** Calculation Tool, by entering the sample mean 58

Formula:
The *z*-score test statistic

as X, the null-hypothesized population mean of 50 as μ, and the standard error 4 as σ. Figure 5.4, a screenshot of the graphic from the Calculation Tool, shows that the resulting z-score of 2.0 falls pretty far over to the right in the **normal** distribution.

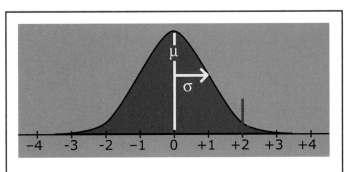

Figure 5.4   Location of $z = 2.0$ in the normal distribution, from the z-Scores Calculation Tool.

As you learned in **Chapter 4,** we know an awful lot about the normal distribution. Most important, we can calculate exactly what proportion of the area of the distribution lies to the left and right of any given z-score. The **Normal Distribution Area** Calculation Tool does just this. Open the Tool, enter our z-score (2.0), and click the radio button to calculate the area beyond z (two-tailed). Figure 5.5 is a screenshot of what you should get. As you can see, only 4.6% of z-

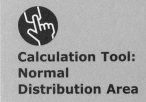

Calculation Tool: Normal Distribution Area

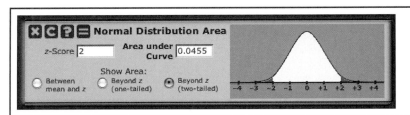

Figure 5.5   Screenshot of the Normal Distribution Area Calculation Tool, illustrating the area beyond $z = +2.0$ and $z = -2.0$ in the two tails.

scores in a normal distribution are ≥ 2.0 or ≤ 2.0. Now we make the following chain of inferences:

- If we assume that the null hypothesis is true, and the mean depression score for the **population** of children with divorced parents really is 50, the **sample** mean Amy obtained (58.0) would correspond to a z-score of 2.0.

- The chances of drawing a z-score with an absolute value this large from a normal distribution are very slim ($\boldsymbol{p} = .046$).

- It thus seems unlikely that our sample mean was actually drawn from the **sampling distribution** we were assuming on the basis of the null hypothesis.

- `Therefore, we probably shouldn't assume that the null hypothesis is true!

In other words, Amy's sample mean of 58.0 constitutes good evidence *against* the **null hypothesis**. It is highly unlikely that random chance would produce a sample mean this large if $\mu = 50$, as proposed by H_0. In statistical parlance, we say that we can *reject the null hypothesis* on the basis of these data.

> If the *p*-value for a test statistic is very low, we *reject the null hypothesis*.

Interactive Page 123

For the example we've been working with, the **probability** that the observed **sample mean** could have come from an experimental **population mean** with $\mu = 50$ was $p = .046$, and we claimed that this was good enough evidence to reject the **null hypothesis.** But what if p had turned out to be .30? Three out of ten are still fairly unlikely odds, but most research scientists agree that a **p-value** this high never constitutes adequate grounds to reject H_0.

The general standard used in most social science forums is that if $p < .05$, the null hypothesis can be rejected, but that if $p \geq .05$, there is not adequate evidence to reject the null. This cutoff value for p is called the **alpha (α) level,** and the traditional alpha level of .05 represents a fundamental degree of **conservatism** on the part of researchers with regard to rejecting the null hypothesis and therefore claiming that an effect exists. The odds of H_0 being correct must be less than 1/20 for the scientific community to accept that an experimental treatment has an effect.

> Glossary Term: alpha level
>
> The *p*-value must generally be below .05 for the null hypothesis to be rejected.

In certain circumstances, an even more conservative alpha level may be adopted. For example, if an experiment is challenging a widely accepted theory that is supported by many other studies, a researcher (and the scientific community of which the researcher is a member) might adopt an alpha level of .01, requiring the odds against H_0 to be 99/100 or greater for it to be rejected.

You may have heard a research result described as "**statistically significant,**" a phrase that is ubiquitous in scientific journals and increasingly common in the nonscientific press. This phrase sounds quite impressive, but it just means that the researcher calculated some test statistic and found that the p-value associated with the statistic was lower than the α level that he or she chose to use. Since the $\alpha = .05$ convention is widely accepted in the behavioral sciences, you can usually assume that an effect described as "statistically significant" beat the .05 level.

Now is a good time to stop and go through the first set of **Review Exercises**. **Exercises 5.1–5.4** guide you through the four steps of hypothesis testing. **Exercise 5.5** and **Exercise 5.6** provide practice with the mechanics of calculating z-score test statistics and p-values. Finally, **Exercise 5.7** and **Exercise 5.8** require you to go through all four steps yourself and draw conclusions on the basis of hypothesis tests. The latter four exercises are all regenerative, so practice them several times each until you're as comfortable as possible with both the logic and mechanics of hypothesis testing.

Interactive
Page 124

5.4 Directional Hypothesis Tests

In **Section 5.3.1,** we stated the **null** and **alternative** hypotheses for Amy's experiment as follows:

$$H_0: \mu = 50 \qquad \text{and} \qquad H_a: \mu \neq 50$$

At the time, though, we noted that this alternative hypothesis does not really capture Amy's prediction about her experiment. Let's review the situation again:

> As part of her doctoral dissertation, Amy gives a questionnaire designed to measure depression to a sample of 10-year-old children whose parents were recently divorced. For the general population of 10-year-olds, scores on the questionnaire form a normal distribution with $\mu = 50$ and $\sigma = 12$. Higher scores indicate greater levels of depression. Amy's hypothesis is that children of divorced parents will be more depressed than the general population of kids. She assesses a small sample of children of divorced parents and obtains depression scores of 55, 79, 60, 66, 44, 62, 66, 51, and 39. Was Amy's hypothesis confirmed?

Amy is really predicting not just that the mean of the experimental population is *different* from the mean of the general population, but that children of divorced parents will show an *increase* in depression scores. Although the **bidirectional** H_a given above is the most common form of the alternative hypothesis, it is also possible to perform a **directional** hypothesis test. For Amy's experiment, the hypotheses for such a test would be:

$$H_0: \mu = 50 \qquad \text{and} \qquad H_a: \mu > 50$$

Figure 5.6 illustrates the contrast between bidirectional and directional tests. The null hypothesis is the same in both cases, predicting that the experimental distribution overlaps the nonexperimental population exactly. In a bidirectional test (Figure 5.6a), the alternative hypothesis states that the mean of the experimental distribution is anything other than the mean for the nonexperimental population. In a directional test (Figure 5.6b), H_a is slightly more specific, predicting that the experimental mean will be either to the right of (greater than) or to the left of (less than) the nonexperimental mean. In the situation depicted in Figure 5.6b, H_a predicts a positive effect—that the experimental distribution will be shifted to the right compared to the nonexperimental distribution.

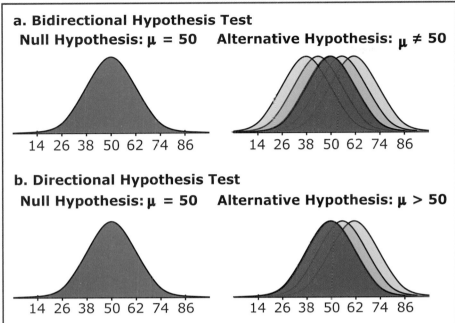

a. Bidirectional Hypothesis Test

Null Hypothesis: μ = 50   Alternative Hypothesis: μ ≠ 50

14 26 38 50 62 74 86   14 26 38 50 62 74 86

b. Directional Hypothesis Test

Null Hypothesis: μ = 50   Alternative Hypothesis: μ > 50

14 26 38 50 62 74 86   14 26 38 50 62 74 86

Figure 5.6 Hypotheses in bidirectional vs. directional tests (this figure is animated on the *Interactive Statistics* website and CD). In a bidirectional test (a), the alternative hypothesis does not specify the direction of effect, while in a directional test (b), the effect is predicted to be either positive or negative. Neither type of test specifies exactly how much the experimental distribution (green) overlaps the nonexperimental distribution (blue).

Interactive Page 125

The procedure for a **directional** hypothesis test is identical to that for a **bidirectional** test except for the calculation of the *p*-value associated with the **test statistic**. In Amy's experiment, for example, the value for **z** is 2.0 in both types of tests. But for a directional test, we calculate the area under the **normal** curve in only one **tail** of the distribution (see Figure 5.7b). For this reason, a directional hypothesis test is often called a *one-tailed test,* while a bidirectional test is known as a *two-tailed test.*

Directional = one-tailed test
Bidirectional = two-tailed test

a. Bidirectional (Two-Tailed) Hypothesis Test   **b. Directional (One-Tailed) Hypothesis Test**

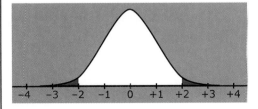
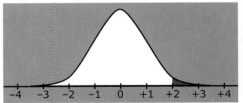

−4 −3 −2 −1 0 +1 +2 +3 +4   −4 −3 −2 −1 0 +1 +2 +3 +4

Figure 5.7 Relevant areas under the standard normal curve for a bidirectional hypothesis test (a) and a directional hypothesis test (b) with *z* = 2.0.

The difference between a one- and a two-tailed test is really quite simple. The goal of the hypothesis testing enterprise is to calculate the probability that a particular sample mean could have come from the distribution of sample means specified by the **null hypothesis.** For Amy's experiment, $\overline{X} = 58.0$, while the null hypothesis specified that the mean of the **sampling distribution** for the experimental population is $\mu = 50$.

If we are using a directional H_a, we want to calculate the probability that a sample mean 58.0 *or higher* could come from a distribution with $\mu = 50$ and $\sigma_{\overline{X}} = 4$. For a bidirectional hypothesis test, however, we want the probability that a sample mean this *extreme*—that is, a sample mean that is two standard errors higher *or lower* than the hypothesized population mean—could come from this sampling distribution. Therefore, we must also consider the probability of randomly drawing a sample mean less than or equal to 42.0 ($= \mu - 2\sigma_{\overline{X}}$).

Based on what you've learned so far, you should realize that there is a simple relationship between the *p*-value for a directional test and the *p*-value for a bidirectional test: the one-tailed *p*-value is always exactly one-half the *p*-value for a two-tailed test.

> The *p*-value for a one-tailed test = 1/2 *p* for a two-tailed test.

This fact has an important implication for the likelihood of rejecting H_0 and concluding that there is a **statistically significant** effect in an experiment. Can you guess what this implication is?

Interactive Page 126

On the previous page, you learned that the **p-value** for a one-tailed (**directional**) hypothesis test is always half the *p*-value for a two-tailed (**bidirectional**) test. One important implication of this fact is that it is always easier to reject the **null hypothesis** with a one-tailed test than with a two-tailed test.

> It is easier to reject the null hypothesis with a one-tailed test than with a two-tailed test.

In Amy's case, for example, $p = .046$ for the two-tailed test, as we saw before, but $p = .023$ for the one-tailed test. For a researcher trying to convince the scientific community that her effect is real, the lower the *p*-value, the stronger an argument she can make. So you might guess that directional (one-tailed) tests would be more common than two-tailed tests. In fact, however, directional tests are rarely used in the behavioral sciences, for at least three reasons:

> Why two-tailed hypothesis tests are more common than one-tailed tests.

- The first reason is the logical consideration noted in **Section 5.2.** Since the null hypothesis specifies that μ is exactly equal to some number (e.g., $\mu = 50$), the most logical **alternative hypothesis** includes all other numbers ($\mu \neq 50$). With a directional hypothesis ($\mu > 50$), a large number of possible results are not covered by either the null or the alternative hypothesis.

- The second reason to use a two-tailed rather than a one-tailed test is that sometimes a researcher doesn't know beforehand which way to predict an effect. And even if the researcher herself strongly believes an effect will go one way, her scientific peers might doubt the certainty of this belief. If there's any doubt at all about which direction to expect, a bidirectional test is required.

- The final, and perhaps most compelling, reason to avoid one-tailed tests is to consider what you would do if you framed an H_a in one direction and the results came out in the other direction. For example, what if Amy had gone with H_a: $\mu > 50$ and had found that $\overline{X} = 34.0$. In this case, z would be extremely large (-4.0), but she would have to conclude that the null hypothesis could not be rejected, with the implication that there was no **significant** effect. Needless to say, this would appear to be a flawed conclusion.

One strategy that you might be thinking about would be to calculate $\overline{X}$ for an experiment and then frame a **directional alternative hypothesis** to fit the data. But this practice is universally considered cheating by the scientific community. To see why, let's consider another example:

> Fiona is a high school guidance counselor. She suspects that minorities in her school score lower than the general population on the verbal portion of the SAT, and decides to test this prediction using a z-score test statistic and a directional hypothesis test. As we've already learned, the distribution of scores for the general population on this test are centered on $\mu = 500$, with $\sigma = 100$. Fiona averages the scores of the 25 minority students currently enrolled (which she considers a sample of all minority students at her school who have or ever will take the SATs) and finds that $\overline{X} = 535$—the minority students actually scored *higher* than average on the SAT. Therefore, Fiona uses H_0: $\mu = 500$ and H_a: $\mu > 500$, even though her original prediction was that μ would be < 500. She calculates that $z = 1.75$, $p = .04$. Should she conclude that minorities at her school score significantly higher than the general population?

Well, according to the one-tailed **p-value,** Fiona's $\overline{X}$ does beat the .05 significance level. But since she waited until after she looked at the data to decide which way H_a should go, she obviously must have been considering both directions of effects as possibilities, so she really needs to be using both tails of the **normal** distribution to calculate the p-value. As we've learned, the two-tailed p-value is always twice the one-tailed value, so in reality $p = .08$, squarely in the gray area where it is difficult to draw any conclusions at all from a hypothesis test.

To use a directional test, you must frame your alternative hypothesis *a priori.* That is, you must specify the direction of effect you're predicting *before the data are collected.* And you'd better have a compelling argument that the effect is not going to go the other way. For this reason, directional hypothesis tests are most often used when one or more prior experimental results strongly imply that the effect in the new experiment, if there is one, will go in a particular direction. See **Exercise 5.9** for a "real-world" example in which a directional test may be justifiable.

The predicted direction of effect must be specified prior to data collection to use a one-tailed test.

Interactive Page 128

5.5 Hypothesis Test Assumptions

Inferential statistics used in **hypothesis** tests each have a set of **assumptions**—conditions that must be met if conclusions based on the statistics are to be valid. For a nonstatistical (albeit somewhat silly) example of assumptions, consider the conditions that must be met if you tell a friend you're going to watch a baseball game on television this weekend. For this plan to be carried out successfully, the game must not be rained out, your TV must be working properly, and you must plant yourself on your couch at the appropriate time.

You probably don't give much thought to these conditions. Rather, as the name given to them suggests, you just *assume* that they will be met. The same is true of assumptions for inferential statistics: even though violating the assumptions will make any conclusions drawn from the statistics invalid (or at least questionable), most researchers and consumers of research don't worry too much about them, because the assumptions for commonly used statistics are not very difficult to meet. Nevertheless, it is important to have a general understanding of the assumptions for each statistic you plan to use.

For the *z*-score test statistic, there are four important assumptions:

1. The sample of subjects in the experiment must be drawn at random from the experimental population. Certain laws of probability only apply to randomly selected samples, and we use these probability laws when we evaluate the *p*-value associated with the test statistic. The consequence of nonrandom sampling would be that the experimental conclusions wouldn't apply to the entire population, and generalization to the population as a whole is usually a primary goal of any research project.

Assumption 1: random sampling

2. We calculate the *p*-value for a hypothesis test using *z* by referring to the **normal** distribution. This assumes that the **sampling distribution** for the experimental population is in fact normal. If the distribution of individual experimental scores is itself normal, the sampling distribution for the scores will also be normal, regardless of the sample size. But even if the distribution of individual scores is not normal, the distribution of samples of scores *will* be normal as long as the sample size is relatively large (i.e., about 20 or

Assumption 2: normal sampling distribution

greater; we can be confident of this fact because of the **Central Limit Theorem**).

3. To calculate the **standard error** of the sampling distribution for the experimental population, we must know the standard deviation of this population. We don't know this value, and the only way to nail down σ exactly would be to measure every member of the experimental population. If we could do this, we wouldn't need to do any inferential statistics in the first place! Therefore, to calculate the *z*-score test statistic, we assume that the experimental population has the same standard deviation as the untreated population.

   This assumption is almost surely violated to some extent nearly every time. Unless the experimental manipulation is trivial, σ for the experimental population will almost never be exactly the same as σ for the nontreated population. This is one reason why *z*-scores are rarely used in practice as test statistics.

   Assumption 3: standard deviation unaffected by experimental treatment

4. However, an even more compelling reason not to use the *z*-score test statistic is that in the vast majority of research situations, σ is not even known for the general (nonexperimental) population. In the next chapter, we will learn about another statistic, the *t* **statistic,** that substitutes for *z* when σ is not known. In fact, most of the research examples in the present chapter are somewhat contrived: In the real world, researchers facing these situations would test their hypotheses using *t*, not *z*.

   Assumption 4: standard deviation is known

5.6 Evaluating Evidence

Experiments are usually designed and carried out in order to demonstrate treatment effects. The experimenter hopes that these effects will sway the scientific community's thinking about some issue. So hypotheses are proposed, data are collected, and some difference is found between the average score of the experimental sample and the average we would expect for untreated individuals. Then **inferential statistics,** such as the **z-score test statistic** we learned about in this chapter, are calculated, resulting in a ***p-value***—the probability that the sample mean could have come from the **sampling distribution** that is specified by the **null hypothesis.** This tells us, roughly speaking, the likelihood that the observed experimental effect is due to chance alone.

To this point (steps 1–3), the hypothesis testing procedure is almost completely objective. The experimenter's peers may quibble over the choice of test statistic, and occasionally calculation mistakes are made, but usually there is no dispute over the calculation of the inferential statistic and associated *p*-value.

However, the interpretation of this *p*-value (step 4) is an extremely subjective matter. For example, say two researchers perform almost exactly the same experiment. Researcher A calculates her inferential statistic and discovers that the p-value for her experiment is .049, while researcher B finds a p-value for his experiment of .051. Assuming that the $\alpha = .05$ convention is adopted by both researchers, how should we view the results of the two experiments?

Interactive Page 130

Some researchers see hypothesis testing strictly as a *decision-making* enterprise. For them, experimental effects get a thumbs-up if the **p-value** is lower than the chosen **alpha level** (usually .05) for the study, or a thumbs-down otherwise. In other words, if $p < .05$, it is decided that an effect exists, while if $p > .05$, it is decided that there is no treatment effect.

Hypothesis tests as decision rules

The advantage of this approach to hypothesis testing is that it makes the decision as to whether or not to accept the null hypothesis very cut-and-dried. If $p < .05$, H_0 is rejected, an effect is claimed, and the experiment is deemed a "success." (This success is trumpeted by that triumphant phrase, "**statistically significant.**") If $p \geq .05$, on the other hand, H_0 is not rejected and it is concluded that the experimental treatment had no effect.

It is much simpler to keep things straight in your mind if you can categorize them. Unfortunately, however, life in general, and experimental data in particular, are not so simple. To see the problem with the hard-line decision-making approach, let's consider Amy's dissertation experiment one more time:

> As part of her doctoral dissertation, Amy gives a questionnaire designed to measure depression to a sample of 10-year-old children whose parents were recently divorced. For the general population of 10-year-olds, scores on the questionnaire form a normal distribution with $\mu = 50$ and $\sigma = 12$. Higher scores indicate greater levels of depression. Amy's hypothesis is that children of divorced parents will be more depressed than the general population of kids. She assesses a small sample of children of divorced parents and obtains depression scores of 55, 79, 60, 66, 44, 62, 66, 51, and 39. Was Amy's hypothesis confirmed?

If her first subject had scored 53 instead of 55 on the depression index, the sample mean for her experiment would have been $\overline{X} = 57.8$ instead of 58.0, z would have come out to 1.95 instead of 2.0, and we would have $p = .051$ instead of .046 (you should confirm these numbers using the appropriate Calculation Tools). Should such a small difference in the size of Amy's effect be enough to change our perception of her experiment from "success" to "failure"?

$$z = \frac{\overline{X} - \mu}{\sigma_{\overline{X}}}$$

where μ is the experimental population mean proposed by the null hypothesis (**Section 5.3.3**).

4. Evaluate the hypotheses. Given a z-score calculated via the above formula, we can use the **Normal Distribution Area** Calculation Tool to determine the probability of obtaining a sample mean this extreme if the experimental population mean is as proposed by the null hypothesis. If this p-value is low, we infer that it is unlikely that the experimental population mean really is what H_0 proposed, so we "reject" the null hypothesis and conclude that H_a is probably correct (**Section 5.3.4**).

- The **alpha (α) level** serves as a guideline for deciding whether or not to reject H_0. Experimenters are free to set any α they choose when evaluating hypotheses, but other scientists may be leery of an experimenter's claims if a too-**liberal** alpha level is adopted. The convention in the behavioral sciences is to use $\alpha = .05$ (**Section 5.3.4**).

- The **alternative hypothesis** can be stated in either of two ways. A **bidirectional** (two-tailed) H_a proposes that a treatment will have some effect on the mean of the experimental population, but does not specify whether scores will be raised or lowered. A **directional** (one-tailed) H_a does specify whether the effect will be to increase or to decrease scores. Even though most researchers plan experiments with an idea about which way an effect will go, bidirectional hypothesis tests are far more common. The main reason for this practice is that if a manipulation has the opposite effect to that predicted by the experimenter, a directional test will fail to reject H_0 even if it is false (**Section 5.4**).

- For the z-score test statistic to be used validly, certain assumptions must be met: the experimental sample must be drawn at random from the population, the **sampling distribution** for the experimental population must be normal, the standard deviation of the general (nonexperimental) population must be known, and this standard deviation must not be affected by the experimental manipulation. The first two of these assumptions must also hold for many other **inferential statistics,** but the last two are unique to z-scores, and severely limit the usefulness of this test statistic (**Section 5.5**).

- The p-value from a hypothesis test should be considered an indicator of the strength of evidence for an experimental effect (the lower the p, the stronger the evidence that the treatment had an effect). If $p < \alpha$, we can conclude that it is very unlikely that the effect is in the opposite direction to that observed, but we should not make the mistake of using the alpha level to make absolute decisions as to whether an effect does or does not exist. Overreliance on alpha levels when drawing conclusions from hypothesis tests leads to oversimplification of experimental results (**Section 5.6**).

Chapter 6
Practical Hypothesis Testing with *t*

6.1 The Problem with *z*

Consider the following experiment:

> For a class project, James performs a small experiment replicating a classic social psychology study on conformity. Subjects in the experiment are asked to judge the length of a wooden dowel that is 12 inches long. But before they make their judgments, the subjects observe four confederates (friends of James's who are acting as if they are also randomly chosen subjects) who all overestimate the dowel's length, claiming that it is between 13 and 15 inches long. Ten subjects give the following estimates (in inches) of the dowel's length:

13.9 13.0 15.1 12.4 13.4 13.3 16.9 13.1 10.7 18.2

The idea behind the experiment is that subjects will alter their judgments to conform to the estimates made by the other people in the group. James reasons that without seeing the confederates make their judgments, the subjects should guess, on average, that the dowel is its true length, 12 inches. Therefore, he plans to perform a bidirectional hypothesis test with H_0: $\mu = 12$.

Can James use the **z-score** test statistic introduced in the previous chapter for his hypothesis test? Let's quickly review the formula for this statistic:

$$z = \frac{\overline{X} - \mu}{\sigma_{\overline{X}}} = \frac{\overline{X} - \mu}{\sigma / \sqrt{n}}$$

In words, we can express this formula as:

$$z = \frac{\text{The sample mean minus the population mean proposed by the null hypothesis}}{\text{The standard deviation of the experimental population divided by the square root of the sample size}}$$

In James's experiment, as in most behavioral science studies, values for three of the four terms in this formula are known. However, the value of the fourth term is not known here (or in most other studies), and this fact severely limits the usefulness of z as a **test statistic.** Which value is impossible to calculate given the information James has available?

6.2 The *t* Statistic

We cannot use **z** as a test statistic here because the **standard deviation** of the experimental population, σ, is not known. In fact, if James is assuming that his experimental result will apply to all humans (or at least all college-age humans), the value of σ is not only unknown, but unknowable—to determine the standard deviation of the entire **population,** every human in the world would have to be tested in James's experimental procedure!

Luckily, there is a way around this difficulty. Even though we may never know the population standard deviation for subjects receiving an experimental treatment, we can always calculate the **sample standard deviation** for the subjects in the experiment. We can then *estimate* the population standard deviation σ using the standard deviation s of the sample.

The sample standard deviation can be used to estimate the population standard deviation when the latter is unknown.

Substituting s for σ in this way forces us to make three changes to our **hypothesis** testing procedure. The first two are mostly semantic. First, the test statistic can no longer be called a z-score (the reason should become clear soon). Instead, it's called a **t statistic.**

Second, the denominator of the test statistic formula can no longer be called the **standard error,** because we're not actually using the population standard deviation. Instead, we're using the sample standard deviation as an estimate of this value, so we call the denominator the **estimated standard error,** and denote it by the symbol $s_{\bar{X}}$. Thus the test statistic formula changes like this:

$$t = \frac{\bar{X} - \mu}{s_{\bar{X}}} = \frac{\bar{X} - \mu}{s / \sqrt{n}}$$

The third change in the hypothesis testing procedure is more practical, and will be described in **Section 6.2.1.** But given the data from James's experiment and the formula above, you're already prepared to calculate your first *t* statistic. Give it a try now (with the help of the **Descriptive Statistics** Calculation Tool if you wish).

Glossary Term: estimated standard error

Formula: The *t* statistic

Calculation Tool: Descriptive Statistics

Interactive Page 136

Here is how James should calculate **t:**

$$t = \frac{\bar{X} - \mu}{s_{\bar{X}}} = \frac{\bar{X} - \mu}{s / \sqrt{n}} = \frac{14.0 - 12}{2.20 / \sqrt{10}} = 2.87$$

(We'll go through this calculation again in more detail later, so don't worry if you got some of it wrong this time.)

As you can see, calculating a *t* test statistic is almost identical to calculating a **z-score** test statistic. But we cannot interpret the *t* statistic in exactly the same way, as we'll see in the next section.

Interactive Page 137

6.2.1 *t* Distributions

To start to understand the difference between interpreting *z* and *t* test statistics, let's go back and review the logic of hypothesis testing with **z-scores** (see **Chapter 5** and Figure 6.1 at right). We assume, for argument's sake, that μ for the experimental population equals the value proposed by the **null hypothesis** (a). If we know σ for the experimental population, the **Central Limit Theorem** (CLT) specifies that the **distribution of sample means** should be **normal** (b). We can therefore compare the observed sample mean $\bar{X}$ to this distribution. To do so, we **standardize** $\bar{X}$ using the *z*-score formula (c) and compare the standardized score to the normal distribution (d). If the *z* statistic falls way out in one of the **tails** of the distribution, we conclude that our assumption was wrong, and that μ is *not* equal to the value proposed by the null hypothesis (e).

Hypothesis Testing with *z*-Scores

a. Assume μ is as proposed by H_0.

b. Central Limit Theorem ensures that distribution of $\bar{X}$'s will be normal, with $\sigma_{\bar{X}} = \sigma / \sqrt{n}$.

c. Calculate *z*-score using observed $\bar{X}$, assumed μ, and $\sigma_{\bar{X}}$.

d. Compare *z*-score to unit normal distribution.

e. If test statistic falls in outer tail of distribution, conclude that μ is **not** as proposed by H_0.

Figure 6.1 Flowchart of the logic involved in hypothesis testing with *z*-scores.

Another way of stating all this is as follows: Suppose we were to take 1000 randomly chosen **samples** of n subjects from an experimental **population.** For each sample, we use $\overline{X}$, the hypothesized value of μ, and the assumed value of σ to calculate a z-score. If μ is specified correctly, the CLT guarantees that the distribution of z-scores will be identical to the normal distribution. Therefore, we can use the **Normal Distribution Area** Calculation Tool to evaluate the probability that any one particular sample is actually drawn from a population with the μ specified by the null hypothesis.

Now, here's where **t statistics** differ from z-score test statistics: *When we use s to estimate σ, the* **distribution** *of test statistics will not be quite normal.* The distribution of t statistics will still be bell-shaped and symmetrical, but it will be somewhat "flatter" than a normal distribution.

If you think about it, this should be fairly intuitive. Since s is an imperfect estimate of σ, it makes sense that there should be more **variability** in **test statistics** calculated with $s_{\overline{x}}$ than with $\sigma_{\overline{X}}$. More variability corresponds to a wider, flatter distribution.

So how much flatter is a **t distribution** than a z distribution? The answer to this question depends on how large the sample is, and this is also why this section is called "t Distributions," not "The t Distribution." Suppose the entire population we were testing in an experiment consisted of exactly 5000 individuals. If we wanted to get the best possible estimate of the population standard deviation σ from a sample standard deviation s, how many subjects would we want to include in the sample? (Warning: This is a bit of a trick question.)

Interactive Page 138

Given a population of 5000, a sample that included every single subject would give the best possible estimate of σ. In fact, the standard deviation of this "sample" would equal σ exactly, since every member of the population was assessed.

What if our **sample size** were slightly smaller, say 4900? Each time we draw a sample of this size, s will be slightly different, and will not equal σ exactly. But you should realize that s will be pretty darn close to σ every time. Even if the population is infinitely large, as in the case of James's experiment, a sample size of 1000 or more will provide a near-perfect estimate of σ (and even with 100 subjects, the sample s will estimate σ quite precisely).

At the opposite extreme, let's say we were drawing samples of only 3 subjects. As examples, we'll take the first 3 and the last 3 subjects from James's experiment. Remember that μ for his null hypothesis is 12. For these two mini-samples, the **descriptive sta-**

Distributions of t statistics are not perfectly normal.

tistics and *t* would be calculated as follows (you should verify these calculations yourself):

	Scores	$\overline{X}$ Mean	St. Dev. *s*	$t = \dfrac{\overline{X} - \mu}{s_{\overline{X}}}$
Sample 1:	13.9, 13.0, 15.1	14.0	1.05	3.29
Sample 2:	13.1, 10.7, 18.2	14.0	3.83	0.90

Here we have two samples with the same means, but we get widely varying *t* statistics (3.29 vs. 0.90). Why? Because the standard deviations we used in the *t* formula varied widely.

Now you should see the crux of the difference between **t distributions** and the distribution of **z-score** test statistics. If we're using σ in the denominator of the test statistic formula, the denominator will be identical for every sample we test. Therefore, the only source of **variability** in test statistics will be in the numerator (because sample means can change from sample to sample). But when we have to estimate σ from the sample standard deviation, both the numerator and the denominator can vary from sample to sample, so there's more variability in the test statistics.

> Both the numerator and denominator vary from sample to sample when calculating *t* statistics.

Interactive
Page 139

On the previous page, we saw that with a large sample size *n*, *s* provides an excellent estimate of σ. In this case, the **t** distribution should look almost identical to the **normal** distribution. But as the sample size shrinks, *s* becomes a less reliable estimator of σ, and the **t distribution** gets flatter and flatter. This relationship between *n* and the shape of the *t* distribution is illustrated in Figure 6.2.

> The *t* distribution spreads out as *n* decreases.

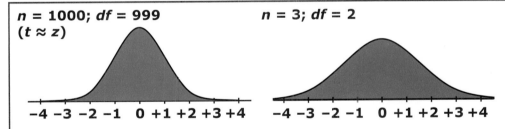

n = 1000; df = 999
(t ≈ z)

n = 3; df = 2

−4 −3 −2 −1 0 +1 +2 +3 +4   −4 −3 −2 −1 0 +1 +2 +3 +4

Figure 6.2 Distributions of *t* statistics for two different sample sizes *n* and degrees of freedom *df* (*df* = *n* − 1) (this figure is interactive, with four parts, on the *Interactive Statistics* website and CD). When *n* = 1000, the *t* distribution is essentially identical to the *z*-score distribution. But when *n* = 3, the "tails" include a larger percentage of the area of the distribution. Note: This figure actually overstates the difference between *t* and *z* distributions. Even with an *n* of only 3, the *t* distribution looks more similar to the *z* distribution than pictured here.

Note that the shape of the *t* distribution does not depend on the value of *s*. Individual *t* statistics will of course be affected by *s*, but the distribution of *t* statistics depends only on the number of subjects in a sample.

The practical significance of all this is that when we're using a *t* statistic to evaluate a null hypothesis, the *t* distribution to which we need to compare our test statistic will vary depending on the number of subjects in the experimental sample. For technical reasons, *t* distributions are indexed not by the actual sample size, but by the **degrees of freedom** (*df*) for the sample mean. The concept of degrees of freedom is a bit tricky, and if you want to understand it better, read the **Box** on degrees of freedom (see p. 55). Practically, though, all you need to know is this:

In **hypothesis** tests involving a single
sample of size *n*, *df = n − 1*.

6.3 Hypothesis Testing for Single Samples with *t*

Now that we've covered the conceptual background for the ***t* statistic,** we're ready to go through the steps involved in conducting **hypothesis** tests with *t*. The logic is very similar to that for hypothesis tests with *z*, except that:

- We use the sample standard deviation *s* in place of σ in the test statistic formula.

- We compare the *t* statistic to a ***t* distribution** instead of the **normal** distribution.

These changes to the logic of hypothesis testing are illustrated in Figure 6.3.

If you wish, use the **Hypothesis Testing Steps** Activity to keep track as we proceed through the four steps in the hypothesis testing procedure. We'll use James's conformity experiment as our example.

> For a class project, James performs an experiment in which subjects are asked to judge the length of a 12-inch-long wooden dowel after observing four confederates who all overestimate the dowel's length, claiming that it is between 13 and 15 inches long. Ten subjects give the following estimates (in inches) of the dowel's length:
>
> 13.9 13.0 15.1 12.4 13.4 13.3 16.9 13.1 10.7 18.2

The first step is to establish the **null** and **alternative** hypotheses, H_0 and H_a. Both are phrased in terms of the mean (μ) of the experimental population. Think of the experimental population as a "theo-

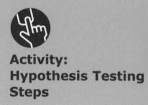

**Box: Degrees of
Freedom**

**Activity:
Hypothesis Testing
Steps**

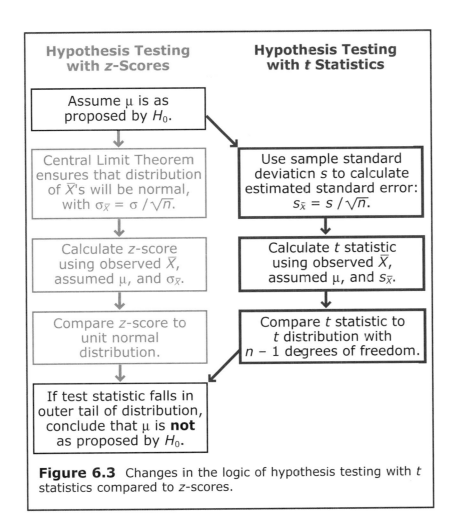

Hypothesis Testing with *z*-Scores	Hypothesis Testing with *t* Statistics
Assume μ is as proposed by H_0.	
Central Limit Theorem ensures that distribution of $\bar{X}$'s will be normal, with $\sigma_{\bar{X}} = \sigma / \sqrt{n}$.	Use sample standard deviation s to calculate estimated standard error: $s_{\bar{X}} = s / \sqrt{n}$.
Calculate *z*-score using observed $\bar{X}$, assumed μ, and $\sigma_{\bar{X}}$.	Calculate *t* statistic using observed $\bar{X}$, assumed μ, and $s_{\bar{X}}$.
Compare *z*-score to unit normal distribution.	Compare *t* statistic to *t* distribution with $n - 1$ degrees of freedom.
If test statistic falls in outer tail of distribution, conclude that μ is **not** as proposed by H_0.	

Figure 6.3 Changes in the logic of hypothesis testing with *t* statistics compared to *z*-scores.

retical population"—it consists of all the potential subjects in the world (for James, all humans) after they have gone through the experimental **treatment.** The subjects in the experiment form one small sample of this population.

James assumes that if his subjects make judgments on their own (without hearing the confederates make their judgments first), their average estimate of the dowel length would be equal to the dowel's actual length, 12 inches.* But prior studies suggest that subjects will conform, to at least some extent, to the confederates' judgments. Since the rules of logic dictate that we can't conclusively

---

*You might question this assumption—can we be sure that people would estimate the dowel length accurately even without seeing the confederates' judgments? What if people have a systematic tendency to overestimate the lengths of all wooden dowels? It is possible to redesign this experiment to take care of this concern, but for now, let's just assume that the 12-inch judgment is a good standard to use as the basis for the null and alternative hypotheses.

prove that people conform, we try to prove that they don't not conform. That is, we make the following hypotheses:

$$H_0: \mu = 12 \qquad\qquad H_a: \mu \neq 12$$

If we can establish that there is a very low probability that the mean judgment is 12 inches, we can conclude that the confederates are having some effect on judgments.

Note that we're using a bidirectional hypothesis here, because it's conceivable that the effect could go in the opposite direction to what James expects. In **Section 6.3.1,** we'll discuss directional tests using t.

The second step in the hypothesis testing procedure is to run the experiment and calculate relevant **descriptive statistics.** For hypothesis testing with the **t statistic,** we need to know the **sample mean,** the **sample standard deviation,** and the **number of subjects** in the experiment. These values, which we calculated earlier, are:

Descriptive Statistic	Value in James's Experiment
Sample mean $\overline{X}$	14.0
Standard deviation s	2.20
Sample size n	10

In the third step of the hypothesis testing procedure, we calculate our test statistic, **t.** We've already seen its formula several times. Plugging in the values above, we get:

$$t = \frac{\overline{X} - \mu}{s_{\overline{X}}} = \frac{\overline{X} - \mu}{s / \sqrt{n}} = \frac{14.0 - 12}{2.20 / \sqrt{10}} = 2.87$$

The numerator of the t statistic formula, $\overline{X} - \mu$, represents the difference between the observed sample mean for the experiment and the center of the distribution of all possible sample means, assuming that the null hypothesis is correct.

The denominator of the formula, $s_{\overline{X}} = s / \sqrt{n}$, represents an estimate of the standard deviation of the distribution of sample means.

By dividing the raw **effect size** by the standard error, we get a **standardized** measure of the estimated distance from the observed sample mean to the center of the distribution of sample means (Figure 6.4). This distribution looks a lot like a **normal** distribution, but it's not quite the same, as discussed in **Section 6.2.1:** It's a bit flatter and more spread out, with the amount of flattening determined by the **degrees of freedom** for the sample (here, $df = n - 1$

= 9). This difference will have implications for step 4 of the hypothesis testing procedure, which is discussed next.

Interactive Page 143

For a class project, James performs an experiment in which subjects are asked to judge the length of a 12-inch-long wooden dowel after observing four confederates who all overestimate the dowel's length, claiming that it is between 13 and 15 inches long. Ten subjects give the following estimates (in inches) of the dowel's length:

13.9 13.0 15.1 12.4 13.4 13.3 16.9 13.1 10.7 18.2

James uses a bidirectional hypothesis test with H_0: μ = 12 to assess whether or not the confederates had a significant effect on subjects' judgments. He calculates the sample mean $\overline{X}$ to be 14.0 inches, with a sample standard deviation s of 2.20 inches. The t statistic for the experiment is thus 2.87.

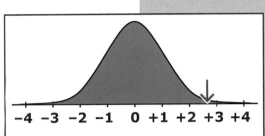

Figure 6.4 Location of the t statistic for James's experiment in the distribution of t statistics with $df = 9$.

In the fourth and final step of the hypothesis testing procedure, we compare our computed **t statistic** to the **t distribution** with $n - 1$ **degrees of freedom.** More specifically, we determine the proportion of the area of this distribution that falls outside (in the tails) of the calculated t value.

Use the **t Distribution Area** Calculation Tool to calculate this p-value. This Calculation Tool works just like the **Normal Distribution Area** Tool, except that you must enter the degrees of freedom (df) as well as the t statistic. Figure 6.5 shows what the Calculation Tool looks like when you enter 2.87 and 9 for t and df, the values we calculated on the previous page.

Calculation Tool: t Distribution Area

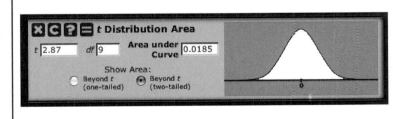

Figure 6.5 Screenshot of the t Distribution Area Calculation Tool, illustrating the area beyond $t = 2.87$ in the two tails of the t distribution for 9 degrees of freedom. (The area is so small that it takes up just a few red pixels in each tail.)

Note that we do not use the **t Distribution Area** Tool to calculate the probability of getting a t value of 2.87 exactly. Rather, we calculate the probability that t is ≥ 2.87 or ≤ -2.87. As you can see, values of t this extreme make up a very small proportion (1.85%) of the t distribution with 9 df (.00924. of the distribution is ≥ 2.87 and .00924 of the distribution is ≤ -2.87; add these two probabilities up and we get the two-tailed p-value we're interested, .0185).

This tells us that if μ were really 12, as assumed under the null hypothesis, the probability of James observing an $\overline{X}$ of 14.0 or higher (or an $\overline{X}$ of 10.0 or lower) is only $p = .0185$. Since this probability is so low, we conclude that the initial assumption of H_0 is wrong: The population mean is probably *not* 12. Hence the experimental manipulation most likely had an effect on subjects' judgments of dowel length.

Looked at another way, our analysis indicates that if the null hypothesis were true, 98.15% (1 – .0185) of all sample means with $n = 9$ would result in t statistics that were less extreme than the one observed by James. It is possible that H_0 really is true, and James happened to draw an anomalous sample. But it is much more likely that H_0 is false, and that the population mean is really some value greater than 12. (So then, what is the exact value of the population mean? We can't know for sure, but **Chapter 9** discusses how we can use James's data to estimate the value of μ.)

If you have to find the p-value for a t statistic and don't have access to a computer, you may need to use a statistical table. The procedure for looking up p-values in a t Distribution Table is illustrated in Figure 3 of the **Statistical Tables** Box (see p. 89).

Box: Statistical Tables

6.3.1 Directional t Tests

In James's experiment, as in most studies, we had to consider the possibility that the experimental effect would be in the opposite direction to that anticipated—that is, that people might in fact *under*estimate the dowel length after seeing other people make overestimations. Therefore, we performed a standard **bidirectional** hypothesis test.

Sometimes, though, we only have to worry about effects in one direction or the other. Consider the following situation:

> The U.S. Food and Drug Administration (FDA) specifies that children should get at least 250 milligrams (mg) of vitamin C in their diets every day. There is no harm in consuming more vitamin C, but children who get less than 250 mg have an increased risk of contracting diseases. Dr. Ceci, a researcher for the FDA, is assigned to study whether or not children are meeting this minimum requirement. She designs a study in which the diets of 500 randomly selected children are carefully monitored by their parents for a single day; the parents report exactly what their kids

ate. Researchers then calculate vitamin C intake for the day, and an average is calculated. Dr. Ceci decides to use a single-sample *t* test to evaluate the data, and chooses a null hypothesis of:

$$H_0: \mu = 300 \text{ mg}$$

reasoning that if children *averaged* only 250 mg, a great many would not be meeting the minimum requirement. She chooses to use a directional alternative hypothesis:

$$H_a: \mu < 300$$

because she will recommend action (an ad campaign to inform parents of health risks) only if the mean is significantly *less* than 300. If the mean is 300 or greater, nothing needs to be done.

The data are collected, and the average vitamin C intake is calculated to be 302 mg, with a standard deviation *s* of 25 mg. What should Dr. Ceci conclude?

Interactive
Page 145

The *t* statistic for this experiment is calculated as follows:

$$t = \frac{\overline{X} - \mu}{s_{\overline{X}}} = \frac{\overline{X} - \mu}{s / \sqrt{n}} = \frac{302 - 300}{25 / \sqrt{500}} = 1.79$$

When you enter this *t* value and the associated **df** (499) in the **t Distribution Area** Calculation Tool and choose to show the area "Beyond *t* (one-tailed)," you should find that the proportion of the **t distribution** that falls beyond *t* = 1.79 is about .037 (Figure 6.6a). This low **p-value** (< .05) leads many students to think that H_0 should be rejected.

However, the correct decision is *not* to reject H_0. Why? *Because the effect is in the opposite direction to that specified by* H_a.

The graphic in the Calculation Tool is misleading in this case because the Tool assumes that you care about the area of the curve greater than a positive *t* value or less than a negative *t* value. However, because H_a stated that μ would be less than 300, we want to know the chances of drawing a sample whose *t* statistic is less than or equal to the one we observed, if μ were really 300. That is, we are interested in the area of the curve *less than* +1.79. This probability is shown in Figure 6.6b. Since the total area under the curve is 1.0, the area to the left of +1.79 is equal to 1.0 − .037 = .96.

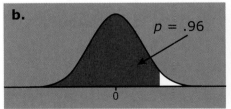

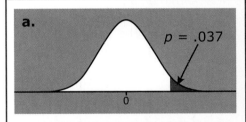

a.   *p* = .037

b.   *p* = .96

Figure 6.6 (a) The area of the *t* distribution falling beyond *t* = 1.79, taken from the *t* Distribution Area Calculation Tool. (b) Area of the *t* distribution that corresponds to the probability value for Dr. Ceci's hypothesis test. This is the inverse area to that displayed by the Calculation Tool.

Remember, the FDA doesn't care if children are getting more than 300 mg of vitamin C. They only need to know if they're getting less than this amount. Since the observed mean is 302, it is clear that the data are not consistent with a significant deficiency in vitamin C intake.

This example illustrates the confusion that can ensue when directional hypothesis tests are employed, especially when research results in an outcome counter to the researcher's expectations. Although bidirectional hypothesis tests may initially seem more complicated than directional tests, the logic of hypothesis testing is more consistent with two-tailed than with one-tailed tests.

Moreover, the goal of Dr. Ceci's experiment was really to estimate the average vitamin C content of kids' diets, not to test any one specific hypothesis (hence the somewhat arbitrary choice of $\mu = 300$ for the null hypothesis). It would be better to analyze Dr. Ceci's results with a confidence interval, discussed in **Chapter 9.**

Now is a good time to work through the first four **Review Exercises,** which challenge you to tackle on your own the four steps of hypothesis testing with t. These exercises will also help you understand some of the nuances of directional and bidirectional hypothesis testing.

Interactive Page 146

6.4 Assumptions for t Tests

In **Section 5.5,** we saw that the following **assumptions** have to be met in order for conclusions based on the **z-score** test statistic to be valid:

- The sample of subjects in the experiment must be drawn at random from the experimental population.

- The **sampling distribution** of the experimental population must be approximately **normal.**

- The standard deviation σ of the nonexperimental population must be known.

- The standard deviation of the experimental population must be equal to the standard deviation of the nonexperimental population.

As we've discussed, the great advantage of hypothesis testing with t **statistics** rather than z-scores is that the last two assumptions in this list don't have to be met. However, the first two assumptions are just as important for the t statistic as for the z-score test statistic. Let's consider these two assumptions more carefully now, since they will also apply to many other **test statistics** that we'll discuss later in the course.

The random-sampling and normality assumptions are as important for t tests as for z tests.

Interactive
Page 147

6.4.1 Random Sampling

If every person in the world behaved exactly the same, we could do experiments with a single subject and be perfectly confident that the result obtained for that subject holds for the entire human **population.** Unfortunately (for behavioral science researchers, anyway), people are different. (And rats, rabbits, monkeys, and other types of subjects used in animal research aren't much more consistent than humans.)

To overcome this difficulty, we use **inferential statistics** to take the results of a **sample** of a variable population and draw reasonable conclusions about the population as a whole. But this generalization only works if the sample is **representative** of the population as a whole. Representative and nonrepresentative (also called **biased**) samples are illustrated in Figures 6.7a and b, respectively.

A sample only generalizes to a population if the sample is representative of the population as a whole.

The population that the researcher assumes she's studying is represented in these figures by a green-shaded normal **distribution** of scores on some behavioral measure. Sampled individuals are represented by red arrows. In Figure 6.7a, the sample fairly represents the population: There are a few individuals from the top and bottom of the population distribution, but most of the people in the sample come from the middle of the distribution.

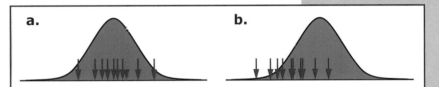

Figure 6.7 Sampled individuals (red arrows) from a population (green curve; this figure is animated on the *Interactive Statistics* website and CD). In (a) the sample is representative of the population as a whole, but in (b) the sample is not representative.

We call this a random sample because it's as if we put all the individuals from the population in a giant bag, jumbled them up, and drew our sample at random from the bag. If we select samples in this way, they'll usually be representative of the population.

A sample of randomly selected individuals will usually be representative of the population as a whole.

In Figure 6.7b, however, a disproportionate part of the sample is drawn from the lower part of the population distribution. It should be clear to you that conclusions drawn from a biased sample such as this will not reflect the population as a whole.

So how do we make sure we have representative samples? Well, occasionally a sample will turn out to be nonrepresentative purely by chance. This is when the randomness of the sampling process, which usually works to our advantage, catches up with us. This situation will be discussed in **Chapter 7,** when we cover hypothesis tests gone bad.

But more often, a sample is biased because of poor planning on the experimenter's part. This is avoidable if we know what to look out for, and this knowledge is best gained by looking at some examples of flawed experimental designs.

Interactive Page 148

Here's our first example of an experimental design in which the random-sampling assumption is violated:

> Dr. Kappa wants to challenge the hypothesis, widely held in his field, that Americans wear hats less often these days than they did in the 1930s and 1940s. He conducts an experiment in which he stations research assistants at each of the eight gates of his local baseball team's stadium and has them count how many people are and are not wearing hats on one particular game day. As it happens, it rains on the day of the experiment. The percentages of hat-wearers at each gate turn out to be 60, 57, 42, 56, 69, 49, 37, and 70, resulting in a grand average of 55%, with a standard deviation of 11.83. For comparison, Dr. Kappa uses a 1938 study that found that 45% of Americans wore hats regularly. Testing his data against this figure, Dr. Kappa calculates that $t = 2.39$ with 7 df. This t value has an associated p-value of .048. "Hah!" he declares at the A.F.P. (Association of Fashion Psychologists) conference, "Americans in the 21st century wear hats more often than they did in 1938."

The problem with this study is that the **sample** has not been drawn from the general **population** of all Americans. (Actually, there are other problems with this study, but here we're just interested in this particular misstep.) Instead, the experiment included only Americans who attend baseball games on rainy days. Dr. Kappa's conclusion may apply to this population, but there are good reasons to suspect that other Americans in other circumstances would wear hats much less frequently.

This example is silly, of course, but subtler versions of Dr. Kappa's mistake are made all the time. For example, many behavioral science researchers draw subjects from their experiments exclusively from introductory-level college courses. These researchers often imply that conclusions drawn on the basis of their experiments are applicable to the general population of all adult humans, but their subjects have a strong tendency to be younger, smarter, and more motivated than the typical adult.

Nonrepresentative samples are usually a result of flawed research designs.

Studies that sample a subset of a population can only draw conclusions about that subset.

Interactive Page 149

Now see if you can find the flaw in the following experimental design:

> Mr. Kidd, a school superintendent, is concerned about the poor performance of students in his district on a standardized science test for sixth graders. He hypothesizes that the problem is due to the textbook the district has chosen for the science classes. He chooses two classes from two different schools, with 20 students in each class, and has the teachers of these 40 students use a new textbook. Encouragingly, Mr. Kidd finds a significant increase in test scores for these students over the average score for the previous year, with *t* (39 degrees of freedom) = 2.54, p = .015.

The problem here is that Mr. Kidd has not drawn his "subjects" at random from all the students in his district. Rather, he has drawn two *classrooms* at random from all the classes in the district. The difference is very important.

In statistical jargon, we say that the observations in Mr. Kidd's experiment are not **independent.** Two events are independent if the outcome of the first event is not related in any way to the outcome of the second event. This is not the case here because the students in the classrooms chosen for the study are affected by all kinds of factors that don't necessarily apply to students in other classes: the particular teaching style of the teachers, the particular interactions among the particular sets of students, the socioeconomic status of students at these schools compared to others in the district, and so on.

Thus Mr. Kidd's **sample size** in this study is not $n = 30$ students, but $n = 2$ classrooms. With $n - 1 = 1$ degree of freedom, a *t* of 2.54 corresponds to a *p*-value of .239, not .015. To test the impact of the new textbook properly, Mr. Kidd either has to have 40 students from 40 different classes use the new textbook (an impractical design, since you can't have different students in the same classroom using different texts), or he has to treat each classroom as a "subject" and include more than two classes in his study. It's probably safe to assume that the average scores of different classrooms are independent of each other (as long as the chosen classes aren't all drawn from the same school).

Interactive Page 150

6.4.2 Normal Sampling Distribution

The second **assumption** that has to be met to use a *t* **test** is that the **sampling distribution** of the experimental population must be

Glossary Term: independence

If experimental observations are not independent of each other, conclusions based on the observations will be flawed.

approximately **normal.** As discussed in **Section 4.3.5,** with large sample sizes, the Central Limit Theorem guarantees that sampling distributions are normal even for underlying populations that are not at all normal. So with $n \geq 40$ or so, this assumption is virtually always met.

With slightly smaller samples, an **asymmetrical** population distribution—one that is **skewed** in one direction or another—will result in a more or less non-normal sampling distribution. Nevertheless, the t statistic is still **robust** to most violations of the normality assumption as long as n is approximately 15 or higher. (A procedure is said to be robust if conclusions based on the procedure hold even when assumptions are violated.)

With small samples, however, asymmetrical population distributions lead to serious problems for the t statistic. So use the following rules when deciding whether or not the use of a t test is appropriate:

- If **$n < 15$,** use a t test only if the data appear roughly normal (i.e., not skewed strongly in one direction or another), and no apparent outliers are present in the data. If these conditions are not met, consider a **nonparametric statistic** as an alternative, or try to collect data from more subjects.

- If **$15 < n < 40$,** use a t test unless the data are extremely skewed. You also need to be cautious if there are significant outliers.

- If **$n > 40$,** you can use a t test even if the data are strongly skewed. A couple of outliers won't hurt things either, although outliers should always be noted when discussing the results.

Guidelines for using t tests with non-normally shaped sample distributions.

Exercise 6.5, Exercise 6.6, and **Exercise 6.7** quiz your knowledge of when the various assumptions discussed in the last few pages are met or violated.

6.5 Reporting Hypothesis Tests

Because of the roundabout logic involved in the **hypothesis** testing procedure, researchers are forced to employ some linguistic gymnastics when they report conclusions drawn with the aid of hypothesis tests. Outside the meeting rooms of a scientific conference, you might overhear researchers talking about experiments that either "worked" or "failed," and discussion sections of journal articles will sometimes include statements that cast experimental results in black and white. But when actually reporting **test statistics,** we are constrained by the fact that the hypothesis testing procedure only gives us shades of gray.

The easiest way to describe how hypothesis tests are reported is through examples. The report of James's class experiment might include the following language:

> The 10 subjects' estimates of the dowel length averaged 14.0 inches, with a standard deviation of 2.2 inches. This sample mean is significantly different from the actual length of the dowel (12 inches), $t(9) = 2.87$, two-tailed $p = .019$.

This description gives readers all the information they need to calculate the test statistic themselves and confirm that James did his statistics properly. This should be the goal in any research report. Note that the **degrees of freedom** for the *t* **statistic** should always be given in parentheses after the letter *t*, as shown here.

A report of a hypothesis test should provide enough information for the reader to recalculate the test statistic.

However, this description is also somewhat verbose, and much of the relevant information (e.g., the sample size and standard deviation for the general population) might be given elsewhere in the report. Also, hypothesis tests are usually assumed to be **bidirectional** unless specified otherwise. To shorten the report a bit, James might write:

> The mean length judgment, 14.0 inches, was significantly greater than the actual length of 12 inches, $t(9) = 2.87$, $p = .019$.

Remember that the phrase "**statistically significant,**" often shortened (as it is here) to "significant," just means that $p < \alpha$, allowing the researcher to reject the null hypothesis.

Note another subtle change in the second statement compared to the first: the subjects' length estimates are said to be "greater than," rather than just "different from" the actual length. Even though the bidirectional **alternative hypothesis** states simply that the experimental population mean is not the same as the general population mean, a successful rejection of the **null hypothesis** allows a researcher to conclude that the effect is in the direction implied by the sample mean.

6.6 When Should You Use *t* vs. *z*?

The answer to the question heading this section is pretty simple: Except in very rare circumstances, always use *t.* Even if you know σ for the nonexperimental population (which is rare in itself), with the *z* test we also have to assume that the standard deviation of the experimental population is perfectly predicted by the standard deviation of the nonexperimental population.

Real-world research designs should almost always be analyzed with *t* tests rather than *z* tests.

In contrast, with the *t* test we use the sample standard deviation *s* to estimate σ for the experimental population. Therefore, it's safer to use *t* than *z* even if σ is known.

The *t* test also has the advantage that it can be used in research designs where there is no nonexperimental population at all, as discussed in the next section.

Interactive Page 153

6.7 What You Can Test with *t*

A *t* **statistic** can be used to conduct hypothesis tests in a large variety of research situations. In the present chapter, we've discussed comparing the mean of a single experimental sample to some standard value. This is called a **single-sample *t* test,** since only one group of experimental scores is involved. The standard against which the sample mean is compared can be any of the following:

- An assumed value for the nonexperimental population. For example, in James's conformity experiment, he simply assumed that without his **experimental treatment,** people would judge the dowel length to be 12 inches on average.

- A value that has been well established by previous research on the population being studied. This was the standard in Dr. Kappa's experiment, in which the data were compared to the percentage of hat wearers from an earlier study.

- Any arbitrary number, such as the FDA's vitamin C recommendation in Dr. Ceci's experiment.

Note that in Kappa's and Ceci's experiments, there was no true experimental manipulation: The researchers simply tested whether the mean of a population differs from a hypothesized value. The logic of hypothesis testing applies equally well to these situations as to true experiments.

Although the single-sample *t* test has its uses, a more common research design involves testing two samples of subjects, one of which has and one of which has not undergone the experimental manipulation. Sometimes the two samples are **related,** as in research designs where one group of subjects receives a pre-test before the manipulation and a post-test afterwards. In other designs, the two samples are **independent** of each other.

These designs can also be tested with the *t* statistic, but the procedures for calculating *t* are somewhat different. We'll cover these types of *t* tests in **Chapter 8.**

Glossary Term: single-sample *t* test

Research designs in which a single-sample *t* test is useful

Related- and independent-samples *t* tests

Interactive
Page 154

6.8 Chapter Summary/Review

This Chapter Summary/Review is interactive on the *Interactive Statistics* website and CD. Also be sure to go through all the **Review Exercises** for this chapter.

- It is not feasible to use **z-scores** in most real-world research situations because σ, the standard deviation of the experimental population, is almost never known. To get around this difficulty, we can use the sample standard deviation *s* from the experimental data as an estimate of σ. When the **standard error** of the distribution of sample means is calculated using s rather than σ, it is called the estimated standard error, and symbolized $s_{\bar{X}}$ (although we usually drop the word "estimated," since we are almost always working with $s_{\bar{X}}$ rather than $\sigma_{\bar{X}}$). A **test statistic** calculated using $s_{\bar{X}}$ is called a ***t* statistic (Section 6.2).**

- Because *s* is never a perfect estimate of σ, distributions of *t* statistics are never perfectly **normal.** The amount of deviation from normality is determined by the degrees of freedom for the test statistic; for a single-sample *t* statistic, ***df*** $= n - 1$. Figure 6.2, reproduced below, shows how the **distribution** of *t* statistics varies with sample size (**Section 6.2.1**).

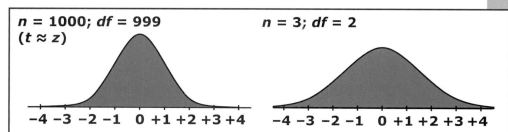

$n = 1000; df = 999$
$(t \approx z)$ $n = 3; df = 2$

−4 −3 −2 −1 0 +1 +2 +3 +4   −4 −3 −2 −1 0 +1 +2 +3 +4

Figure 6.2 Distributions of *t* statistics for four different sample sizes *n* and degrees of freedom *df* ($df = n - 1$).

- The hypothesis testing procedure for comparing a mean from a single sample to a specified value using *t* is almost identical to the procedure using **z** (**Section 6.3**). The main differences are:

  ○ We substitute *s* for σ in the test statistic formula (**Section 6.3**):

$$t = \frac{\bar{X} - \mu}{s_{\bar{X}}} = \frac{\bar{X} - \mu}{s / \sqrt{n}}$$

  ○ The probability value associated with the calculated *t* statistic must be determined based on the *t* distribution with $n - 1$ degrees of freedom, rather than the normal distribution. The ***t* Distribution Area** Calculation Tool will provide the appropriate *p*-values (**Section 6.3**).

- **Directional** hypothesis tests work the same way with *t* as described in the last chapter for *z*. In **Section 6.3.1**, we illustrated the difficulty in interpreting the outcome of a directional test when the experimental effect is in the opposite direction to that predicted by the alternative hypothesis.

- The hypothesis testing procedure will only work properly if the subjects in an experiment are sampled randomly from the population about which the experimenter would like to draw conclusions (**Section 6.4.1**). The observations in an experimental sample must also be **independent:** The score for any one subject must not be influenced by any other subject in the experiment (**Section 6.4.1**). The other major **assumption** for *t* tests is that the sampling distribution must be normally distributed. This assumption will almost always be met as long as there are at least 15 subjects in an experiment and the data are not extremely skewed. More detailed guidelines are given in **Section 6.4.2**.

- When hypothesis tests are reported, the most common convention is to use the phrase "statistically significant" when $p < \alpha$ for a treatment effect. This phrase does not imply that the effect was theoretically important or even particularly large. It just means that there was enough evidence to reject the hypothesis that the treatment had no effect whatsoever (**Section 6.5**).

- In choosing between using the *z* or *t* statistic for a hypothesis, there is really no choice: The *t* statistic should almost always be used, even if σ for the nonexperimental population is known (**Section 6.6**). The **single-sample *t* test** described in this chapter can be used any time a mean from a single sample is compared to some fixed value. If the comparison value is the mean of a different condition in the same experiment, a slightly different procedure must be used, as described in upcoming chapters (**Section 6.7**).

Chapter 7
Making and Avoiding Hypothesis Test Errors

7.1 When Hypothesis Tests Fail

Most researchers conduct experiments because they expect (or at least hope) to find certain effects. Ideally, an experimental manipulation really does cause an effect (i.e., a deviation from the **null hypothesis**), and a **test statistic** confirms that the null hypothesis can be rejected. Unfortunately, however, this is not the only possible outcome of an experiment and accompanying hypothesis test. In fact, there are four potential outcomes, which are described below and depicted graphically in Figure 7.1:

1. The experiment may "work": here, there truly is an experimental effect, and the researcher rejects the null hypothesis. This is the only outcome we have discussed in any detail up to now.

Possible combinations of experiment and hypothesis test results

2. No experimental effect exists, and the researcher correctly does not reject the null hypothesis. In such a situation, the experiment is usually judged to have "failed" (we usually would not bother doing an experiment if we didn't expect to find some effect of the experimental manipulation), but the hypothesis test correctly serves its purpose.

3. The researcher rejects the null hypothesis even though there really is no experimental effect. Here, the hypothesis test has caused an error in judgment—we are incorrectly led to believe that an effect exists. This situation is called a **Type I error.**

4. The researcher does not reject the null hypothesis when there really *is* an effect. Again, this is an error (more specifically, a **Type II error**), since we are led to abandon support for an effect that in reality does exist.

In the first part of this chapter we will explore the three undesirable outcomes. Two are hypothesis testing errors and one is a correct—though unfortunate and ambiguous—outcome of the testing procedure. Then in the second part of the chapter, we'll examine a technique called **power** analysis, which allows us to assess the **proba-bility** of one type of error, and sometimes do something about lowering this probability, before the experiment is conducted.

		Actual Treatment Effect	
		No effect: H_0 true	Effect exists: H_0 false
Conclusion of Hypothesis Test	Reject H_0	Type I error	Experiment "works"
	Fail to reject H_0	Experiment "fails"	Type II error

Figure 7.1 The four possible outcomes of an experiment and subsequent hypothesis test. If the experimental treatment really does cause an effect and the null hypothesis is rejected, the experiment is generally considered a success (green-shaded cell).

Before we get started with this discussion, you should be warned that the subject of hypothesis test errors is confusing and frustrating for students and researchers alike. When a researcher conducts an experiment and tests a null hypothesis, she must make a decision to reject H_0 or not. As you can see in Figure 7.1, either decision may lead to an error. The problem is that we *never* know with absolute certainty whether or not a treatment actually has an effect, so we can never say with absolute certainty whether or not an error has been made!

The best we can do is to try to understand the circumstances that can lead to hypothesis test errors and the steps we can take to minimize errors. To properly address these topics in this chapter, we will generally need to assume that we actually *do* know whether or not H_0 is true in the research examples discussed. As you read on, keep in mind the plight of the poor researcher, and consumers of research, who must carry on without the benefit of this omniscient knowledge.

Interactive
Page 157

7.1.1 Rejecting a True Null Hypothesis

Consider the following research scenario:

> A group of anthropologists working in Egypt discovers a necklace bearing an ancient inscription that claims the wearer of the necklace will be granted extrasensory perception (ESP). Most of the scientists are skeptical of this claim, but one, Bob, decides to test the powers of the necklace. He gathers a group of volunteers and has each one predict the outcomes of coin flips while wearing the necklace. For each person, he flips a coin 25 times and has them predict heads or tails for each flip. By chance alone, each person should get about 50% of his or her predictions correct, and Bob reasons that anyone who predicts correctly more than half the time is showing evidence of ESP. In fact, Bob knows from his graduate training that one can perform a hypothesis test for things like coin flipping with a z-score test statistic: the null hypothesis is $\mu = .50$, the sample size is the number of coin flips (25 in this case), and the standard deviation is $\sigma = .25$.

> Bob tests 20 volunteers and finds that 19 of them get between 11 and 13 out of 25 predictions correct, revealing no evidence at all of ESP. One person, however, gets 15/25 predictions, or 60%, correct while wearing the necklace, and Bob calculates that for this participant in his study, $z = 2.0$, $p = .046$. He whips up a paper excitedly claiming that he has the statistical evidence to prove that the necklace really does confer ESP, at least to certain people. However, his article is rejected by *Anthropology Today* with an angry letter from the editor demanding that he never submit to her journal again. Where did Bob go wrong in his analysis?

To appreciate Bob's mistake, we have to keep in mind the logic of **hypothesis** testing: The **z-score test statistic** locates the mean of one particular experimental sample in the overall **sampling distribution** for the given sample size. A sample mean that computes to a z score of 2.0 is unlikely ($p = .046$), but by chance alone, we will get a sample mean this extreme (i.e., ≥ 2.0 or ≤ -2.0) occasionally—about once out of every 20 samples ($1 / 20 = .05$). And this is exactly what happened in Bob's project: He tested 20 people (i.e., he collected 20 samples of coin-flip predictions), and one of them produced a "significant" result.

Over the years, thousands of studies have tested for ESP, and almost every scientifically sound experiment has failed to find any evidence of the paranormal. Thus we can be fairly sure that in Bob's project, *the null hypothesis is actually true:* people wearing the necklace will average exactly the same proportion of correct coin-flip predictions, 50%, as people not wearing the necklace. Therefore, the hypothesis test that rejected this null hypothesis probably produced an incorrect conclusion—an error.

Interactive Page 158

When a test statistic is used to reject a **null hypothesis** that is actually true, we say that a **Type I error** has occurred (see Figure 7.2). There is no way to completely guard against Type I errors, because samples that produce extreme means will occur randomly from time to time, and (by definition) we cannot predict **random events.**

		Actual Treatment Effect	
		No effect: H_0 true	Effect exists: H_0 false
Conclusion of Hypothesis Test	Reject H_0	Type I error	Experiment "works"
	Fail to reject H_0	Experiment "fails"	Type II error

Figure 7.2 The four possible outcomes of an experiment and subsequent hypothesis test. If the experimental treatment actually has no effect but the null hypothesis is still rejected, a Type I error has occurred (green-shaded cell).

We can try to minimize Type I errors, however, by avoiding research designs that require too many hypothesis tests.* If we adopt an **alpha** level of .05, only trusting the results of experiments that produce ***p*-values** of .05 or lower, we will get a Type I error at most 1/20th of the time. That is, out of every 20 experiments in which the null hypothesis is actually true (experiments where the experimental population mean is equal to the μ specified by the null hypothesis), null will be rejected (incorrectly) only once. As noted above, however, if we run 20 experiments all at once it actually becomes fairly likely that the null hypothesis will be rejected once in a while because of random chance factors.

---

*How many is "too many?" Unfortunately, there is no way to give a single general-purpose answer to this question; this is a judgment call on the part of the researcher (and those who evaluate the researcher's work).

Glossary Term: Type I error

Type I errors can be minimized by limiting the number of hypothesis tests in a research project.

Interactive
Page 159

The only way to decrease the likelihood of Type I errors is to reduce the **alpha level** used in evaluating **test statistics.** In fact, the **alpha** level can be defined as the **probability** that a **Type I error** will occur. As discussed on the previous page, with $\alpha = .05$, there is a 1/20 chance that a Type I error will occur if the **null hypothesis** is true. But if we reduce α to .01, the odds go up to 1/100 that a sample mean will be large enough to be declared **significant** when the null hypothesis is true.

Adopting a smaller alpha level will lower the chances of a Type I error.

However, the downside of decreasing α is that the scientific community may dismiss experimental treatments that produce effects that are real, but too small or unstable to produce test statistics with very small *p*-values. The psychology literature, for example, is full of experiments that found important results with $p > .01$ but $< .05$.

The better way to deal with Type I errors is to make sure that experimental results are replicable. If an apparently significant result is spurious, an attempt to replicate the effect in a new experiment will usually fail to beat the .05 level. But if an effect is real, multiple experiments that test for the effect should *consistently* produce *p*-values less than .05. Real effects stand the test of time. Type I errors will always be weeded out, as long as someone is willing to put in the effort to try to replicate them.

Experimental replications will weed out Type I errors.

7.1.2 Accepting a False Null Hypothesis

Interactive
Page 160

The flip side of a **Type I error** is a **Type II error,** which occurs when an effect does actually exist (i.e., the **null hypothesis** is false), but the ***p*-value** of our **test statistic** is so high that we fail to reject the null (Figure 7.3).

Glossary Term:
Type II error

		Actual Treatment Effect	
		No effect: H_0 true	Effect exists: H_0 false
Conclusion of Hypothesis Test	Reject H_0	Type I error	Experiment "works"
	Fail to reject H_0	Experiment "fails"	Type II error

Figure 7.3 The four possible outcomes of an experiment and subsequent hypothesis test. If the experimental treatment actually does have an effect but the null hypothesis fails to be rejected, a Type II error has occurred (green-shaded cell).

Consider the following experiment:

Carl, a biological psychologist, is studying the effect of paternal care on infant well-being, using a rabbit model. It is known that 20-day-old rabbits raised by both parents have an average body weight of $\mu = 60$ grams (g). Based on previous studies, Carl has good reason to believe that depriving the infant rabbits of their fathers will reduce their average body weight by 5–10 g. He performs an experiment with $n = 10$ rabbits and finds, sure enough, that $\bar{X} = 54$ (a decrease of 6 g from the normal infant body weight), with a standard deviation of $s = 10$ g. However, when he calculates his test statistic, he finds that $t(9) = -1.90$, $p = .09$.

Should Carl conclude that lack of paternal care really doesn't affect rabbit development? Not necessarily. To see why, imagine that in a second experiment he tested 25, rather than 10, infant rabbits and found exactly the same sample mean, 54, and exactly the same standard deviation, 10. Calculate t and use the **t Distribution Area** Calculation Tool to find **p** for this new experiment.

Interactive Page 161

On the previous page, we saw an experimental situation with the following facts:

- H_0: $\mu = 60$

- H_a: $\mu \neq 60$

- $n = 10$

- $\bar{X} = 54$

- $s = 10$

- $t = -1.90$

- $df = 9$

- $p = .09$

Despite the fact that the **null hypothesis** could not be rejected, the experimenter, Carl, had good reason to believe that H_0 was in fact false. And when Carl ran Experiment 2 with 25 rabbits instead of 10 and obtained the same sample mean and standard deviation, he found **$t(24) = -3.00$, $p = .006$**.

Thus it appears that in Experiment 1, a **Type II error** occurred: The experimental treatment did in reality have an effect, but the null hypothesis was not rejected. In statistical lingo, we say that the experiment was not **powerful** enough to reveal the treatment effect. With a standard deviation of 10 and an **effect size** of 54 – 60 = –6, the null hypothesis cannot be rejected with a sample size of 10. With a sample of 25 subjects, however, Experiment 2 was powerful enough to reject H_0 for an **effect size** this large.

In **Section 7.4,** we will learn a procedure called power analysis that can help experimenters avoid Type II errors. For now, note that it is usually very difficult to distinguish between a Type II error and a situation where the null hypothesis is actually true. In other words, for experiments that fail to reject H_0, it is virtually impossible to know whether there really was no experimental **treatment** effect, or whether there was a treatment effect that was too small to be revealed by the hypothesis test.

It is difficult to distinguish between Type II errors and actual null effects.

Interactive
Page 162

As we've already seen, **power** can be increased, and the likelihood of **Type II errors** decreased, by increasing the sample size of an experiment. This observation sometimes leads researchers to employ the tactic described in the following anecdote:

The larger the sample size, the smaller the chances of a Type II error.

> Diane, an enthusiastic graduate student in education psychology, has developed a new one-week training program that she is sure will increase fifth graders' problem-solving abilities. She recruits 30 volunteer fifth graders, puts them through the training course, and administers a standard problem-solving test, for which it is known that the general population of fifth graders averages $\mu = 50$. She obtains a sample mean $\overline{X} = 54$ with $s = 12$, from which she calculates $t(29) = 1.83$, $p = .078$. Diane decides that she really needs this experiment to work in order to nail down her master's thesis, so she tests three more subjects, and the sample mean goes up to 54.5, while the standard deviation stays constant at 12. Now, with $n = 33$, she calculates that $t(32) = 2.15$, $p = .039$! Should Diane get her master's?

Though Diane's intentions may have been good, her actions were not statistically sound. It might appear that the second, more favorable, hypothesis test resulted from the increase in power that was obtained by running three more subjects. In reality, however, Diane simply benefitted from a lucky break.

Since her training program is brand new, Diane did not know before the experiment began how large an effect to expect. Based on the original sample of 30 subjects, the **effect size** of her treatment (the problem-solving training program) appeared to be $54 - 50 = 4.0$. The extra three subjects tacked on at the end of the experiment raised this effect size to 4.5, but this increase might have been due to one or two students who were particularly good problem solvers to begin with (or who were particularly good guessers on the test). What if instead she happened to have recruited a particularly dull fifth grader as her 33rd subject, and the sample mean with $n = 33$ turned out to be 53.5? In this case, t would have gone down to 1.68 and p would have gone up to .10.

If an experiment fails to reject H_0 and the researcher suspects a power problem, the correct response is to design a new experiment that addresses the lack of power (raising the sample size is the primary means of raising power, as discussed in the next chapter). If Experiment 2 reveals a more compelling result (i.e., a smaller p-value), Experiment 1 can be deemed a pilot experiment, and the results can be published with confidence. Adding subjects a few at a time and hoping that p eventually dips below .05 is intellectually dishonest, and it rarely works in practice anyway.

As with Type I errors, the best way to remedy a Type II error is to do more research.

Interactive Page 163

7.1.3 Is the Null Hypothesis Ever Really True?

We've now covered three of the four possible hypothesis testing outcomes. The fourth and final possibility is that we correctly fail to reject the **null hypothesis** (Figure 7.4). Since most experiments are designed to reveal a **treatment** effect of some kind, this result usually means that the experiment has "failed," even though the hypothesis testing procedure has worked properly.

		Actual Treatment Effect	
		No effect: H_0 true	Effect exists: H_0 false
Conclusion of Hypothesis Test	Reject H_0	Type I error	Experiment "works"
	Fail to reject H_0	Experiment "fails"	Type II error

Figure 7.4 The four possible outcomes of an experiment and subsequent hypothesis test. If the experimental treatment has no effect and the null hypothesis is not rejected, the hypothesis testing procedure has worked correctly, but the experiment is usually deemed a failure (green-shaded cell).

However, we must be extremely careful in interpreting failures to reject H_0. Let's consider an example of this type of situation:

> Ernest works for the marketing department of a large chain of coffee houses. The chain has just started getting its beans from a new supplier, and Ernest is asked to find out if people prefer the new beans over those from the company's old supplier. Based on extensive prior company testing, Ernest knows that on a scale of 1 to 10, the general population of coffeeholics gave the old variety of beans an average rating of $\mu = 7.85$. He recruits 100 volunteers to taste

the new varieties of beans and finds that their average rating, using exactly the same testing procedures as before, is $\overline{X}$ = 7.80 (with s = 0.50), an apparent drop in satisfaction for the new beans. What should Ernest report to his superiors?

The first thing Ernest must do, of course, is perform the appropriate inferential test, which in this case is a **single-sample t test.**

In the absence of any compelling reason to suspect a **Type II error,** and given that his sample size (n = 100) was large enough to presumably provide sufficient statistical **power,** Ernest must conclude that the **null hypothesis** cannot be rejected, since $t(99)$ = 1.00, p = .32. In fact, since p is so high, you might be tempted to think (and Ernest might be tempted to report to his superiors) that this hypothesis test proves that the null hypothesis is actually true—that people like the new variety of beans *exactly* as much as the old variety.

But think about it: In an infinite universe, how likely is it that the two types of beans are so similar that there is absolutely no reason to prefer one over the other? Surely if every single coffee drinker in the world were tested, some degree of preference for one bean or the other would emerge, even if the difference came out to something as small as 7.8500000 for one bean and 7.8499999 for the other.

Strictly speaking, the null hypothesis is almost *never* true (except in the case of ESP experiments). And given a large enough sample size, even the smallest of effects will come out to be **significant.** In Ernest's case, if he had included 2000 rather than 100 taste-testers in his study and gotten the same **effect size** (a difference of 0.05 in taste ratings) and standard deviation, he would have found that t = −4.47, p < .0001, and he would have been forced to tell his boss that there was a significant preference for the old beans over the new ones.

> The null hypothesis is (almost) never true.

7.1.4 What *Can* We Conclude?

At this point, you may be feeling somewhat flummoxed. When you were first introduced to the hypothesis testing procedure in Chapter 5, you were told that due to limitations of logic, we can't prove that an experiment has an effect; instead, we are forced to try to prove that the experiment did not have no effect (i.e., we try to prove the **null hypothesis** false; see **Section 5.2**). We learned a procedure for doing this, but in this chapter you were told that:

- When we reject H_0, we can never rule out a **Type I error,** in which the null hypothesis is really true and by random chance we got an extreme sample that is unrepresentative of the experimental population as a whole.

- When we fail to reject H_0, we can never rule out a **Type II error,** in which the null hypothesis is really false but the experiment didn't have enough **power** to reject it.

- To top it off, the null hypothesis is almost never true to begin with, so even when a statistically powerful experiment fails to reject it, we can't claim that there was no experimental effect at all.

So just what *can* we conclude from a hypothesis test?

The answer to this question is that we can make some educated guesses about experimental results, but we have to be willing to live with a degree of uncertainty. The best thing that can happen is that a hypothesis test produces a very small **p-value,** giving us confidence that H_0 is truly false and that an effect exists. To seal the deal and be sure that a Type I error did not occur, another experiment should then be done that replicates the treatment effect from Experiment 1 and also produces a small *p*-value. This is why behavioral science journal articles usually include multiple experiments.

> When $p < .05$, we can state with some confidence that an effect probably exists.

When a hypothesis test produces a large *p*-value (say, .20 or greater), the researcher generally concludes that the experiment did not work. This does not mean, however, that the experimental **treatment** had absolutely no effect. Rather, the proper conclusion from such a result is that the treatment effect was so small that it could not be distinguished from a null effect on the basis of the hypothesis testing procedure.

> When $p > .20$, we can state that the effect is so small that it can't be distinguished from the null hypothesis.

Occasionally, a researcher may be pleased with a null result. For example, if a well-established theory strongly predicts that an effect *should* be found for a given experimental manipulation, it may be intriguing to find that an experiment that tests the manipulation fails to reject H_0. But any number of factors (faulty equipment, inattentive subjects, and dumb luck are some of the most common) can dilute an effect that would otherwise manifest itself in an experiment, so it is generally difficult for a researcher to convince the scientific community that a null result is meaningful.

The worst result of a hypothesis test is when *p* falls somewhere between .05 and .20. In this case, there appears to be some evidence against the null hypothesis, but not enough to allow us to claim an effect with any confidence. The only way to resolve such a situation is to do more research. However, we can try to avoid this situation in the first place by doing a power analysis before the experiment is even run, as discussed in **Section 7.4.2.**

> When $.05 < p < .20$, we can draw no conclusions at all about the experimental effect.

Before moving on, you should go through **Exercise 7.1, Exercise 7.2,** and **Exercise 7.3** to make sure you have a basic understanding of all the possible ways hypothesis tests can fail.

7.2 Fishing for Significance

Let's revisit the case of Bob the anthropologist (from **Section 7.1.1**) for a moment:

> A group of anthropologists working in Egypt discovers a necklace bearing an ancient inscription that claims the wearer of the necklace will be granted extrasensory perception (ESP). Most of the scientists are skeptical of this claim, but one, Bob, decides to test the powers of the necklace. He gathers a group of volunteers and has each one predict the outcomes of coin flips while wearing the necklace. For each person, he flips a coin 25 times and has them predict heads or tails for each flip. By chance alone, each person should get about 50% of his or her predictions correct, and Bob reasons that anyone who predicts correctly more than half the time is showing evidence of ESP. In fact, Bob knows from his graduate training that one can perform a hypothesis test for things like coin flipping with a z-score test statistic: the null hypothesis is $\mu = .50$, the sample size is the number of coin flips (25 in this case), and the standard deviation is $\sigma = .25$.

> Bob tests 20 volunteers and finds that 19 of them get between 11 and 13 out of 25 predictions correct, revealing no evidence at all of ESP. One person, however, gets 15/25 predictions, or 60%, correct while wearing the necklace, and Bob calculates that for this participant in his study, $z = 2.0$, $p = .046$. He whips up a paper excitedly claiming that he has the statistical evidence to prove that the necklace really does confer ESP, at least to certain people. However, his article is rejected by *Anthropology Today* with an angry letter from the editor demanding that he never submit to her journal again. Where did Bob go wrong in his analysis?

Bob performed 20 hypothesis tests, found one result **significant** at the .05 level, and tried to claim he'd found a psychic. The technical description of Bob's gaffe is an inflation of the **Type I error** rate (remember that a Type I error occurs when one rejects a null hypothesis that was actually correct). Less technically, we can say that Bob went "significance fishing"—if you cast your hook (the hypothesis testing procedure) into the stream enough times, you're bound to catch something (a "significant" result) eventually.

No one is issued a license to significance-fish, but researchers sometimes find themselves doing it unintentionally, often in ways more subtle than Bob's multiple casts for a single significant ESP result.

The thing we want to avoid here is making a small **p-value** the object of the scientific enterprise. A good researcher starts with a theoretical prediction, designs an experiment to test the prediction, then uses the hypothesis testing procedure as an aid in interpreting the results of the experiment. Imagine the following scenario:

> George is a student in a one-year master's program. He spent his first semester designing an experiment to test his advisor's latest theory, then spends the first half of the spring semester running the experiment. He analyzes the data, calculates the most straightforward test statistic, and finds that $p = .12$. Panic sets in: How is he going to get his degree with a measly p of .12?
>
> So he sits himself in front of the computer one Friday night and starts up his favorite statistics program. He reanalyzes his data, tries a new test statistic, and gets p down to .073. Emboldened, he does yet another reanalysis, then finds a dialog box that offers to calculate an **inferential statistic** that he's never even heard of. He clicks "OK," waits a few seconds while the computer churns away, and scrolls down to the p-value. Eureka—.004! "I don't even have to settle for $p < .05$," he thinks, "*my* experiment beats the .01 level." George's thesis is saved!

Or is it? The fact is that even a string of random digits will contain some kind of meaningful-looking pattern, and if you search long enough, especially with the aid of a computer, you'll find it eventually.

This is not to say that George's data are nothing more than a string of random digits. The pattern he found on his third analysis may very well be an important finding, worthy of reporting and maybe even worthy of a master's degree.

The exercise George engaged in that Friday night is called **exploratory data analysis**, or EDA for short. EDA is something of an art form that all researchers would do well to learn (see this **Box** for a quick look at the exploratory approach to data analysis). The problem is that when a pattern is found through EDA, the regular rules for interpreting hypothesis tests no longer apply. No matter how low the p-value was in his exploratory analysis, the effect George stumbled across must remain questionable until it is confirmed by a second experiment that is expressly designed to test for it.

A small *p*-value should be a means toward accomplishing a scientific goal, not the goal itself.

Box: Exploratory Data Analysis

Box: Exploratory Data Analysis

The hypothesis testing procedure emphasized in this text is known formally as "model fitting." We can see why it is called this by considering the following example. Say a researcher gives a standard problem-solving test to a set of 93 elementary school students enrolled in a "gifted" program, then uses a single-sample t test to decide whether or not the average score of the gifted students differs from the average score for other students of the same age. The model that is implicitly assumed in this analysis is that a single statistic (the sample mean) provides a good indicator of the typical score for the entire population being studied (in this case, the population would be all students that are, were, or will be enrolled in the gifted program).

But what if the distribution of scores for the sample were as pictured below? Although a single mean can be computed for these scores, the histogram clearly suggests that this one number would not adequately describe everyone in the sample, and by extension, in the population.

A better model would describe the population with two means, one for the left peak in the distribution and one for the peak on the right. The appropriate hypothesis test would then compare these two means, rather than comparing the single mean of all scores to some standard value.

Furthermore, by looking closely at which students scored lower on the test and which scored higher, we might be able to determine why we need two means instead of just one. For example, perhaps some students were selected for the gifted program on the basis of their artistic ability and others on the basis of their science aptitude. If the artsy students are scoring low on the test while the technically minded students are scoring high, the people administering the program might want to break the students up when doing exercises designed to improve problem-solving skills, so that the former aren't lost and the latter aren't bored.

This very useful conclusion would never be drawn if we stuck to the traditional model-fitting approach to analyzing data. Exploratory data analysis (EDA) emphasizes the study of graphical displays such as histograms to uncover unexpected results like the one found here. Instead of deciding on a model before the data are even collected, as we do in traditional data analysis, EDA practitioners try to discover a model by sifting through the data after they have been collected.

Although they seem like competitors, traditional and exploratory analyses are best used in tandem. Researchers almost always approach an experiment with some kind of model in mind, and it is sensible to test this model first once the data have been collected. But regardless of whether or not this first analysis reveals "significant" results, a sharp researcher will then use EDA techniques to see if any other model fits the data better than the model she started with. If so, this new model should form the basis of a new experiment, to confirm that the new model is really better than the old one.

One lesson of this section, as well as this chapter as a whole, is that no single experiment ever has the last word about an experimental effect. No matter how compelling the results of Experiment 1 look, you must wait until after Experiment 2 (and possibly until after Experiments 3 and 4 as well) is completed to be able to make strong claims. This is especially true if the effect from Experiment 1 was not the one you originally set out to look for.

7.3 Reporting Failed Hypothesis Tests

In **Section 6.5,** we discussed how hypothesis tests are reported in the happy event that the null hypothesis is rejected. Researchers must get more creative when they wish to report results for which the null was *not* rejected. Take Ernest's report to his superiors about coffee taste ratings. He might write:

> Previous research tells us that consumers gave the old variety of beans an average rating of 7.85. A new sample of 100 testers gave the new beans an average rating of 7.81, a difference that does not approach statistical significance, $t(99) = 1.00$, $p = .32$.

Note the word "statistical" near the end of this description, a favorite qualifier of researchers that allows them to make claims that seem stronger than they actually are. If Ernest was even more bold, he might say that the sample mean was "statistically indistinguishable" from the **null hypothesis,** implying that the two values are the same but technically only saying that there wasn't enough **power** to detect a difference. This is not to say that Ernest is being dishonest, but consumers of research articles need to know that "statistically identical" is not *really* identical.

The ultimate in flowery statistical language appears when the **p-value** comes out in the marginal range between about .05 and .20. Remember Carl's initial experiment with fatherless rabbits? If he were going to report this experiment, probably as a precursor to another experiment with more power and a more convincing result, he might write:

> The 10 infant rabbits in Experiment 1 showed a trend toward lower body weights, averaging 54.0 g as compared to 60 g for rabbits raised with their fathers. However, this statistical trend did not reach the traditional level of significance, $t(9) = -1.90$, $p = .09$. Therefore, we designed Experiment 2 with a larger sample size. . .

The phrases "statistical trend" and "marginally significant" should be taken to mean something like "my experiment almost worked, so please believe that the experimental manipulation does have an effect!" You should be wary of such claims, given what you've learned in this chapter about the difficulty in distinguishing "true" null hypotheses from **Type II errors.** As long as they're just used as supporting evidence for a more robust finding in a later experiment, though, it is perfectly legitimate to report such marginal effects.

One final note about reporting hypothesis tests: Always remember that the **descriptive statistics,** not the **inferential statistics,** tell the story of what happened in an experiment. Never report a *t* statistic (or any of the other **test statistics** you'll learn about later) without also reporting the mean(s) from which the test statistic was calculated.

Never report an inferential statistic without the descriptive statistic(s) on which it is based.

Interactive Page 169

7.4 Power

As noted in **Section 7.1.1,** the **probability** of committing a **Type I error** is determined by the **alpha** level an experimenter chooses when evaluating the *p*-value of a hypothesis test. Typically, α is set at .05, meaning that if the **null hypothesis** is true, there is a 1/20 chance of incorrectly rejecting it.

But what about the likelihood of committing a **Type II error,** incorrectly accepting a false null hypothesis? This probability goes by the name "**beta**" (β). Turning this assessment around, $1 - \beta$ gives the probability of correctly rejecting the null hypothesis when it is false. This latter value goes by yet another name that we've already mentioned several times: **power.** (From here on, we'll largely ignore β, which assesses the probability of the bad outcome, and focus instead on power, which more optimistically assesses the positive outcome.)

Glossary Term: power

Power is a relatively simple concept in the abstract: The power of an experiment is the probability that it will "work" (i.e., show a significant effect, assuming there is one). But power is extremely tricky to understand in detail. Fasten your seatbelts and do your best to hold on as we go through the next few sections, because once you have a firm grasp of experimental power, you will understand the whole hypothesis testing enterprise better.

Interactive Page 170

The best way to comprehend statistical **power** is to visualize the **probabilities** of rejecting and accepting an incorrect **null hypothesis.** In this section, we'll build up a graphic illustrating these probabilities in a series of steps.

We begin the hypothesis testing procedure by assuming that the null hypothesis (H_0) is true—that the experimental population has a **distribution** centered on μ_0. Given this assumption, we infer that the **distribution of sample means** for the population will have a mean μ_0 and a standard deviation $\sigma_{\bar{x}}$, as shown in blue in Figure 7.5c.

(Only parts c, e, and f of Figure 7.5 are shown in the *Interactive Statistics* printed text. The complete interactive figure, with all six parts, is on the *Interactive Statistics* website and CD.)

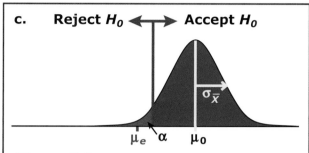

Figure 7.5c Graphical illustration of the power of a hypothesis test to reject an incorrect null hypothesis. Here, the sampling distribution proposed by the null hypothesis is shown along with the critical value of *t*, the "rejection region," and the estimated value for μ_0, μ_e.

If the sample mean falls far enough into one of the **tails** of this distribution, we will reject our initial assumption and conclude that the null hypothesis is wrong. The probability of the sample mean falling far enough out to reject H_0 is α, which is set by the experimenter.

The purple line in Figure 7.5c is the divider between sample mean values that will lead us to accept or reject H_0, a point known as the **"critical value."** Alpha is equal to the proportion of the area of the null hypothesis sampling distribution that falls beyond the critical value—the area shaded purple in the figure.

The alpha level establishes ranges of sample means that lead us to accept or reject the null hypothesis.

Note that in the figure, we show only one "rejection region," in the left-hand tail of the blue **sampling distribution.** However, you shouldn't take this to imply that we're assessing power for a one-tailed hypothesis test. For reasons we don't need to get into, the other tail of the sampling distribution can be ignored when assessing power, regardless of whether the test is **directional** or **bidirectional.** So assume throughout this chapter that we're using a two-tailed α level of .05. This means that 2.5% of the area of the blue sampling distribution in Figure 7.5 falls beyond the critical value.

Experimenters usually don't really think that the null hypothesis is true. Let's say we estimate the mean of the experimental population to be μ_e, as shown in Figure 7.5c. This value is pretty far out in the tail of the sampling distribution predicted by the null hypothesis, well into the rejection region defined by α. So you might think that if μ_e is an accurate estimate of the experimental mean, we will always be able to reject H_0.

Interactive
Page **171**

On the previous page, we established the following:

- If the **null hypothesis** is correct, the **sampling distribution** for an experimental population will be as shown in the blue distribution in Figure 7.5a.

- If the sample mean from the experiment were to fall in the area shown in purple in Figure 7.5b, we would consider it too extreme to have come from the blue distribution, so we would reject the null hypothesis.

- If the experimental population mean is actually μ_e, rather than μ_0, as shown in Figure 7.5c, it would seem very likely that null will be rejected, since μ_e falls well beyond the cutoff value defined by α.

But remember that the **sample mean** you obtain in an experiment will not always be exactly the same as the actual experimental mean. Even if we know the **population mean** is μ_e, the sample mean could be any value in the sampling distribution centered on this mean, shown in green in Figure 7.5d (click the "Add predicted sampling distribution" radio button on the interactive figure).

> Even when the null hypothesis is false, a sample mean could appear consistent with its sampling distribution.

Figure 7.5e (click the "Add some possible sample means" radio button on the interactive figure) illustrates why **Type II errors** occur: There is considerable overlap between the sampling distribution we would expect if μ_0 is correct and the sampling distribution we would expect if μ_e is correct. And keep in mind that before the experiment starts, we don't know which of these values is the true population mean. So any of the sample means indicated by the dancing red arrow in Figure 7.5e are consistent with both hypothesized means. (Of course, the sample mean could end up falling somewhere else—either farther to the left or farther to the right than the ones shown. The point of Figure 7.5e is to show that many of the possible sample means would lead to ambiguity about which of the two hypotheses to believe.)

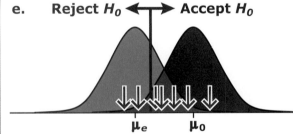

Figure 7.5e Graphical illustration of the power of a hypothesis test to reject an incorrect null hypothesis, with the predicted sampling distribution and some possible sample means added to Figure 7.5c. The complete interactive figure, with all six parts, is on the *Interactive Statistics* website and CD.

However, if we have a good estimate for μ_e, we can at least determine the **power** of the hypothesis test—the likelihood of rejecting the null hypothesis if the experimental population mean is actually μ_e rather than μ_0. Can you guess which of the following areas corresponds to this probability?

Interactive Page 172

The **power** of our hypothesis test is indicated by the striped area in Figure 7.5f: the portion of the **sampling distribution** assuming μ_e is true (shown in green) that falls to the left of the **critical value** for rejecting H_0.

Graphical definition of power

If you're having trouble following this section, it may help to remember that the two sampling distributions shown in Figure 7.5 are both hypothetical. We don't know what the experimental population mean is before the experiment begins. Indeed, the purpose of doing the experiment is to estimate this value! The **null hypothesis** makes one guess, μ_0, and μ_e is a second estimate, a value for μ that would be consistent with the **alternative hypothesis**.

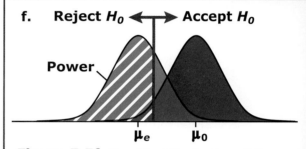

Figure 7.5f Graphical illustration of the power of a hypothesis test to reject an incorrect null hypothesis. The complete interactive figure, with all six parts, is on the Interactive Statistics website and CD.

The essential point of our discussion is this: Even if μ_e rather than μ_0 is the true population mean, there will still be a pretty high probability that we will fail to reject H_0. Why? Because samples are inherently variable, so the sample mean for an experiment will not always be exactly equal to the true population mean. The power of an experiment—the likelihood that we *will* reject H_0 if μ_e is the true population mean—is given by the proportion of the sampling distribution for μ_e that falls to the left of the critical value for the hypothesis test (the striped area in Figure 7.5f).

Given an estimate for μ_e, it is possible to calculate power—the likelihood of a successful rejection of H_0—exactly. We'll discuss this calculation in **Section 7.4.2** and discuss ways to increase power in **Section 7.4.3**. First, however, we'll look more closely at some factors that affect statistical power. (And before getting to these factors, you might want to try **Exercise 7.4** and **Exercise 7.5,** which review the symbols and terms used in power analysis.)

Interactive Page 173

7.4.1 Factors Affecting Power

Consider the following research proposal:

It has long been known that young people (i.e., the college students employed as subjects in a large proportion of behavioral science experiments) can keep about seven pieces of information in short-term memory at a time. For example, one study established that under normal circumstances, college students can memorize a list containing an average of 7.0 nonsense syllables (e.g., "jik," "lom," and "yud") and repeat the syllables back 30 seconds later. Dr. Durley, our gerontologist from Chapter 1, wants to assess the extent to which short-term memory declines in old age. He plans to give the nonsense syllables test to a group of 65-year-old subjects and compare their mean to that of the college students from the earlier study.

In **Section 7.1.3,** we asserted that the **null hypothesis** is almost never true—nearly every reasonable experimental manipulation has at least *some* effect. In the present example, we would certainly expect 65-year-olds to suffer at least a slight loss in short-term memory capacity compared to college students.

But we've also learned that even when H_0 is false, we will sometimes fail to reject it. What factors contribute to the likelihood of such an unhappy event (a **Type II error**)? Or, turning the question around to focus on the positive, what factors increase the likelihood of correctly rejecting H_0 (power)?

To begin answering this question, consider the formula for the test statistic that Dr. Durley will use, the single-sample **t statistic**:

$$t = \frac{\bar{X} - \mu}{s_{\bar{X}}} = \frac{\bar{X} - \mu}{s / \sqrt{n}}$$

The null hypothesis will be rejected if t is large enough to produce a **p-value** less than .05. Logically, then, any factor that works to increase t will increase power. An examination of the t formula reveals that:

- Power will increase as the **effect size** ($\bar{X} - \mu$) increases.

- Power will increase as the **standard deviation** (s, which is an estimate of σ) decreases.

- Power will increase as the **sample size** (n) increases.

For reasons we will discuss below, there is one more factor that affects power:

- Power will increase as the **alpha** (α) level increases.

Let's start by looking at the effect of effect size on power.

Factors that increase the value of t also increase power.

Interactive Page 174

Figure 7.6 illustrates how the size of the experimental effect in Dr. Durley's study will affect **power.** Let's say that a decrease in short-term memory capacity of one nonsense syllable would constitute a "medium" **effect size,** as shown in Figure 7.6b. The experimental power in this case would be as shown by the striped area in the figure.

But what if the effect size were actually smaller? Figure 7.6a shows what would happen if the population mean for elderly adults is 6.5, only 0.5 nonsense syllables lower than that for college students.

The larger the effect size, the greater the power.

Here, the **sampling distribution** for μ_e (6.5) overlaps much more with the sampling distribution for μ_0 (7.0). As a result, a smaller proportion of the former distribution falls to the left of the **critical value** (represented by the purple line in the figure), and power is lower.

Conversely, a larger effect size, as shown in Figure 7.6c, would lead to an increase in power, because the two sampling distributions would overlap less than with a medium effect size.

Note that in Figures 7.6a, b, and c, the sampling distribution proposed by the null hypothesis (shown in blue) and the critical cutoff value defined by α remain constant. The only thing that changes is our estimate of the center of the sampling distribution for the actual population mean (shown in green). The less the two distributions overlap, the greater the power, because there is a smaller chance that a sample drawn from the green distribution will fall to the right of the critical value.

Interactive Page 175

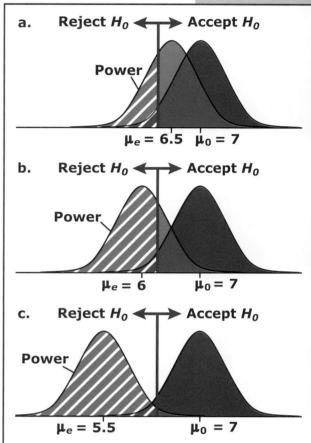

Figure 7.6   The effect of effect size on experimental power. A larger effect size (c) results in greater power than a smaller effect size (a).

Figure 7.7 shows how the standard deviation (σ, which is estimated by s in a t test) and sample size (n) affect **power.** Together, these two factors determine the **standard error** of the **sampling distribution** ($\sigma_{\bar{X}} = \sigma / \sqrt{n}$) for the hypothesis test. In Figure 7.7b, a medium-sized standard deviation and sample size lead to a moderate amount of power.

In Figure 7.7a, we see how a decrease in σ and/or an increase in n causes the standard error to shrink, thereby squeezing the sampling distributions centered on μ_0 and μ_e. Conversely, Figure 7.7c shows how increasing σ and/or decreasing n enlarges $\sigma_{\bar{X}}$, expanding the sampling distributions. The narrower the sampling distributions, the less they will overlap, and therefore the greater the power. Note that the critical value for rejecting H_0 also shifts with changes in $\sigma_{\bar{X}}$, further decreasing power when the standard error increases and vice versa.

The smaller the standard deviation, the greater the power.

The larger the sample size, the greater the power.

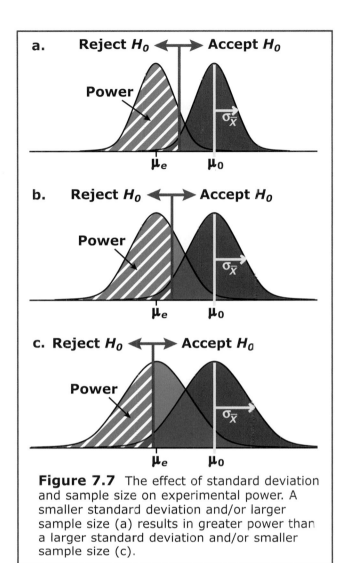

Figure 7.7 The effect of standard deviation and sample size on experimental power. A smaller standard deviation and/or larger sample size (a) results in greater power than a larger standard deviation and/or smaller sample size (c).

Interactive Page 176

The final factor affecting **power** is the **alpha** level, as shown in Figure 7.8. As α increases from a medium (Figure 7.8b) to a large (Figure 7.8c) value, a greater proportion of sample means from the distribution centered on μ_e are taken as strong enough evidence to reject H_0, so power increases. Conversely, if α is set very low (e.g., if the experimenter decides that **p** must be less than .01 for H_0 to be rejected; Figure 7.8a), fewer sample means will count as evidence against the **null hypothesis,** so power will decrease.

The relationship between power and alpha illustrated in Figure 7.8 should be intuitive if you think about it a bit. As α is decreased (say, from .05 to .01), the likelihood of a **Type I error**—claiming an effect when in fact there was none—decreases. At the same time, however, the likelihood of a **Type II error**—claiming a manipulation

The larger the alpha level, the greater the power.

Decreasing the chances of a Type I error increases the chances of a Type II error, and vice versa.

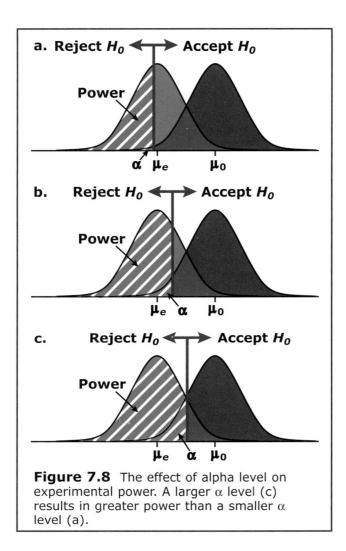

Figure 7.8 The effect of alpha level on experimental power. A larger α level (c) results in greater power than a smaller α level (a).

has no effect when in fact it does have one—*increases.* The opposite is also true: If you increase the likelihood of a Type I error by increasing α, you will concurrently decrease the chances of a Type II error.

7.4.2 Assessing Power

Power should be of great concern to all experimenters. If the power of an experiment is high, there is a very good chance that it will successfully reveal an effect. But if an experiment has low power, chances are that the **null hypothesis** will *not* be rejected *even if it is false.*

Unfortunately, however, researchers often plan studies with no idea about how much power their experiments will have. In part, this sad state of affairs is due to a lack of understanding on the part of many researchers about issues surrounding experimental power.

But an accurate assessment of the likelihood of **Type II errors** is difficult even for power-conscious scientists. The reason is that many of the factors noted in the previous section that affect power are unknown before an experiment begins.

Power is difficult to estimate because most factors that affect power are initially unknown.

Interactive Page 178

Of the four factors affecting experimental **power,** researchers have control over only two as they plan experiments: the α level and the sample size n.

However, if the researcher can come up with a reasonable estimate of the two unknown factors (**effect size** and **standard deviation**), it is possible to do an *a priori* **power analysis** of a planned experiment. Although the estimate of power one gets from such an analysis is never perfect, it's better than planning an experiment with no idea whatsoever of the likelihood of a **Type II error**!

Power can be estimated beforehand if the effect size and standard deviation of an experiment are estimated.

Calculation Tool: *t*-Test Power

The calculations involved in power analyses are fairly complex and won't be detailed here. But we do provide a Calculation Tool for calculating *t*-Test Power. To use the Tool, enter the known values for μ_0, n, and α, along with estimates for μ_e and σ, then click the "calculate" button.

For an example, let's go back to Dr. Durley's short-term memory experiment:

> It has long been known that young people can keep about seven pieces of information in short-term memory at a time. For example, one study established that under normal circumstances, college students can memorize a list containing an average of 7.0 nonsense syllables (e.g., "jik," "lom," and "yud") and repeat the syllables back 30 seconds later. Dr. Durley wants to assess the extent to which short-term memory declines in old age. He plans to give the nonsense syllables test to a group of 65-year-old subjects and compare their mean to that of the college students from the earlier study.

Let's say that Dr. Durley expects a 1.5-syllable deficit for his elderly subjects (that is, he expects them to be able to memorize an average of 5.5 syllables). He predicts a standard deviation of 3.0 syllables, and has 12 subjects lined up for the experiment. Since he will be comparing the 65-year-old mean to the college-student mean of 7 syllables found in the earlier research, he will use a **single-sample *t* test** (described in **Chapter 6**), with a **null hypothesis** of $\mu = 7.0$.

Enter the following values into the *t*-Test Power Calculation Tool to determine how much power Dr. Durley will have in this experiment:

- $\mu_0 = 7.0$

- $\mu_e = 5.5$

- $\sigma = 3.0$

- $n = 12$

- $\alpha = .05$

Make sure that the "Single/Related Samples" radio button is selected (we'll learn about **related**- and **independent**-samples tests later) and click the "calculate" button. You should get a power estimate of .32.

A **power** estimate of .32 means that there is a 32% chance the **null hypothesis** will be rejected, assuming the experimental population mean is as predicted. Note that this does *not* mean that there is a 32% chance of the null hypothesis being rejected if it is wrong, because it is possible that the population mean is neither μ_0 nor μ_e. If μ is actually 6.9, for example, we would have an even more difficult time rejecting H_0, even though it would still be incorrect. Our power estimate is only as good as our estimate of μ_e.

Looking at our power estimate from the other direction, there is a 68% chance that Dr. Durley will fail to reject the null hypothesis, *even if the elderly subjects really do remember 1.5 syllables fewer than college-age subjects!*

This statement is difficult for students (and many researchers) to grasp: If the experimental manipulation has a relatively substantial effect (here, we're talking about a 21% drop in memory capacity), how can the hypothesis test have such a high likelihood of failing?

The answer lies in the vagaries of **random events.** If you flip 10 coins, how many will come up heads? We know that on average, heads will come up half the time, so $\mu = 5$. But sometimes, by chance, you'll get 8, 9, or even 10 heads when you sample 10 flips.

Similarly, even if the average number of syllables a 65-year-old can remember is 5.5, there will be times when a sample of 65-year-olds will produce a mean $\bar{X}$ of 6, 6.5, or even 7 syllables. These sample means would all be statistically indistinguishable from the null hypothesis, given this sample size.

Obviously, Dr. Durley would like to reject the null hypothesis and claim a significant effect, especially if the effect really is as big as he thinks it is! So is there anything we can do to help him?

A power estimate gives the probability that the null hypothesis will be rejected if the population mean and standard deviation are as estimated.

Interactive
Page 180

7.4.3 Increasing Power

How can Dr. Durley increase the **power** of his experiment? Well, we've already noted that two of the four factors that influence experimental power are outside of the experimenter's control: The mean and standard deviation of the experimental population are what they are, and there isn't much we can do to change them. Theoretically, one could raise the **alpha** level to increase power, but practically speaking, the $p < .05$ cutoff for declaring results **significant** is almost universal in behavioral science research. Dr. Durley would be unlikely to get his research published if he used an α rate that was any higher.

But Dr. Durley does have control over the one remaining factor, **sample size**. As shown in Figure 7.7, by increasing n we decrease the **standard error** of the **sampling distribution**. Reducing the standard error in turn reduces the amount of overlap between the distributions predicted by the null and alternative hypotheses, thus increasing power.

The only practical way to increase power is to increase sample size.

It may help here to reconsider the coin flipping example introduced on the previous page. If you flip 10 coins, the likelihood of getting 70% (7/10) heads is pretty good. But if you flip 1000 coins, it's much less likely that you'll get heads 70% (700/1000) of the time. In the same way, increasing the sample size of elderly subjects will decrease Dr. Durley's chances of getting a sample mean that falls to the right of the **critical value** in Figure 7.7.

To see just how much power is affected by the sample size, go back to the *t*-**Test Power** Calculation Tool and enter the assumptions for Dr. Durley's experiment, changing the sample size from 12 to 20. You should find that power now jumps up to .56. So with 20 subjects instead of 12, the experiment will at least be more likely to reject than to accept the null hypothesis.

Interactive
Page 181

7.4.3 Increasing Power (2/3)

It is also possible for an experimenter to turn the **power analysis** calculation around and ask how many subjects are necessary to achieve a given amount of **power**. For example, suppose Dr. Durley wants at least an 80% chance of obtaining a **statistically significant** result. (This is considered a satisfactory amount of power by most researchers.)

Working in the *t*-**Test Power** Calculation Tool again, enter μ_0 (7.0), the estimates of μ_e (5.5) and σ (3.0), and the α level (.05) as before. Then leave the "Sample Size" text box blank and enter .80 in the "Power" text box. Now when you click the "calculate" button, the Tool should fill in the "Sample Size" box with the number 34. This tells us that to have an 80% chance of rejecting the **null hypothesis,** Dr. Durley will need to find 22 more subjects than he was planning to use.

Calculating the necessary sample size to achieve a given amount of power

Analyzing power during the planning stages of a study, as we've done here, can be a valuable exercise. If an experiment has such low power that there is less than a 50% chance it will show a significant effect, why would a researcher want to run the experiment in the first place? Clearly, it will be in the researcher's best interests to run more subjects and boost power to a more acceptable level. If it's not practical to run more subjects, it may be better not to conduct the experiment at all.

However, the problem with *a priori* sample-size analysis is that it is often difficult to come up with accurate estimates of μ_e and σ—as noted earlier, if we knew what μ_e was beforehand, there would be no need to do the experiment. And changes in these estimates can have dramatic effects on the estimated sample size needed to achieve a given amount of power.

For example, how would Dr. Durley's sample-size assessment change if he had estimated μ_e to be 6.0 instead of 5.5, or if he had estimated the standard deviation to be 2.0 instead of 3.0? Use the Calculation Tool to assess the number of subjects he would need for these situations:

(Note: you need to make sure the "Sample Size" text box is cleared each time before you click the "calculate" button in the ***t*-Test Power** Tool; otherwise, the Tool will assume you want to calculate power for the given sample size.)

Below are the results of the alternative sample-size analyses for Dr. Durley's experiment.

Subjects needed for .80 power with μ_e = 6.0:   73

Subjects needed for .80 power with σ = 2.0:   16

As these calculations make clear, the sample-size estimate will vary widely depending on the estimates for μ_e and σ. Often researchers will calculate the necessary sample size given a range of different estimates and use the average of these calculations in planning their experiments. (**Exercise 7.6, Exercise 7.7, Exercise 7.8,** and **Exercise 7.9** take you through the process of power analysis for another research study.)

An even better way to estimate **power** is to run a pilot experiment before beginning an ambitious research project. The pilot experiment might include 5 to 10 subjects and test only some of the experimental conditions planned for the main study. From the results of these first few subjects, reasonable estimates of μ_e and σ

A pilot experiment can greatly increase the accuracy of a sample-size analysis.

can be made, and the number of subjects needed in later experiments can be calculated with a greater degree of confidence.

Interactive
Page 183

7.5 Chapter Summary/Review

This Chapter Summary/Review is interactive on the *Interactive Statistics* website and CD. Also be sure to go through all the **Review Exercises** for this chapter.

- Experimenters usually hope that their experimental treatment has an effect, and they hope that a hypothesis test allows them to reject the **null hypothesis.** However, in addition to this happy combination of circumstances, there are three other potential outcomes of the hypothesis testing procedure (Figure 7.9; see below) (**Section 7.1**):

  - A **Type I error** occurs when the null hypothesis is actually true (there is no treatment effect), but we reject the null hypothesis. Studies that require many hypothesis tests are particularly prone to Type I errors, since by chance alone, true null hypotheses will be rejected 1/20 of the time if **alpha** is set at .05 (**Section 7.1.1**). (Type I errors are very common in extrasensory perception experiments, since the null hypothesis in such experiments is almost certainly true, so as a memory aid, you can remember that experiments testing for an "inner eye (I)" will lead to Type I errors.)

  - A **Type II error** occurs when the null hypothesis is actually false (there is a treatment effect), but we fail to reject the null hypothesis (**Section 7.1.2**).

  - If the null hypothesis is actually false (there is no treatment effect) and we fail to reject it, the hypothesis testing procedure has worked properly, but the outcome is usually considered disappointing. In truth, the null hypothesis is almost never true (nearly every experimental manipulation has *some* impact), so all we can really say if H_0 is not rejected is that any effect that the manipulation did have was minimal (**Section 7.1.3**).

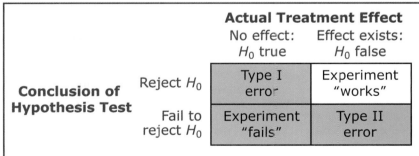

		Actual Treatment Effect	
		No effect: H_0 true	Effect exists: H_0 false
Conclusion of Hypothesis Test	Reject H_0	Type I error	Experiment "works"
	Fail to reject H_0	Experiment "fails"	Type II error

Figure 7.9 The four possible outcomes of an experiment and subsequent hypothesis test. The three green-shaded cells are all undesirable outcomes.

- Since we never know when a Type I or Type II error might occur, we can never place complete confidence in the outcome of any one experiment. The scientific community will usually not trust an effect until it is confirmed by multiple studies (**Section 7.1.4**).

- "Significance fishing" refers to the practice of making a low probability value the sole object of a research project. Hypothesis tests can and should be used to help us interpret results, but a small **p** does not in itself indicate that an experiment is meaningful; likewise, a large *p* does not mean that an experiment or its thesis is meritless (**Section 7.2**).

- When the *p*-value of a hypothesis test is greater than the α level set by the experimenter (usually .05), a researcher will often report that the result did not reach "statistical significance," or that the conditions in the experiment were "statistically indistinguishable." However, because of the possibility of a Type II error, we can never categorically state that an experimental manipulation has absolutely no effect. A *p*-value between .05 and .10 is often reported as a "trend" (**Section 7.3**).

- The **probability** of committing a Type II error (incorrectly accepting a false **null hypothesis**) goes by the symbol β. Statistical power is defined as the likelihood of correctly rejecting a false null hypothesis, and is equal to $1 - \beta$ (**Section 7.4**).

- **Power** can be graphically defined as the portion of the **sampling distribution** centered on an estimated population mean me that falls to the left of the **critical value** for rejecting H0 (where this critical value is determined by the alpha level chosen by the experimenter). See Figure 7.5, reproduced in part below (Section 7.4):

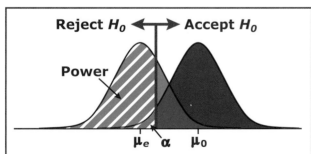

Figure 7.5 Graphical illustration of the power of a hypothesis test to reject an incorrect null hypothesis.

- Four factors affect the amount of power in an experiment (**Section 7.4.1**):

○ Power increases as the size of the experimental effect ($\mu_e - \mu_0$) increases (**Section 7.4.1**).

○ Power increases as the standard deviation (σ) of the experimental population decreases (**Section 7.4.1**).

○ Power increases as the sample size (n) increases (**Section 7.4.1**).

○ Power increases as α increases (**Section 7.4.1**).

• If estimates are made for μ_e and σ, power can be assessed during the planning stages of an experiment (the **_t_-Test Power** Calculation Tool will help with this assessment; see **Section 7.4.2**). If such an **analysis** reveals that the experiment will not have a satisfactory chance of rejecting the null hypothesis, power can be increased by increasing the sample size. The **_t_-Test Power** Tool can also be used to determine the sample size required to achieve a given amount of power (**Section 7.4.3**).

Chapter 8
t Tests for Two Means

8.1 Two-Condition Experimental Designs

Let's review our progression of hypothesis testing knowledge up to now.

- In **Chapter 5**, we learned how to use *z*-scores to test hypotheses about a population mean on the basis of a sample mean.

- However, to use a *z*-score as a test statistic, we have to know the standard deviation σ of the population in question. Since we almost never know this value, *z* tests are almost never used in practice.

- So in **Chapter 6,** we learned to use t statistics, in which we estimate σ with the standard deviation s of our sample, to test hypotheses.

But the form of t test introduced in **Chapter 6** is, like the z test, also rarely used in practice. The reason is that most research projects produce at least two sets of numbers, and the single-sample t test we learned about is only designed to handle, as its name implies, a single set of data.

To make this discussion more concrete, let's return to James's research project from **Chapter 6.** Here is our original description of the project:

> For a class project, James performs a small experiment replicating a classic social psychology study on conformity. Subjects in the experiment are asked to judge the length of a wooden dowel that is 12 inches long. But before they make their judgments, the subjects observe four confederates (friends of James's who are acting as if they are also randomly chosen subjects) who all overestimate the dowel's length, claiming that it is between 13 and 15 inches long. Ten subjects give the following estimates (in inches) of the dowel's length:
>
> 13.9 13.0 15.1 12.4 13.4 13.3 16.9 13.1 10.7 18.2
>
> The idea behind the experiment is that subjects will alter their judgments to conform to the estimates made by the other people in the group. *James reasons that without seeing the confederates make their judgments, the subjects should guess, on average, that the dowel is its true length, 12 inches.* Therefore, he plans to perform a bidirectional hypothesis test with H_0: $\mu = 12$.

In **Section 6.3** we noted (in a footnote) that the assumption in italics above is questionable: We cannot be completely sure that the average dowel length estimate without peer pressure would be exactly 12 inches. Researchers hate questionable assumptions, because they open up opportunities for critics. Consider the following hypothetical debate between James and a critic of his project:

> **James:** "My sample of 10 subjects estimated the dowel to be an average of 14 inches long, two inches longer than the actual length of 12 inches. Therefore, peer pressure must have caused them to alter their judgments in order to conform."
>
> **Critic:** "You might have just gotten lucky—if you put 1000 people in your experimental situation, perhaps

Most experiments include multiple sets of data that must be compared to each other.

the mean would come out to 12 inches. In other words, the population mean might actually be 12, but your sample included, by chance, a bunch of overestimators."

James: "Ah, but I conducted a *t* test, which indicates that the probability of drawing a sample of 10 people whose mean is $\bar{X} = 14$ from a population with mean $\mu_0 = 12$ (given an estimated standard deviation of 2.20) is only $p = .019$."

Critic: "Good point. OK, I concede that the population mean for peer-pressured dowel estimators is not 12. In fact, I'll grant that the mean for this population might even be 14. But, what if the population mean for *non*-peer-pressured estimators is also 14? That is, perhaps your data simply show that people have a general tendency to overestimate the length of dowels. Peer pressure may have nothing to do with it."

James: "Ummm. . ."

Critic: "Or, what if subjects making the judgment alone would actually estimate the dowel length at an average of 16 inches? If this were the case, your data would actually show that people alter their judgments in order to go *against* their peers."

James: "Ahhh. . ."

Instead of ummming and ahhhing, what James should do is re-run his experiment, this time using two conditions: a **control condition** and an **experimental condition.**

The experimental condition is usually the one in which the interesting manipulation occurs. Here, the experimental condition can be identical to James's original experiment: subjects estimating the length of the dowel after seeing other people make estimates of between 13 and 15 inches.

In the control condition, the researcher tries to keep things as close as possible to the experimental condition, but omits the crucial experimental manipulation. So James should use the same dowel, position it and the subjects in exactly the same places, and use the same instructions when asking subjects to make their judgments. The only difference should be that in the control condition, subjects don't see other people making judgments before they are asked to judge the dowel's length.

Control conditions are used to counter potential criticisms about experimental conditions.

If the mean dowel-length judgment in the experimental condition is significantly longer than the mean judgment in the control condition, James will have an effective counter to his critic's arguments.

Since James will have two means to deal with in this new experiment, he will no longer be able to use the single-sample *t* test to evaluate the significance of his results. He can still use a *t* statistic, however. The exact procedure he will use to conduct his *t* test will depend on the details of how he designs his experiment:

- James could have each subject make the dowel judgment twice, first before seeing the confederates make their judgments and then a second time after witnessing the other judgments. This would be a **repeated-measures** experimental design, and he would evaluate the results using a **related-samples *t* test**. This design and test are covered in **Section 8.2** of this chapter.

- Alternatively, James could employ two different groups of subjects, one making judgments alone and one making judgments under peer pressure. Here, he would be using an independent-samples design, which is evaluated using a **independent-samples *t* test**. This design and test are covered in **Section 8.3.**

We'll use new examples to cover these two types of tests, then we'll come back to James in **Section 8.5** and compare the pros and cons of the two experimental designs.

> Repeated measures and independent samples experimental designs

Interactive Page 187

8.2 Repeated Measures

Consider the following example:

> Sergeant Thursday is a dog trainer for the K-9 division of the Springfield Police Department. He develops a special olfactory training program to increase a dog's ability to sniff out illegal drugs in closed car trunks. To make sure the program works, he takes a sample of 7 dogs and has them "inspect" 50 cars, half of which have drugs planted in the trunks and half of which are drug free. He then puts these 7 dogs through the training program, then runs them through another test with the 50 cars (mixing up which ones have the drugs in them and which don't). The dogs' scores on the sniff tests are:

Dog's name	Rex	Bo	Tiger	Lassie	Smokey	Dexter	Shadow
Initial score	23	18	25	28	33	30	22
Score after training	35	16	30	34	42	29	30

(Here, a "score" is the number of cars that a dog correctly identified as either containing or not containing drugs. There are better and more complicated ways of calculating scores in this type of experiment, but we won't go into them here.)

Can Sergeant Thursday claim that his training program helps dogs learn to sniff out drugs better?

As discussed in the previous section, there is a crucial difference between the research design in this experiment and those in the experiments described in Chapters 5 and 6. Until now, we have been comparing the **mean** of a set of scores collected in an experiment to some established standard. Now we're comparing two different sets of scores that were both collected in the same experiment.

More specifically, Sergeant Thursday's experiment utilizes a **repeated-measures** design. Each subject is tested once (the dogs sniff the cars), then the experimental manipulation is applied (they undergo the training program), then the same kind of test is repeated (the dogs sniff the cars again). We end up with a pair of observations for each subject.

There are other types of repeated-measures designs, which you can read about in this **Box.** For Thursday's design, we usually call the assessment before the experimental manipulation a pre-test and the final assessment a post-test. He hopes that his training program will allow dogs to correctly identify more drug-laden cars in the post-test than in the pre-test. In other words, he's betting that the mean score for the post-test will be greater than the mean score for the pre-test. Was his prediction confirmed? Calculate the two means and try making your own assessment.

Glossary Term: repeated-measures design

Box: Repeated-Measures Experimental Designs

Box: Repeated-Measures Experimental Designs

The primary example of a repeated-measures design given in the text is the pre-test/post-test design, in which a group of subjects is assessed on some behavioral measure both before and after an experimental manipulation.

However, any study in which the same subjects are assessed multiple times on a similar measure qualifies as a repeated-measures design. Here are a few more examples:

1. In a memory experiment, subjects are asked to remember a list of words including 10 concrete nouns (e.g., "dog," "brick," "pencil") and 10 abstract nouns ("love," "mind," "war"). Twenty minutes after learning the list, the subjects are asked to recall all the words. The research

Box: Repeated-Measures Experimental Designs *(continued)*

question is which class of words is easiest to remember. Each subject contributes one score for concrete nouns and one score for abstract nouns.

2. In an experiment on family dynamics, college students answer questionnaires designed to assess how much they confide in their mothers and fathers about their romantic relationships. The mean "confide-ence" score for mothers is then compared to the mean score for fathers.

3. In a developmental psychology experiment, the size of infant boys' and girls' vocabularies are tested at ages 18, 24, 30, and 36 months, to determine whether one gender acquires words faster than the other.

The essence of a repeated-measures experimental design is that multiple observations are collected for each subject, so subjects act as their own controls. To see why this is advantageous, consider what would happen if the memory experiment described above were conducted using a between-subjects, instead of a repeated-measures, design.

Instead of having all subjects memorize both types of words, the experimenter could have had subjects in Group A memorize concrete nouns and subjects in Group B memorize abstract nouns. But what if, by the luck of the draw, Group B had better overall memories than Group A? In this case, the experimenter might observe a higher mean memory score for abstract nouns than for concrete nouns, even though countless previous experiments have demonstrated that concrete nouns are actually easier to remember.

In the repeated-measures design, variability between different subjects is much less of an issue. Subject A might have a better overall memory than subject B, but if concrete nouns are easier to remember than abstract nouns, both subjects should remember more of the former than the latter.

The third example above makes two additional points about repeated measures-designs. First, the measure may be repeated more than twice (here, vocabulary is assessed at four different ages). When we have more than two observations per subject, we can no longer use a *t* statistic to analyze the data; instead, we must use an analysis of variance (ANOVA), covered in **Chapter 10.**

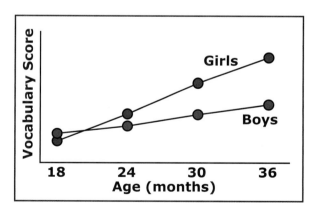

The second important point is that repeated-measures variables (age) can be combined with between-subjects variables (gender) in the same experiment. This experimental design might result in the graph above, which indicates that while boys have vocabularies that are equal to or better than girls' at 18 months, girls' vocabularies subsequently increase at a faster rate than boys' vocabularies.

For this group of dogs, Sergeant Thursday's training program worked as planned: Sniff scores rose for the post-test (for which the mean was 30.86) compared to the pre-test (for which the mean was 25.57). But when evaluating research results, we always have to keep in mind the first question a critic will ask: Couldn't the observed effect have arisen solely from chance processes? Perhaps the training program doesn't really have any effect at all, and the pre- and post-test means only differ because of random factors.

In **Chapter 5,** we learned that the best way to deal with this criticism is to restate it in the form of a **null hypothesis** (H_0) and try to prove this hypothesis false. If we can do this, then the **alternative hypothesis** (H_a), which says that the experimental manipulation does have an effect, is indirectly supported.

For **repeated-measures** designs, we state the null and alternative hypotheses in terms of individual subjects' **difference scores.** As its name implies, a difference score D represents the difference between the pre- and post-test scores. By convention, we subtract the pre-test score (X_1) from the post-test score (X_2):

$$\text{Difference score} = D = X_2 - X_1$$

A D score indicates the experimental effect on an individual subject. Subtracting the post-test score from the pre-test score would work just as well for the hypothesis testing procedure, as long as you're consistent and as long as you're using a **bidirectional** hypothesis test.* But when calculated as shown above, D scores reveal the effect of the experimental manipulation on each individual subject. A positive D score indicates that the effect of the training program was to raise sniff scores. A negative D score suggests that a dog's drug-sniffing ability actually suffered as a result of the training. And a D score of 0 implies that the training had no effect at all.

Now let's look at how we conduct a *t* test on these difference scores.

8.2.1 The Related-Samples *t* Test

By computing **difference scores,** we've converted two numbers into one for each subject. Now we can conduct a *t* test in exactly the same way we did for single samples of scores in **Chapter 6.** The only difference is in terminology, since we're working now with D's instead of X's.

You may sometimes see this type of hypothesis test called a repeated-measures *t* test in a journal article. But for reasons that will

Glossary Term: difference score

Glossary Term: related-samples t test

---

*If you subtract post-test scores from pre-test scores instead of vice versa, the sign of the **test statistic** will change, but this will have no effect on the outcome of a bidirectional hypothesis test, since the *p*-value for bidirectional tests is determined solely by the magnitude (i.e., absolute value) of the test statistic.

become clear later in the chapter, we'll refer to this procedure by the more general title **related-samples *t* test.**

Just as we did with the **z** test and the **single-sample *t* test,** we'll go through each of the four steps in this hypothesis testing procedure in turn using the example we've been working with:

> Sergeant Thursday takes a sample of 7 dogs and has them "inspect" 50 cars, half of which have drugs planted in the trunks and half of which are drug free. He then puts these 7 dogs through a special olfactory training program he's developed, then runs them through another test with the 50 cars (mixing up which ones have the drugs in them and which don't). The dogs' scores on the sniff tests are:

Dog's name	Rex	Bo	Tiger	Lassie	Smokey	Dexter	Shadow
Initial score	23	18	25	28	33	30	22
Score after training	35	16	30	34	42	29	30

> Can Sergeant Thursday claim that his training program helps dogs learn to sniff out drugs better?

Our first job is to establish the **null** and **alternative** hypotheses, H_0 and H_a. We've already informally discussed what these will be: The unseen critic claims that the population mean for the post-test (μ_2) is identical to the population mean for the pre-test (μ_1), while Sergeant Thursday claims that there is a difference between these two population means.

Remember, though, that in a **repeated-measures** design we test the difference scores, rather than the raw X_1's and X_2's. So we need to state our two hypotheses in terms of μ_D, the population mean of the differences between X_1 and X_2. What value for μ_D would be proposed by critic who believes there is no effect?

Interactive Page 190

The **null hypothesis** for a **related-samples *t* test** is almost always $\mu_D = 0$. If an experimental manipulation has no effect, then the **distribution** of difference scores will be centered on 0, as shown in Figure 8.1. The **bidirectional alternative hypothesis** states that μ_D is some value other than 0, because the experimental manipulation has the effect of changing post-test scores relative to pre-test scores. So we have:

$$H_0: \mu_D = 0 \qquad\qquad H_a: \mu_D \neq 0$$

The null hypothesis for a related-samples *t* test states that the mean of the difference scores is 0.

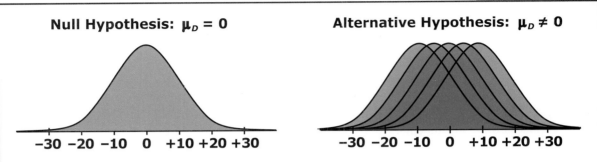

Figure 8.1 Graphical illustration of the null and alternative hypotheses for a related-samples *t* test (this figure is interactive on the *Interactive Statistics* website and CD). H_0 states that the mean of the population of difference scores is exactly equal to 0. The bidirectional H_a states that this mean is something other than 0. The latter hypothesis makes no claim about the magnitude or direction of the effect.

(As with single-sample *t* tests and *z* tests, we can also frame our alternative hypothesis in a **directional** way. Look back at **Section 5.4** to review the benefits and limitations of directional tests.)

Now that the **null** and **alternative** hypotheses have been stated (thus completing step 1 of the hypothesis testing procedure), our next task (step 2) is to collect and describe the sample data. Since you're reading about this experiment in a textbook, the (fictional) data have already been collected, but remember that in the real world, this is the most time-consuming and important part of the research process.

Describing the data in this case means calculating **D** scores for each subject and finding the mean and standard deviation of these differences. The following table illustrates this procedure:

Dog	X_1	X_2	D	$(D - \bar{D})^2$
Rex	23	35	12	45.08
Bo	18	16	−2	53.08
Tiger	25	30	5	0.08
Lassie	28	34	6	0.51
Smokey	33	42	9	13.80
Dexter	30	29	−1	39.51
Shadow	22	30	8	7.37
			$\Sigma D = 37$	$SS = 159.43$
			$\bar{D} = 5.29$	$s_D = 5.15$

In practice, no one in the age of computers actually computes difference scores manually. Statistics packages such as SPSS allow you to enter the data for the pre- and post-tests in two columns, and you direct the program to compute the related-samples *t* statistic for the two columns of data. The program does all the intermediate steps for you, including the calculation of *D* scores.

While the Calculation Tools accompanying this textbook are no match for SPSS, the **Data Transformations** Tool will at least release you from the tedium of calculating *D* scores individually. To use the Tool for this purpose, enter the post-test scores in the *X* text box and the pre-test scores in the *Y* text box (as with other Tools, it's often easiest to copy and paste sets of scores into these text boxes). Make sure the first *X* and the first *Y* correspond to your first pair of scores, the second *X* and second *Y* correspond to your second pair of scores, and so on. Then select "*X* – *Y*" in the "Transformation" pull-down menu and click the "calculate" button. The difference scores will be calculated and placed in the "Results" text box on the right side of the Tool.

Once you've calculated the *D* scores, you can copy and paste them into the **Descriptive Statistics** Tool to calculate their mean and standard deviation. Do this now for Sergeant Thursday's results and confirm that the mean and standard deviation you get match those in the table above.

Once we have the mean and standard deviation of the **difference scores,** we can calculate *t* (step 3 in the hypothesis testing procedure). The formula for the related-samples *t* statistic is identical to that for the single-sample *t*, except that we substitute *D*'s for *X*'s:

$$ t = \frac{\bar{D} - \mu_D}{s_{\bar{D}}} = \frac{\bar{D} - 0}{s_{\bar{D}}} = \frac{\bar{D}}{s_D / \sqrt{n}} $$

In words, this formula says to take the observed sample mean $\bar{D}$, subtract the population mean μ_D predicted by the null hypothesis, and divide by the estimated standard error $s_{\bar{D}}$. Since H_0 almost always predicts that μ_D will be 0, we can drop this term out of the equation, and we can expand the denominator of the equation from $s_{\bar{D}}$ to $s_D / \sqrt{n}$.

Try using this formula now to calculate *t* for Sergeant Thursday's experiment (recall that we've already calculated $\bar{D}$ = 5.29 and s_D = 5.15). Enter the *t* statistic, along with the appropriate *df*, below.

For Sergeant Thursday's experiment, *t* is calculated as follows:

$$ t = (\bar{D} - 0) / (s_D / \sqrt{n}) $$
$$ = (5.29 - 0) / (5.15 / 2.65) $$
$$ = 2.72 $$

The **related-samples *t* statistic** has *n* – 1 **degrees of freedom**, where *n* is the number of *pairs* of scores (i.e., the number of subjects in the study). Sergeant Thursday tested seven dogs, so his *t* statistic has 7 − 1 *df.*

The final step (step 4) in the hypothesis testing procedure is to evaluate the **null** and **alternative** hypotheses in light of the **test statistic.** In practice, this means finding the appropriate ***p*-value** and deciding whether or not it is low enough to reject H_0. Computing and evaluating test statistics and *p*'s may have become rote to you by now, but it is important to remind yourself periodically of the underlying logic:

- We assumed, for argument's sake, that μ_D for the experimental population is 0.

  Reviewing the logic of hypothesis testing

- We used the standard deviation of our sample of *D* scores to estimate σ_D for the population of *D* scores. A population with mean 0 and standard deviation σ_D will have a **sampling distribution** with mean 0 and standard deviation $\sigma_D / \sqrt{n}$. This value is called the **standard error**, but because we're using s_D to estimate σ_D, we're working here with an **estimated standard error**.

- Using the estimated standard error $s_{\bar{D}}$, the observed sample mean $\bar{D}$, and the hypothesized population mean 0, we calculated a standardized statistic called **t**. In this case, *t* = 2.72.

- If one were to take an infinite number of samples of size *n* from a population with mean 0, the resulting distribution of *t* statistics would look something like the one depicted in Figure 8.2a. Recall that the **t distribution** is similar to the distribution of **z-score** test statistics, except that the *t*'s are slightly more spread out because of the **variability** introduced by the estimation of σ. The more **degrees of freedom** we have (where *df* = *n* − 1), the closer the *t* distribution is to being exactly normal.

- The location of our *t* in the distribution is indicated by the red arrow in Figure 8.2b. The likelihood of getting a *t* value this high is given by the proportion of the area of the *t* distribution that falls to the right of this arrow. Since we're conducting a **bidirectional** hypothesis test here, we add in the area that falls to the left of *t* = −2.72. All told, the **probability** of getting a *t* value this extreme, if the population mean of *D* scores is as predicted by H_0, is .0346.

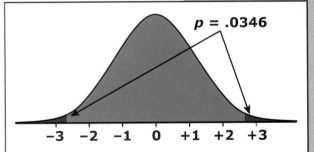

Figure 8.2 The logic of hypothesis testing with *t* for two related samples. This figure is interactive, with four parts, on the *Interactive Statistics* website and CD.

- Since this probability is so small, we surmise that the initial assumption we started out with, that μ_D was 0, is probably wrong. We reject H_0 in favor of H_a, and conclude that Sergeant Thursday's training program most likely does have the effect of raising sniff scores.

To calculate *p*-values, use the ***t* Distribution Area** Calculation Tool in exactly the same way as in **Chapter 6:** Enter the *t* statistic and *df* (the number of subjects minus one), choose whether you want the one-tailed or two-tailed probability, and click the "calculate" button. Do this now and make sure you get the *p*-value given above. (Or use the *t* Distribution Table if you're away from your computer, as described in the **Box** on statistical tables on p. 89.) Then try **Exercise 8.1** and **Exercise 8.2** to test your new related-samples *t* testing skills.

Calculation Tool: *t* Distribution Area

Box: Statistical Tables

Interactive Page 193

8.2.2 Matched-Subject Pairs

The **related-samples *t* test** is a very **powerful** way of comparing two means. As we learned in **Chapter 7,** "powerful" in this context means that if there is a difference between the two means—if the experimental manipulation has an effect—the **null hypothesis** is quite likely to be rejected.

Greater power is always desirable, but sometimes it is difficult or impossible to have each subject in an experiment perform both the pre-test and the post-test. In cases such as these, you can always use an **independent-samples *t* test,** covered in the next section, to assess the difference between the two means. But if an experimenter wants to take advantage of the greater power of the related-samples *t* procedure, it is sometimes possible to use a **matched-pairs** design. We can illustrate this type of design with an example:

> Professor Meene, who teaches a statistics course using this interactive textbook, wants to assess the value of requiring her students to complete and submit the Review Exercises for each chapter. At the beginning of a semester, she has all 20 students in the class perform a test measuring their basic math skills and how much they already know about statistics. She then divides the class in two via the following procedure. First, she makes a single list of the 20 students in order of their scores on the skills test. She then takes the first two students on the list and randomly assigns one to group A and the other to group B. Then she takes the next pair of students and randomly assigns them to groups A and B. After dividing the other 16 students in the same way, she has 10 matched pairs of students:

	Group A		**Group B**	
Pair	**Student**	**Skills Score**	**Student**	**Skills Score**
1	Taylor	95	Roberts	98
2	Gordon	93	Little	91
3	Wang	87	Brown	89
4	Bernstein	85	Evans	83
5	Peters	77	King	78
6	Nielson	75	Melnick	74
7	Albus	64	Curtis	67
8	Green	61	Olson	62
9	Inger	59	Tong	58
10	Lowell	51	Furman	50

Now that Professor Meene has a set of matched pairs, she can perform the experimental manipulation (here, the requirement to complete and submit Review Exercises) on one member of each pair and use the other member as a control. In other words, she treats each matched pair as if it represents the pre- and post-test data from a single subject. Note, however, that both the "pre-test" and the "post-test" are administered at the same time. For this "experiment," the assessment is simply the students' final grades for the semester:

Professor Meene requires the students in Group B to complete and submit Review Exercises, but makes no such requirement for the students in Group A. At the end of the semester, she compares the final grades of the two groups:

	Group A		**Group B**	
Pair	**Student**	**Final Grade**	**Student**	**Final Grade**
1	Taylor	100	Roberts	96
2	Gordon	94	Little	97
3	Wang	83	Brown	81
4	Bernstein	92	Evans	98
5	Peters	78	King	88
6	Nielson	95	Melnick	99
7	Albus	78	Curtis	87
8	Green	84	Olson	96
9	Inger	64	Tong	72
10	Lowell	72	Furman	73

Does performing the Review Exercises appear to make a significant difference in students' grades?

In a matched-pairs design, each pair is treated as the pre- and post-test score from one subject.

To perform the hypothesis test for a matched-pairs design, we calculate the difference between each matched pair, then calculate t and evaluate p for these D scores in exactly the same way as with a repeated-measures design. Again, for bidirectional hypothesis tests it doesn't matter which score you subtract from which, as long as you're consistent, but by convention we arrange things so that a positive D score indicates that the experimental condition produced a higher score than the control condition. The **degrees of freedom** for the hypothesis test are equal to the number of *pairs* – 1. Go ahead and give this calculation a try for Professor Meene's data.

Here are the final grades of Professor Meene's students again, along with the appropriate D scores for the matched pairs:

Pair	Group A Student	Group A Final Grade	Group B Student	Group B Final Grade	Difference
1	Taylor	100	Roberts	96	−4
2	Gordon	94	Little	97	+3
3	Wang	83	Brown	81	−2
4	Bernstein	92	Evans	98	+6
5	Peters	78	King	88	+10
6	Nielson	95	Melnick	99	+4
7	Albus	78	Curtis	87	+9
8	Green	84	Olson	96	+12
9	Inger	64	Tong	72	+8
10	Lowell	72	Furman	73	+1

The mean of these 10 difference scores is 4.70, with a standard deviation of 5.27. Plugging these values into the t formula, we get:

$$t = \frac{\overline{D} - \mu_D}{s_{\overline{D}}} = \frac{\overline{D} - \mu_D}{s_D / \sqrt{n}} = \frac{4.7 - 0}{5.27 / \sqrt{10}} = 2.82$$

And plugging this t value, 2.82, into the t **Distribution Area Calculation Tool** along with 9 **degrees of freedom,** we find that .020 of the area of this t **distribution** falls to the right of 2.82 or to the left of −2.82. This p**-value** should be low enough to convince Professor Meene that it's worth it to assign students the Review Exercises: The students who did the Exercises did better in the course than those who did not, and it is very unlikely that this difference is due to chance.

Matched-pairs designs are relatively rare in the behavioral science literature, because their use is only justified to the extent that the individuals in each pair are truly matched—and since no two people are identical, the matches will never be perfect. In Professor Meene's study, each matched pair of subjects had almost identical skills scores entering the semester. This matching is obviously important, but many other factors also contribute to students' final grades (how many other classes they're taking, how many nights they stay up late partying, how important the course is to them, etc.). We have no idea how well the pairs of subjects match up on these other factors.

For Professor Meene's purposes, the matched-pairs design is fine. Her study provides justification for assigning the Review Exercises, which is probably what she was looking for, and it is highly unlikely that there will be any negative consequences if the assignments actually don't improve grades. But in scientific research, a researcher has to be able to make a strong case that factors other than those on which subjects were matched won't have a meaningful effect on the experimental outcome. If this case can't be made, it is safer to use an **independent-samples *t* test,** which we turn to in the next section.

Exercise 8.3 and **Exercise 8.4** go through the calculations involved in analyzing another matched-pairs experimental design.

If unmatched factors might affect the experimental outcome, it is safer to use an independent- samples t test.

Interactive Page 195

8.3 Independent Samples

We begin this section with a new example:

> Kimberly and John are social psychology graduate students who share a research interest in the effectiveness of different media for disseminating news. Kimberly believes that the in-depth reporting of National Public Radio (NPR) makes it a better way of delivering information. John posits instead that delivery via all-news television networks such as the Cable News Network (CNN) is more effective, since visual images help to make the information more concrete.
>
> To test their theories, Kimberly and John conduct an experiment in which they ask 24 subjects to get their news exclusively from one source for two months. Each subject is randomly assigned to either the NPR group, in which subjects are instructed to listen to NPR as much as they wish but not to watch any television news source, or the CNN group, in which subjects are instructed to watch CNN as much as they want but not listen to any news radio.
>
> After the two months are up, all subjects take the same current-events test, on which scores range

from 0–50, with higher scores indicating greater awareness of the major news stories from the two-month period of the study. The data look like this:

NPR group:	Subj.#	1	4	5	6	11	12	16	17	18	20	21	24
	Score	48	42	25	33	37	37	37	30	20	46	39	37

CNN group:	Subj.#	2	3	7	8	9	10	13	14	15	19	22	23
	Score	30	41	41	25	22	29	22	16	27	18	35	39

So who was right, Kimberly or John? Well, the mean current-events knowledge score for the NPR listeners was 35.92, higher than the mean score of 28.75 for the CNN watchers. But while these means show unequivocally that this group of listeners scored higher than this group of watchers, we must, as always, be careful when generalizing from samples to populations. Before Kimberly can attach that highly prized adjective "**significant**" to the advantage of NPR over CNN, she will have to wait to see the result of a hypothesis test.

Like Sergeant Thursday's test from **Section 8.2.1,** Kimberly's test will have to compare two means. Another similarity between Kimberly's hypothesis test and the earlier **related-samples test** is that the null hypothesis H_0 will again be that the difference between the two means is 0. That is, a critic of this result (e.g., John) will claim that if the entire population (all humans) could be tested, the current-events knowledge of NPR listeners and CNN watchers would not differ at all. The fact that there happened to be a difference between the present samples could be due to chance. (Originally, of course, John claimed that NPR listeners would be *less* well-informed than CNN watchers, but given the results of the experiment, the best John can really hope for at this point is to fend off the claim that radio is better than TV for delivering news.)

Given these similarities, you might think that Kimberly would conduct her **hypothesis** test in the same way Sergeant Thursday did. The first step for Sergeant Thursday was to calculate **difference scores** for each subject in his experiment. These *D* scores came out like this:

Initial score	23	18	25	28	33	30	22
Score after training	35	16	30	34	42	29	30
Difference score	12	–2	5	6	9	–1	8

Kimberly might calculate difference scores for her data like so:

NPR group	48	42	25	33	37	37	37	30	20	46	39	37
CNN group	30	41	41	25	22	29	22	16	27	18	35	39
***D* score**	18	1	–16	8	15	8	15	14	–7	24	4	–2

Does it seem to you that Kimberly is on the right track here?

Results from samples cannot be automatically generalized to populations.

Interactive Page 196

Like earlier experiments in this chapter, Kimberly and John's experiment involves two conditions. But in previous experiments such as Sergeant Thursday's (**Section 8.2**), every subject participated in both conditions. Here, in contrast, each subject participated in only one of the two conditions. In statistical jargon, we say that the experiment involved two **independent samples**.

> Glossary Term: independent samples

Therefore, pairing scores in the NPR condition with scores in the CNN condition, as Kimberly tried to do, makes no sense. Since we cannot calculate *D* scores, we must frame our **hypotheses** directly in terms of the difference between the means of the two **populations**. So whereas in the case of a **related-samples** design we stated:

$$H_0: \mu_D = 0$$

we now state, for an **independent-samples** design:

> The null hypothesis for independent-samples designs

$$H_0: \mu_1 - \mu_2 = 0$$

where μ_1 stands for the population mean for one condition in our experiment (the NPR group) and μ_2 stands for the population mean for the other condition (the CNN group; it doesn't matter which group we assign as condition 1 and which as condition 2). As usual, we can use a **directional** or **bidirectional alternative hypothesis.** Since Kimberly and John had no reason, other than their personal beliefs, to assume before the experiment started that the difference would be in one direction or the other, a bidirectional H_a is appropriate here. So we have:

> The bidirectional alternative hypothesis for independent-samples designs

$$H_a: \mu_1 - \mu_2 \neq 0$$

And so, without even realizing it, we've stated our null and alternative hypotheses, completing step 1 of the **hypothesis** testing procedure for independent samples. As illustrated in Figure 8.3, H_0 states that the distribution of current-event scores for NPR listeners is identical to the distribution for CNN watchers, whereas H_a states that the two distributions are centered on different points.

This figure may be somewhat confusing at first, reflecting the fact that the null and alternative hypotheses in an **independent-samples *t* test** are fairly nebulous. Neither hypothesis specifies exactly what the mean of either population **distribution** is; in the figure, the possible means range from 10–40. But the null hypothesis states that the distributions for NPR listeners and CNN watchers will overlap precisely (regardless of where the distributions are centered), whereas the alternative hypothesis specifies that there will be some separation between the two distributions.

Interactive Page 197

Step 2 of the **hypothesis** testing procedure is to collect and describe the **sample** data. We need the **mean, standard deviation,** and **size** of each sample. We'll continue using the subscripts 1

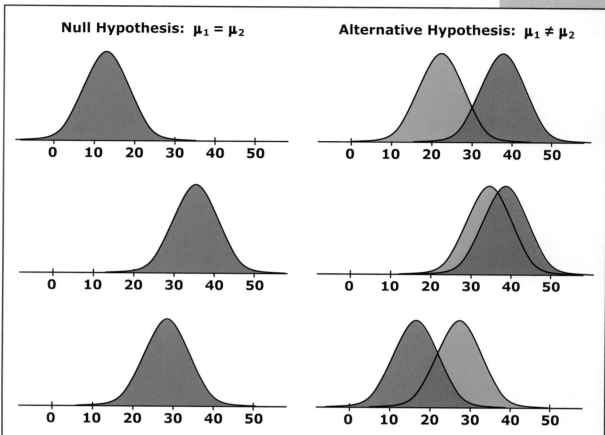

Figure 8.3 Graphical illustration of the null and alternative hypotheses for an independent-samples *t* test (this figure is animated on the *Interactive Statistics* website and CD). The null hypothesis states that the mean of population 1 is exactly equal to the mean of population 2 (since the distributions overlap precisely, they appear to be a single distribution in the figure). The bidirectional H_a states that there is some difference between the means. Neither hypothesis makes any claim about the absolute magnitude of μ_1 or μ_2, and H_a makes no claim about the magnitude or direction of the effect.

and 2 to refer to the NPR and CNN conditions, respectively. Here are the **descriptive statistics** for Kimberly and John's data (you can confirm these numbers using the **Descriptive Statistics** Calculation Tool if you wish):

NPR	Subj.#	1	4	5	6	11	12	16	17	18	20	21	24
group:	Score	48	42	25	33	37	37	37	30	20	46	39	37

CNN	Subj.#	2	3	7	8	9	10	13	14	15	19	22	23
group:	Score	30	41	41	25	22	29	22	16	27	18	35	39

Statistics for NPR group: $\overline{X}_1 = 35.92$, $s_1 = 8.06$, $n_1 = 12$
Statistics for CNN group: $\overline{X}_2 = 28.75$, $s_2 = 8.70$, $n_2 = 12$

In step 3 of the hypothesis testing procedure, we calculate the **test statistic,** which in this case will be a **t** statistic. The *t* formulas we used for single samples and related samples both had the following general form:

$$t = \frac{\text{sample statistic} - \text{parameter value proposed by the null hypothesis}}{\text{estimated standard deviation of the sampling distribution for the statistic (estimated standard error)}}$$

For a single-sample *t* test:

1. The sample statistic is $\overline{X}$.

2. The parameter value proposed by H_0 is the **population mean** (μ). The experimenter can choose any value he or she wishes for μ_0.

3. And the **estimated standard error** $s_{\overline{X}}$ is $s / \sqrt{n}$, as specified by the **Central Limit Theorem (Exercise 8.7)**.

For a related-samples *t* test:

1. The sample statistic is $\overline{D}$, the mean of the **differences** between the two conditions for each subject.

2. The parameter specified by H_0 is the population mean of the difference scores, μ_D. The null hypothesis almost always proposes that $\mu_D = 0$.

3. And the estimated standard error $s_{\overline{D}}$ is $s_D / \sqrt{n}$.

Now, for our **independent-samples t test,** we've already established the two values that go in the numerator of the *t*-statistic formula, but we haven't addressed the value of the denominator:

1. The sample statistic is $\overline{X}_1 - \overline{X}_2$, the difference between the means of the two samples we've collected.

2. The parameter specified by H_0 is the difference between the two population means ($\mu_1 - \mu_2$), which is again almost always proposed to be 0.

3. The denominator should be the estimated standard error of the difference between the two sample means, a value symbolized by $s_{\overline{X}_1 - \overline{X}_2}$. Calculating this value is tricky, and we'll spend the entire next section covering this calculation.

Once we have $s_{\overline{X}_1 - \overline{X}_2}$, we will calculate our *t* value, which will have ($n_1 - 1$) + ($n_2 - 1$) = $n_1 + n_2 - 2$ **degrees of freedom.** Armed with a *t* statistic and the appropriate *df*, we will be able to evaluate our hypotheses (step 4 of the hypothesis testing procedure) in exactly the same way we did for single-sample and related-samples *t* tests.

Now let's figure out what to use for $s_{\overline{X}_1 - \overline{X}_2}$.

Interactive
Page 198

8.3.1 The Standard Error for Independent-Samples *t* Tests

Figure 8.4 is designed to give you a sense of what the **standard error** of $X_1 - X_2$ will look like.

Suppose that listening to NPR really does result in greater awareness of world events (or at least higher scores on current-events quizzes) than watching CNN. The two distributions of quiz scores might then look like the ones shown in Figure 8.4a.

Now, what will happen when we take **samples** from these two **population distributions**? Most likely, we will get sample means close to the two population means. And if we took lots of samples and built up two distributions of sample means, the **Central Limit Theorem** tells us that these two sampling distributions would look like the ones shown in Figure 8.4b: they would be centered on the population means, but squished toward these centers in proportion to the square root of the sample size.

But we're not actually interested in the **sampling distributions** of the two populations individually; what we really want to know about is the distribution of differences between the two sample means. The sample mean for NPR listeners will come from somewhere in the distribution on the right, and the sample mean for CNN watchers will come from the distribution on the left.

On average, the difference between the two sample means will be about the same as the difference between the means of the two sampling distributions. But sometimes the sample mean for CNN watchers will be lower than normal and the sample mean for NPR listeners will be higher than normal, resulting in a larger than normal difference. Other times, the CNN mean will be high and the NPR mean will be low, resulting in a smaller than normal difference.

a. Population distributions for raw quiz scores of NPR listeners (blue) and CNN watchers (green)

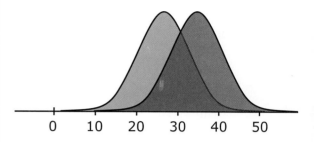

b. Sampling distributions for mean quiz scores of NPR listeners (blue) and CNN watchers (green)

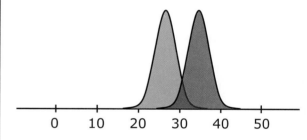

d. Distribution of differences between sample means

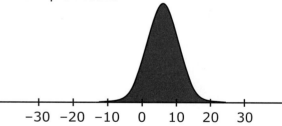

Figure 8.4 Possible experimental results of Kimberly and John's experiment, assuming that NPR listeners average higher scores than CNN watchers on the current events quiz. Part c, not shown here, is animated on the *Interactive Statistics* website and CD.

The upshot of all this is that the distribution of differences between the sample means, shown in Figure 8.4d, will be more **variable** than either sampling distribution alone. We won't work through the math, but it turns out that if the standard deviations of the two samples are exactly the same and the sample sizes are also equivalent, the **estimated standard error** of the differences between the sample means will be greater than the estimated standard error of the individual sample means by a factor of $\sqrt{2}$. That is:

$$\text{If } s_1 = s_2 = s \text{ and } n_1 = n_2 = n,$$
$$s_{\bar{X}_1 - \bar{X}_2} = \sqrt{2}\,(s / \sqrt{n})$$

Unfortunately, many experiments include different numbers of subjects in the two conditions (that is, n_1 is often not equal to n_2), and we can never count on s_1 being identical to s_2. Therefore, calculating the estimated standard error for an independent-samples *t* test is a bit more complicated than the above formula suggests. We'll work out the formulas (yes, there will be more than one) we need on the next page.

Calculating $s_{\bar{X}_1 - \bar{X}_2}$ for an independent-samples *t* test involves a two-step procedure:

1. First, we calculate the average amount of **variability** (the "pooled" **standard deviation**) in the two samples of scores we've collected.

2. Then we use this average to calculate the **estimated standard error.**

Now, you might think that the first step could be accomplished by simply averaging the standard deviations of the two samples. Unfortunately, however, it's not this simple, for two reasons:

- For mathematical reasons we won't go into, we have to average the sample **variances** of the two samples (s_1^2 and s_2^2), not the sample standard deviations.

- The average must reflect the fact that one sample may be bigger than the other. For example, if n_1 is 25 and n_2 is 10, we would want the variability in sample 1 to have a larger influence on the average than the variability in sample 2. So we weight each sample variance according to the degrees of freedom in the sample (more *df* = more weight).

And so, without further ado, we present the formula for s_p^2, the **pooled variance** from two samples:

The distribution of differences between two sample means will be more variable than either sampling distribution alone.

Steps for calculating the standard error for an independent-samples *t* test

Glossary Term: pooled variance

$$s_p^2 = \frac{(n_1 - 1)\, s_1^2 + (n_2 - 1)\, s_2^2}{(n_1 - 1) + (n_2 - 1)}$$

The pooled standard deviation, s_p, is the square root of the pooled variance:

$$s_p = \sqrt{s_p^2}$$

Once you have s_p^2 or s_p, you can proceed to compute the estimated standard error, which can be calculated using either of the following formulas:

$$s_{\bar{X}_1 - \bar{X}_2} = \sqrt{\frac{s_p^2}{n_1} + \frac{s_p^2}{n_2}} = s_p \sqrt{\frac{1}{n_1} + \frac{1}{n_2}}$$

If you're planning an experiment with an independent-samples design, you should strive to have an equal number of subjects in each group, because the independent-samples t test is most reliable with equal n's. The calculations for s_p and $s_{\bar{X}_1 - \bar{X}_2}$ are also simplified in this case, as so:

When $n_1 = n_2 = n$,

$$s_p = \sqrt{\frac{s_1^2 + s_2^2}{2}} \qquad s_{\bar{X}_1 - \bar{X}_2} = s_p \sqrt{\frac{2}{n}} = \sqrt{2}\,(s_p / \sqrt{n})$$

If you check back on the previous page, you'll see that this last formula is identical to the one we originally derived for $s_{\bar{X}_1 - \bar{X}_2}$, with the substitution of s_p for s.

Exercise 8.5 provides practice calculating s_p and $s_{\bar{X}_1 - \bar{X}_2}$.

 Interactive Page 200

8.3.2 Final Formulas for Independent-Samples t Tests

You've been subjected to quite a few formulas on the previous few pages, so here we'll sum up the ones you need to conduct an **independent-samples t test.** We'll highlight the formulas that use s_p (rather than s_p^2), since they're most similar to the formulas for **single-sample** and **related-samples** t's, which both use s:

$$s_p = \sqrt{\frac{(n_1 - 1)s_1^2 + (n_2 - 1)s_2^2}{(n_1 - 1) + (n_2 - 1)}}$$

$$s_{\bar{X}_1 - \bar{X}_2} = s_p \sqrt{\frac{1}{n_1} + \frac{1}{n_2}}$$

$$t = \frac{(\bar{X}_1 - \bar{X}_2) - (\mu_1 - \mu_2)}{s_{\bar{X}_1 - \bar{X}_2}}$$

$$df = n_1 + n_2 - 2$$

If $n_1 = n_2$, the formulas for s_p, $s_{\bar{X}_1-\bar{X}_2}$, and *df* can be simplified to:

$$s_p = \sqrt{\frac{s_1^2 + s_2^2}{2}}$$

$$s_{\bar{X}_1-\bar{X}_2} = s_p\sqrt{\frac{2}{n}}$$

$$df = 2(n-1)$$

In this second set of formulas, *n* is the number of subjects in either group alone; that is, $n = n_1 = n_2$.

Remember that $\mu_1 - \mu_2$, the difference in population means postulated by H_0, is almost always 0.

Here are the statistics for Kimberly and John's experiment again:

NPR group: $\bar{X}_1 = 35.92$, $s_1 = 8.05$, $n_1 = 12$
CNN group: $\bar{X}_2 = 28.75$, $s_2 = 8.70$, $n_2 = 12$

Use the above formulas to calculate s_p, $s_{\bar{X}_1-\bar{X}_2}$, **t,** and **df,** then use the ***t* Distribution Area** Calculation Tool to find the two-tailed **p-value** associated with the *t* value.

Here are the appropriate calculations for Kimberly and John's **independent-samples *t* test:**

$$s_p = \sqrt{\frac{s_1^2 + s_2^2}{2}} = \sqrt{\frac{64.99 + 75.66}{2}} = 8.39$$

$$s_{\bar{X}_1-\bar{X}_2} = s_p\sqrt{\frac{2}{n}} = 8.39\sqrt{\frac{2}{12}} = 3.43$$

$$t = \frac{(\bar{X}_1 - \bar{X}_2) - (\mu_1 - \mu_2)}{s_{\bar{X}_1-\bar{X}_2}} = \frac{35.92 - 28.75}{8.39} = 2.09$$

$$df = 2(12 - 1) = 22$$

$$p = .048$$

Since *p* is (just barely) under the .05 level, Kimberly can, after all, claim a statistically significant advantage for NPR over CNN as a news delivery mechanism.

You can get more practice making these calculations in **Exercise 8.6, Exercise 8.7,** and **Exercise 8.8.**

8.4 Assumptions for Two-Condition t Tests

As originally discussed in **Section 5.5,** all **inferential statistics** require that we meet certain basic requirements in order for the statistics to make any sense. We call these requirements "**assumptions,**" because we assume them to be true even though we usually can't prove that they hold.

The two basic assumptions noted previously for **single-sample *t* tests** hold as well for both **related-samples** and **independent-samples *t* tests.** First, the subjects must be drawn at random from the **population** we're trying to generalize to (**Section 6.4.1**). Proper experimental designs ensure that this assumption is met.

The random sampling assumption

The second assumption for *t* tests is that the sampling distribution in question must be approximately **normal** (**Section 6.4.2**). For related-samples designs, we're concerned with the distribution of *D* scores, whereas for independent-samples designs, we're concerned with the distribution of differences between sample means for condition 1 and sample means for condition 2.

The normal sampling distribution assumption

In both cases, if the **sampling distributions** of each of the conditions separately are themselves normal, the sampling distribution of the *D* scores or the differences between the conditions will also be normal. So if the underlying population distributions are approximately normal, and/or if you have at least 20 or so subjects (for an independent-samples design, this means that both $n_1 > 20$ and $n_2 > 20$), you don't need to worry about this assumption.

There is one additional **assumption** for **independent-samples *t* tests** that is not applicable to either **single-sample** or **related-samples** tests. Recall that s_p^2, the **pooled variance** for the two conditions, is a weighted average of the variances of the conditions separately. In calculating s_p^2 (and s_p), we are assuming that both **sample** variances are approximating an underlying **population** variance that is common to the two conditions. That is, we are assuming that the population variances of the two conditions, σ_1^2 and σ_2^2, are identical.

The equal variances assumption for independent-samples tests

In practice, we'll be OK as long as the two population variances, which you can estimate using the sample variances (s_1^2 and s_2^2), aren't too far off from each other. How far is too far? Here are some rough guidelines:

- If n_1 and n_2 are both **less than 10,** you can use the formulas involving s_p as long as the larger of the two variances is less than four times the smaller of the two variances. To check this, divide the larger s^2 by the smaller s^2. If the result is less than 4.0, you're good to go with s_p.

- If n_1 and n_2 are both **greater than 10,** we can use s_p as long as the larger s^2 divided by the smaller s^2 is less than 2.0. (Note: you have to check the ratio of variances, not the ratio of standard deviations.)

If the variances are too unequal, the **p-value** we derive from the **t statistic** that is calculated using s_p will not be accurate—it may be too small or it may be too large. In other words, the *t* test is meaningless if this assumption is not met (just as any *t* test is meaningless if either of the other two assumptions is violated).

Generally speaking, this assumption will only be violated if the comparison being made doesn't make sense in the first place. For example, it's probably all right to *t* test the difference in weight between apples and oranges, but it doesn't make any sense to compare the weights of cherries and watermelons using the *t* statistic. But do we really need a *t* test to tell us which of the latter two fruits weighs more?

If you ever come across a situation in which this assumption is violated but you're convinced that the comparison in question is meaningful, there are alternative ways to test the **significance** of the comparison. One recourse is to use an alternate form of the *t* procedure for independent samples. This alternate form uses the **sample variances** themselves, instead of s_p^2, in the calculation of the **estimated standard error,** but it requires a hairy formula for calculating the **degrees of freedom,** and you often end up with a *df* that is not a whole number (for example, in Kimberly and John's experiment, the *t* statistic using this alternative method would have 21.88 *df*). Statistics packages will often give you the option of using either the pooled or the nonpooled version of the independent-samples *t* test.

A second option is to use a "nonparametric" inferential statistic, as discussed in the next section. **Exercise 8.9** quizzes you on assumption violations in research examples.

8.4.1 Nonparametric Alternatives to *t* Tests

As we saw in the previous section, in order to use the *t* statistic in a hypothesis test we must make certain assumptions about population

[margin note] How to proceed if the equal variances assumption is violated

[margin note] Glossary Term: non-parametric statistic

parameters (e.g., that population variances are equal in an independent-samples *t* test). For this reason, *t* is sometimes called a **parametric statistic.** When one or more of these assumptions is violated, researchers often use a **nonparametric statistic,** which avoids parameter assumptions, as a substitute for *t*.

Two such nonparametric alternatives are the Mann–Whitney *U* statistic and the Wilcoxon *T* statistic, used for **independent-samples** and **related-samples** designs, respectively. We will not cover the *U* and *T* tests in any detail—you will not learn enough here to be able to conduct one of these tests on your own—but we will give you a flavor of how they work.

In the Mann–Whitney procedure, we start by combining the data from two independent samples and rank ordering the datapoints. For example, suppose we started with the two samples of IQ scores in the following table:

Sample	IQ
A	117
A	109
A	125
A	93
A	107
B	84
B	111
B	76
B	94
B	95

The logic of the Mann–Whitney test is fairly simple: If one **population** has a lower median than the other, then the scores in one **sample** should have consistently lower ranks than the scores in the other sample. To test whether or not this is so, we go through the scores of sample A one by one and award "points" to each one based on how many scores in sample B have higher ranks. Then we do the same thing for sample B. We add up all the points for each sample, and the lower of the two point totals is the *U* statistic (as shown at the bottom of the table below). We then compare our *U* value to a distribution of *U* statistics to arrive at a **probability value,** which is evaluated just like the *p*-values in *t* tests. The ranks and calculations are given in the following table:

Rank ordering datapoints

Mann–Whitney *U* test

Sample	IQ	Rank	Points for A	Points for B
B	76	1	—	5
B	84	2	—	5
A	93	3	3	—
B	94	4	—	4
B	95	5	—	4
A	107	6	1	—
A	109	7	1	—
B	111	8	—	2
A	117	9	—	—
A	125	10	—	—
Total points for A = 3 + 1 + 1 + 0 + 0 = 5				
Total points for B = 5 + 5 + 4 + 4 + 2 = 20				
$U(5, 5) = 5$; $p = .15$				

The Wilcoxon procedure starts by rank ordering the absolute values of the difference scores from a related-samples experiment. We then sum the ranks for the positive D scores and sum the ranks for the negative D scores. The smaller of the two sums is T (not to be confused with t), which can be compared to a T distribution to give us a p-value. The table below illustrates the procedure for related samples of response times.

Wilcoxon *T* test

Response Time 1	Response Time 2	D	Rank
942	935	−7	1
1100	1128	28	2
961	931	−30	3
1106	1138	32	4
1153	1212	59	5
1013	1092	79	6
1152	1254	102	7
943	1056	113	8
Sum of negative ranks = 1 + 3 = 4			
Sum of positive ranks = 2 + 4 + 5 + 6 + 7 + 8 = 32			
$T(8) = 4$; $p = .055$			

The main thing you should take away from these very brief descriptions of the U and T tests is that both tests assess whether or not the **median** of one population is significantly lower or higher than the median of another population. In contrast, t tests test hypotheses about population **means.**

The U and T statistics are used to test hypotheses about medians, not means.

Another thing you should know is that U and T tests are usually less **powerful** than their corresponding T tests. That is, a real effect is less likely to be deemed "**statistically significant**" with a U or a T test than with a t test. For example, the p-value for a t test on the IQ data above is .048, while the p-value for the corresponding U test is only .15.

In **Chapter 12,** we will discuss two additional nonparametric alternatives, the sign test (which can substitute for a related-samples t test) and the median test (which fills in for an independent-samples test). These tests work by converting interval data into categorical data and are somewhat easier to understand conceptually than the T and U tests. However, they are also even less powerful than T and U tests, so if you're faced with the prospect of choosing a nonparametric test to use on your own data, consider learning more about T and/or U.

8.5 Comparing Two-Condition Designs

In this chapter, we've learned about two different t tests for analyzing the difference between two sets of numbers. You may be wondering at this point when to use which. Once the data from an experiment have been collected, you have no choice:

- If the two sets of numbers came from a single group of subjects who each contributed one datapoint to condition 1 and one datapoint to condition 2, you must use the **related-samples t test** to analyze the results.

- If the two sets of numbers came from two different groups of subjects, with each subject contributing a datapoint to one condition or the other, you must use the **independent-samples t test.**

However, in the planning stages of an experiment, researchers often do have the opportunity to choose which experimental design they wish to use, and therefore which t test they will end up employing. In this section we'll discuss some of the factors that lead researchers to choose one design or the other.

One important consideration when designing an experiment, discussed in detail in **Section 7.4,** is the statistical **power** one can expect to obtain. To review, power is the likelihood of successfully rejecting the null hypothesis if, in fact, the **null hypothesis** is false. Since researchers almost always hope to reject H_0, power is essentially the **probability** that an experiment will "work."

Based on your knowledge of the 2 two-condition t tests, do you think one is inherently more powerful than the other?

The design of an experiment dictates which t test procedure to use.

8.5.1 Power

Statistically speaking, a **related-samples *t* test** will almost always be more powerful than an **independent-samples *t* test.** We can prove this with an example, using the ***t*-Test Power** Calculation Tool to compute **power** estimates for related- and independent-samples *t* tests. As described in **Section 7.4.2,** to make such a computation you need to know or estimate the following values, each of which is entered into the Calculation Tool:

1. The **alpha level** you plan to use in evaluating the experiment: This is the ***p*-value** below which you will declare the results of the experiment "**significant.**" In most behavioral science studies, α is set at .05. The Calculation Tool assumes a **bidirectional** α; to calculate power for a **directional** test, enter α × 2 into the Tool.

2. The value predicted by the **null hypothesis:** For both related-samples and independent-samples designs, this is 0, since H_0 predicts no difference between the two conditions in the experiment.

3. Your estimate for what the **effect size** really is: For related- and independent-samples designs, this will be the difference you expect to find between the two conditions.

4. The estimated **standard deviation** of the effect size: For related-samples designs, this is your guess about the standard deviation of the *D* scores. For independent-samples designs, you need to estimate the standard deviation of the X_1 and X_2 variables. Remember that in the form of the independent-samples *t* test presented in this book, we are assuming that the two conditions have identical σ's. Therefore, you only need one estimated value for σ to enter in the Calculation Tool. (We assume that $\sigma_1 = \sigma_2 = \sigma$.)

5. The **size** of the sample you plan to use: For related-samples designs, this is the total number of subjects. For independent-samples designs, enter the *n* for each group, assuming equal sample sizes. If you are going to use differently sized samples, enter the average of n_1 and n_2 (the power calculation in this case will not be precise, but the value we obtain will only be an estimate, anyway).

For our example, we'll return to James, our protagonist from Chapter 6 and earlier in this chapter. When we last left him, he had already completed an experiment in which he obtained a mean dowel-length estimate of 14.0 with standard deviation 2.20. A single-sample *t* test indicated that estimates made under the conditions he tested—in which subjects first saw confederates consistently overestimate dowel length— were significantly greater than the actual length of the dowel (12 inches).

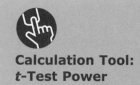

**Calculation Tool:
t-Test Power**

Review of estimates we need to make to calculate power

However, a critic pointed out that this experimental design requires the questionable assumption that without peer pressure, subjects would consistently estimate the dowel to be its true length (12 inches). To counter this criticism, James must run a new experiment using a two-condition design.

Interactive Page 207

As noted in **Section 8.1,** James could choose either a **related-samples** or an **independent-samples** design. In the latter, he would use the same procedure he used in his first experiment for one group of subjects, but the other group would make dowel-length estimates without being exposed to any peer pressure. If he wanted to use a **repeated-measures design,** he would have each subject make the length judgment twice: once before the confederates made their judgments and once after seeing the confederates.

To evaluate the **power** James can expect from these two potential experiments, we'll assume an α level of .05 and a μ_0 of 0 (the **null hypothesis** predicts that length judgments in both conditions will be equivalent). It seems reasonable to guess that the mean judgment in the experimental condition (the condition in which subjects are exposed to peer pressure) will be about 14 inches, as in the previous experiment. It also seems reasonable to assume that in the control condition (with no peer pressure), the mean judgment will be about 12 inches. (Of course, we don't know that this will be the case; this is why we're doing the experiment in the first place. But to estimate power, we have to make some guess, and this is a reasonable one to make.) Therefore, the estimated **effect size** is 14 – 12 = 2.

To estimate σ, we'll again go with the observed value from the first experiment, 2.20. Assume James only has the resources to run a total of 20 subjects in his new experiment. For the repeated-measures design, we will therefore enter 20 for n. For the independent-samples design, half the subjects will go into each group, so $n_1 = n_2 = 10$.

For independent-samples designs, enter the n for one condition into the Calculation Tool.

Now that we've made all our estimates, we can enter them in the **_t_-Test Power** Calculation Tool and see what comes out. Try it yourself and you will find that power for the related-samples design (click the "Single/Related Samples" radio button in the Tool for this analysis) is .98, meaning that the null hypothesis will almost surely be rejected using this design (assuming that all of our estimates are on the mark). Figure 8.5 shows a screenshot of what the Calculation Tool will look like once you've entered the parameters and clicked the "calculate" button.

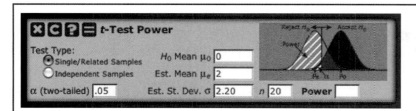

Figure 8.5 Screenshot of the *t*-Test Power Calculation Tool filled in to calculate power for the related-samples version of James's second experiment.

In contrast, power for the independent-samples design (click the "Independent Samples" radio button, and make sure to change the value in the "*n*" text box from 20 to 10, as described above) is only .47—there is better than a 50–50 chance that the null hypothesis will fail to be rejected if our estimates are correct and James uses this design.

The related-samples *t* procedure is generally more powerful than the independent-samples procedure.

Interactive
Page **208**

For this example, then, the **repeated-measures design** is twice as powerful as the **independent-samples** design, given the same number of subjects. One reason for this **power** difference is the discrepancy in the number of observations per condition in the two designs. The repeated-measures design results in 20 observations per condition (since each of the 20 subjects participates in both conditions), whereas the independent-samples design only yields 10 observations per condition (each subject only contributes to one or the other). The more observations we have, the closer the **sample means** will approximate the **population means,** and the more likely we will be to successfully reject the **null hypothesis.**

Related-samples designs gain some power by producing more observations per subject.

However, a little experimentation with the *t*-**Test Power** Calculation Tool will show that another factor is also at work here. Make sure the "Independent Samples" radio button is selected in the Tool, enter the same values as above for α, μ_0, μ_e, and σ (0, 2, and 2.2, respectively), leave the "*n*" text box blank, and enter .98 into the Power text box (Figure 8.6 shows what the Calculation Tool looks like with these values entered). If you now click the "calculate" button, the Tool will tell you how many subjects James will need per condition to achieve .98 power with an independent-samples design—the *n* text box will be filled in with the value 40.

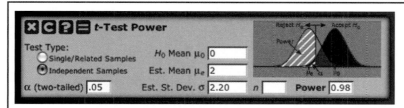

Figure 8.6 Screenshot of the *t*-Test Power Calculation Tool filled in to calculate the number of subjects needed to achieve .98 power for the independent-samples version of James's second experiment. After the "calculate" button is clicked, the "*n*" text box will be filled in by the Tool with the value 40.

Thus James will need twice as many observations in each condition to achieve as much power in the independent-samples design as he would get in the repeated-measures design. Why? The short answer is that by using different subjects in the two conditions, the independent-samples design (as compared to the repeated-measures design) introduces an extra source of **variability** into the data. This extra variability has the ultimate effect of lowering our *t* **statistic,** and therefore raising the significance level of the effect of an experimental manipulation.

This explanation may not be very satisfying, but it's as far as we'll go for now. **Chapter 10** will be all about analyzing the various sources of variability in experiments, and towards the end of that chapter, we'll revisit the issues raised here and discuss them in greater detail.

For now, just take it as a matter of faith that, all other factors being equal, related-samples *t* tests are more powerful than independent-samples *t* tests. **Exercise 2.10** asks you to make power calculations for all three types of *t* tests we've covered in this textbook (single-sample, related-samples, and independent samples).

Interactive
Page **209**

8.5.2 Other Considerations

Given that **repeated-measures** experimental designs are generally more **powerful** than **independent-samples** designs, you might wonder why anyone would ever choose to use independent samples.

One reason is that some research questions just can't be addressed using repeated measures. For example, any hypothesis positing a gender difference—that is, a higher or lower mean score for men compared to women—requires two groups of subjects, one sample of men and one sample of women.

The other important limitation to repeated-measures designs is that effects of the first measurement may **carry over** to the second measurement, compromising the validity of the second measurement. For example, say we're interested in the effect of alcohol on driving ability. We set up a complex "virtual obstacle course" using a driving simulator and plan to have subjects try to navigate the course twice, first while sober and then later after imbibing several beers.

The problem with this experimental design is that subjects may be aided in their second run through the virtual obstacle course by the experience they gained in their first run through the course. That is, there may be a "practice effect" in the experiment, which could at least partially negate any adverse effect due to alcohol consumption. Other repeated-measures designs are compromised by "fatigue effects," in which completing the first measurement causes a decrease in performance on the second measurement.

Related-samples designs also gain power by reducing overall variability.

Many research hypotheses do not lend themselves to repeated-measures designs.

It is sometimes possible to minimize carry-over effects by counter-balancing the order in which the measurements are taken. For example, in the driving simulation experiment, we could use the procedure described above for half of the subjects and have the other half run through the simulator first while under the influence of alcohol, and then a second time several hours later, after they've sobered up.

The goal of counterbalancing is not to eliminate carry-over effects per se, but to eliminate their influence on the overall results of the experiment: Any carry-over effects that might exist will (hopefully) affect both conditions equally (e.g., the extra practice in the simulator will boost scores in the sober condition for half the subjects and scores in the drunk condition for the other half).

Most often, the choice between repeated-measures and independent-samples designs is fairly clear. That is, common sense will usually tell you which design is preferable. **Exercise 8.11** asks you to choose designs for a number of research situations.

8.6 Chapter Summary/Review

This Chapter Summary/Review is interactive on the *Interactive Statistics* website and CD. Also be sure to go through all the **Review Exercises** for this chapter.

- Many experimental designs include two conditions: a control condition that serves as a baseline measure of the variable in question and an experimental condition that includes some interesting change in the procedure of the control condition. To test the significance of differences between the means of the two conditions, we can employ a related-samples or an independent-samples *t* test, depending on the details of the research design (**Section 8.1**).

- In experimental designs with related samples, the mean of a set of scores is compared to the mean of another set of scores taken from the same individuals, or from a matched set of subjects. For example, one common **repeated-measures design** compares a set of pre-test scores with a set of post-test scores, where both tests are taken by the same group of subjects. Usually, the hypothesis testing procedure is used to evaluate a **null hypothesis** that claims the mean of the subjects' **difference (*D*) scores** is 0 (where *D* equals a post-test score minus a pre-test score) (**Section 8.2**).

- The procedure for conducting a **related-samples *t* test** is virtually identical to that for a **single-sample *t* test,** except that the mean of the difference scores is tested instead of the mean of a set of raw scores (**Section 8.2.1**). As stated above, the null

Side notes:

Counterbalancing conditions can help minimize the negative consequences of carry-over effects.

Common sense is the best guide to deciding on a research design.

Interactive Page 210

hypothesis almost always claims that $\mu_D = 0$. The standard deviation used in the t formula should be the standard deviation of the D scores, and $\textbf{df} = n - 1$, where n is the number of subjects in the experiment (not the number of scores). The final formula for the related-samples t statistic is:

$$t = \frac{\overline{D} - \mu_D}{s_{\overline{D}}} = \frac{\overline{D} - 0}{s_{\overline{D}}} = \frac{\overline{D}}{s_D / \sqrt{n}}$$

- If the same subjects cannot be evaluated in both conditions of a repeated-measures design, a **matched-pairs design** can be used. Here, pairs of subjects are matched on as many factors as possible, then one member of each is placed in the control condition and the other member is placed in the experimental condition. Difference scores are calculated by subtracting the score of one member of each pair from the score of the other member of the pair, and the t test is conducted the same way as with repeated-measures designs (**Section 8.2.2**).

- If two separate, nonmatched groups of subjects are tested in an experiment, an **independent-samples t statistic** must be computed to evaluate the difference between the conditions (**Section 8.3**). The null hypothesis for an independent-samples t test usually claims that the difference between the two means, $\mu_1 - \mu_2$, is equal to 0—that is, that the means of the two conditions in the experiment are identical. So the sample statistic we're testing is the difference between the two sample means, $X_1 - X_2$ (**Section 8.3**).

- The sampling **variability** in the difference between the two means is greater than the variability of either sampling distribution alone. So to calculate $s_{\overline{X}_1 - \overline{X}_2}$, we must employ a two-step procedure (**Section 8.3.1**). First, we calculate the pooled standard deviation s_p from the variances of the two samples. Then we use this value to calculate $s_{\overline{X}_1 - \overline{X}_2}$. The latter is then inserted into the formula for t. The distribution of t statistics for an independent-samples test has degrees of freedom equal to the sum of the two sample sizes minus two. The formulas are:

$$s_p = \sqrt{\frac{(n_1 - 1)s_1^2 + (n_2 - 1)s_2^2}{(n_1 - 1) + (n_2 - 1)}}$$

$$s_{\overline{X}_1 - \overline{X}_2} = s_p \sqrt{\frac{1}{n_1} + \frac{1}{n_2}}$$

$$t = \frac{(\overline{X}_1 - \overline{X}_2) - (\mu_1 - \mu_2)}{s_{\overline{X}_1 - \overline{X}_2}}$$

$$df = n_1 + n_2 - 2$$

- There are other versions of the above formulas. Most usefully, they can be simplified when $n_1 = n_2$, as shown in **Section 8.3.2**.

- The assumptions for two-sample *t* tests are mostly the same as those for single-sample *t* tests. First, the subjects must be drawn at random from the population in question. Second, the distributions of scores must be approximately **normally** distributed, or there must be at least 20 subjects in each condition, so that the sampling distribution is normal. For the independent-samples *t* test described in this book, we must additionally assume that σ_1^2 and σ_2^2 are identical. In practice, this last assumption demands that we check to make sure that the larger of the two sample variances (s_1^2 or s_2^2) is no more than twice as great as the smaller of the two sample variances, if n_1 and n_2 are both greater than 10 (**Section 8.4**).

- Related-samples *t* tests are generally more **powerful** than independent-samples *t* tests (see Section 8.5.1 for instructions on calculating power for two-condition *t* tests). That is, a researcher is more likely to find a significant effect with a repeated-measures test than with an independent-samples test. However, many experimental questions just can't be addressed with related-samples designs. Furthermore, users of repeated-measures designs often risk **carry-over** effects that can bring the experimental results into question (**Section 8.5.2**). The best tool for deciding between related-samples and independent-samples designs is usually common sense.

Chapter 9
Confidence Intervals

9.1 Estimation

Consider the following scenario:

> Kidstuff, a company that currently manufactures child safety equipment, is considering getting into the stuffed animal market. To investigate how profitable this move is likely to be, Steve, the head of marketing for Kidstuff, is asked to determine how many stuffed animals the typical 4-year-old American child has. Steve checks the company records and finds that nine current Kidstuff employees have 4-year-old children. To start his investigation, Steve decides to ask these nine employees to count and report the number of stuffed animals their children have.

The research task Steve is undertaking here is similar in many ways to ones we have considered previously, in that he will be collecting data from a sample in hopes of generalizing to a larger population.

But Steve's research question also differs in a fundamental way from previous questions we've looked at. Steve is not interested in comparing his data to some standard value, nor is he comparing data from two different conditions. Indeed, he is not trying to test any hypothesis at all.

Estimation as a research goal

Instead, his goal is estimation of the **population parameter** (mean number of stuffed animals owned by all 4-year-old children in the United States) from the **sample** data (the number of stuffed animals owned by the children of his nine employees).

Suppose Steve collects the following data:

18 9 16 13 15 19 21 19 23

On the basis of these data, what value do you think Steve should estimate for the **mean** number μ of stuffed animals owned by all American 4-year-old children?

Interactive Page 213

The best estimate Steve can give for μ, the mean of the population he's studying, is $\overline{X}$, the mean of the sample he's collected. For the data above, $\overline{X} = 17.0$, so this should be Steve's estimate.

The best estimate of the population mean is the sample mean.

But how good of an estimate for μ should we consider $\overline{X}$ to be? Put another way, how likely do you think it is that μ really is exactly 17.0?

Interactive Page 214

Given the millions of 4-year-old American children, there is almost no chance at all that the mean of Steve's **sample** of nine children is *exactly* the same as the whole **population.**

The sample mean will almost never be exactly the same as the population mean.

This conclusion, that $\overline{X}$ is almost certainly not identical to μ, might seem inconsistent with the conclusion drawn on the previous page, that $\overline{X}$ is the best estimator we have for μ. But in fact, both are correct. The Central Limit Theorem tells us that on average, sample means tend to be similar to the mean of the population from which the sample was drawn. So $\overline{X}$ is likely to be *close* to μ, and no other sample statistic can give us a better estimate.

The sample mean $\overline{X}$ provides what can be called a **point estimate** of μ. When reporting this point estimate to his superiors, Steve could say something like "My initial research indicates that the average American 4-year-old has 17.0 stuffed animals." But Steve might be somewhat nervous making this statement: since his point estimate is based on only nine subjects, he might think that it sounds a bit more precise than it really is.

Glossary Term: point estimate

When we don't feel confident enough to make an exact estimate of some value, we often revert to speculating about the range in which the value probably falls. For example, at what age did you start reading? You probably don't know the exact answer to this question, but you might be comfortable saying that you started reading when you were somewhere between three and four years old. In statistics, this is called an **interval estimate.**

Now, here's another question: How sure are you that you were between three and four years old when you started reading? Would you be more sure if we expanded the range to between two and five years old?

This same question might come up for Steve when he presents his findings. Say he chooses an arbitrary range for his interval estimate. For example, he might report that "The average American 4-year-old appears to have between 15 and 19 stuffed animals." His superiors might reasonably ask, "How sure are you that the true average is really within this range?"

In this chapter, we will adapt the **inferential** techniques covered in the last few chapters to the task of making interval estimates. More specifically, we will address the concern raised in the previous paragraph by developing a procedure in which we decide how much confidence we want to have in our estimate and then determine the interval that provides exactly this much confidence. The resulting range will be called a **confidence interval.**

As we did with hypothesis testing, we will begin our study of confidence intervals in **Section 9.2** with examples in which σ, the **standard deviation** of the population we're interested in, is known. This allows us to use **z-score test statistics** and the standard **normal** distribution, but in most research settings σ is unknown. In **Section 9.2.1** and **Section 9.2.2,** we'll transition to more useful procedures that use **t statistics** and distributions. Finally, in **Section 9.3** we'll discuss some important factors to consider when using and interpreting confidence intervals.

9.2 Constructing Confidence Intervals

As promised on the previous page, we will devote this section to developing the procedure for constructing **confidence intervals** using **z-scores** and the standard **normal distribution.** To do so, we need an example in which we know the value of σ for the **population** we're trying to describe:

> The Springfield school board has decided to begin next year giving all 7th-grade students a standardized test of science knowledge. The test has been previously designed and tested on a very large group of seventh graders, and scores on the test are scaled such that the population standard deviation σ is exactly 15.

To get a baseline measure of how well Springfielders can expect to do on the test next year, the school board decides to conduct a small study in which 25 of this year's seventh graders are given the test. The average score $\overline{X}$ for this sample comes out to 70.

As we saw in the previous section, $\overline{X}$ is the best possible point estimator of μ, so we expect the mean of all Springfield seventh graders to be close to 70. But suppose the school board wants an **interval estimate** of μ, and wants to be 95% sure of this estimate. In other words, we want the **probability** to be .95 that μ falls within the interval.

Before we begin going through the logic behind confidence intervals, take a guess at what you think the 95% confidence interval will be. (There's no reason that you should know how to calculate it at this point; just give the question a bit of thought and take a wild guess.)

Interactive Page 216

To construct the 95% confidence interval for Springfield seventh graders' test scores, we use the following logic, illustrated in Figure 9.1 (parts a and d of Figure 9.1, not shown here, are interactive and animated on the *Interactive Statistics* website and CD):

Logic of confidence interval construction

a. The value of μ, the **population mean** of all Springfield seventh graders, is unknown (see Figure 9.1a on the *Interactive Statistics* website and CD).

b. However, we assume that the value of σ, the **population standard deviation** of this **distribution,** is 15 points, the value determined by the designers of the test (Figure 9.1b).

c. We have at our disposal a sample of 25 children from the distribution. The **Central Limit Theorem** tells us that the distribution of all the possible means of such samples will be centered on μ and will have a standard deviation $\sigma_{\overline{x}}$ of $\sigma / \sqrt{n} = 15 / \sqrt{25} = 3$ (Figure 9.1c).

d. We know from the 68-95-99.7 rule (**Section 4.2.2**) that about 95% of the values in a normal distribution fall within two standard deviations of the **mean** of the distribution.* Therefore, 95% of samples with $n = 25$ will have means between $\mu - 2(\sigma_{\overline{x}}) = \mu - 6$ and $\mu + 2(\sigma_{\overline{x}}) = \mu + 6$. In other words, 95% of the sample means will have means that fall within the range $\mu \pm 6$ (see part d on the *Interactive Statistics* website and CD).

---

*To be more precise, 95% of the values in a normal distribution fall between $z = -1.96$ and $z = +1.96$. Later on we'll use more precision when constructing confidence intervals, but for now we'll simplify things and round our critical z value up to 2.

b. But we do know the standard deviation.

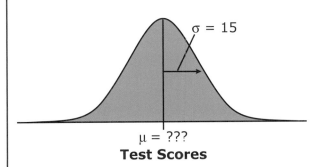

σ = 15

μ = ???
Test Scores

c. Therefore, we also know what the sampling distribution for n = 25 looks like.

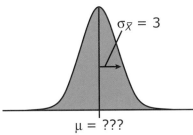

$\sigma_{\bar{X}}$ = 3

μ = ???
**Sample Means of
Test Scores for *n* = 25**

e. Therefore, a 95% confidence interval should be $4\sigma_{\bar{X}}$ wide.

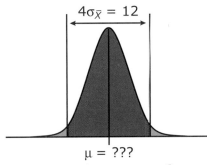

$4\sigma_{\bar{X}}$ = 12

μ = ???
**Sample Means of
Test Scores for *n* = 25**

f. 95% of all possible samples will produce confidence intervals that include μ.

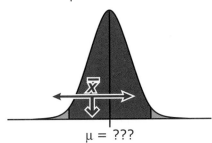

$\bar{X}$

μ = ???
**Sample Means of
Test Scores for *n* = 25**

g. 5% of all possible samples will produce confidence intervals that do *not* include μ.

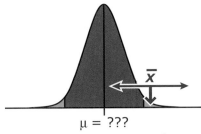

$\bar{X}$

μ = ???
**Sample Means of
Test Scores for *n* = 25**

Figure 9.1 The logic behind construction of a 95% confidence interval for Springfield seventh graders' test scores. This figure is interactive and animated, and includes parts (a) and (d), on the *Interactive Statistics* website and CD.

e. If we know that 95% of $\bar{X}$'s lie within two **standard errors** (6 points) of μ, we can turn this statement around and say that μ will lie within two standard errors of 95% of the $\bar{X}$'s. Therefore, the 95% confidence interval should be $2 \times 2\sigma_{\bar{X}} = 4 \times 3 = 12$ points wide (Figure 9.1e).

Pay special attention to parts e and f.

f. Now, we don't know for sure where the mean for the sample we've collected falls in the **sampling distribution.** But if $\overline{X}$ falls anywhere within the purple-shaded area in the figure, we can be sure that the range $\overline{X} - 6$ to $\overline{X} + 6$ includes μ (Figure 9.1f). And the probability that our $\overline{X}$ falls in the purple-shaded area is, again, .95.

g. Therefore, we can be 95% confident that the true population mean falls with the range 70 – 6 = 64 to 70 + 6 = 76. Thus, our 95% **confidence interval** for the mean test score for all Springfield seventh graders is 64 to 76. We cannot know whether the interval derived from our particular sample is one of the ones that includes μ or one of the unlucky ones that does not (Figure 9.1g). But we do know that the probability is only .05 that we were unfortunate enough to get such a "bad" confidence interval.

If you were hoping that the logic of confidence intervals would be more straightforward than the logic of **hypothesis testing,** we're sorry to disappoint you. The biggest logical leap comes in the steps in parts (e) and (f) above, where we go from the premise that 95% of sample means will be within two standard errors of the population mean, to the conclusion that there is a 95% chance that μ will be within two standard errors of any randomly selected $\overline{X}$.

But careful study of Figure 9.1f and Figure 9.1g should convince you that the logic is sound: When an interval estimate is constructed as described here, *95% of all samples will produce a confidence interval that "captures" μ, while only 5% of the samples will produce a confidence interval that fails to include μ. Therefore, we can be 95% confident that the interval based on the sample we happen to obtain in an experiment will include μ.*

Interactive Page 217

What if we wanted to be even more confident that our **interval estimate** includes μ? For example, what if we wanted a 99% **confidence interval,** rather than the 95% confidence interval calculated on the previous page?

To adapt the logic of interval estimates so that we can specify any amount of confidence, we just have to make a slight alteration to the step in part (d) of the logical sequence presented on the previous page. In that step, we noted that 95% of the **samples** drawn from a **population** with mean μ will have means between $\mu - 2(\sigma_{\overline{X}})$ and $\mu + 2(\sigma_{\overline{X}})$. The value "2" in these calculations was derived from the 68-95-99.7 rule, which tells us that about 95% of the values in a normal distribution fall within 2 standard deviations of the mean. In other words, if we translated all the possible sample means into z-scores, 95% of the z-scores would be between –2.0 and +2.0.

Creating intervals for confidence levels other than 95%

If we want a 99% confidence interval, we need to use some value other than 2 when constructing the interval. More specifically, we need to use the value of z for which 99%, rather than 95%, of sample means fall between −z and +z. Raising the confidence level of an interval estimate is roughly analogous to lowering the **alpha level** in a hypothesis test, and we use the α terminology in confidence intervals, too. Specifically, we speak of 1 − α confidence intervals, where an α level of .05 provides 1 − .05 = 95% confidence, an α level of .01 provides 1 − .01 = 99% confidence, et cetera. Note that the alpha level here provides the inverse **probability** of the confidence level: α = the probability that the interval *fails* to contain the true value of μ.

The general formula for an interval providing a confidence level of 1 − α, when σ for the population is known, is:

$$\bar{X} \pm z_{\alpha/2}(\sigma_{\bar{X}})$$

where $z_{\alpha/2}$ is the value of z for which α / 2 of the area of the normal distribution falls beyond z (e.g., 2.0 for α = .05).

The following table gives values of $z_{\alpha/2}$ to use when constructing other confidence intervals:

Confidence level	α	$z_{\alpha/2}$
80%	.20	1.3
90%	.10	1.6
95%	.05	2.0
99%	.01	2.6

For the example we're working with in this section, $\bar{X}$ = 70 and $\sigma_{\bar{X}}$ = 3, so we can construct the following confidence intervals:

- 80% confidence interval: 70 ± 1.3(3) = 66.1 to 73.9

- 90% confidence interval: 70 ± 1.6(3) = 65.2 to 74.8

- 95% confidence interval: 70 ± 2.0(3) = 64.0 to 76.0

- 99% confidence interval: 70 ± 2.6(3) = 62.2 to 77.8

Note the tradeoff between the amount of confidence we have in an interval estimate and the size of the interval: If we want more confidence, we have to content ourselves with a larger interval.

9.2.1 Confidence Intervals for Single Samples

Interactive
Page 218

Now that we've covered the logic behind constructing **confidence intervals** using **z-score** test statistics, we can go on to discuss confidence interval construction using **t statistics.** We'll start with the

> The alpha level for a confidence interval indicates the likelihood that the population mean falls outside the interval.

> The greater the confidence, the larger the interval.

simplest type of experimental design, in which we collect data from a single sample of subjects. Steve's small study from **Section 9.1** will do nicely as an example to work with:

> Steve, the head of marketing for Kidstuff, is asked to determine how many stuffed animals the typical 4-year-old American child has. He asks nine Kidstuff employees who have 4-year-olds to count and report the number of stuffed animals owned by their children. They provide the following set of data:
>
> 18 9 16 13 15 19 21 19 23
>
> The mean of this sample is 17.0, with a standard deviation of 4.27.

We don't know for sure what the mean number of stuffed animals is for all the 4-year-olds in the whole country. Steve's sample mean, 17.0, provides a **point estimate** of this **population mean.** But we know that μ is almost certainly not exactly 17.0. In other words, we have very little confidence that μ is exactly equal to the point estimate. To achieve greater confidence, we need to construct an **interval estimate.**

Although we assume that μ is probably not exactly 17.0, we still use this value as the basis for our confidence interval. That is, our interval will be:

$$(17.0 - \text{some value}) \text{ to } (17.0 + \text{some value})$$

The question is, what number should we use in place of "some value" in the equation above? In the previous section, we reasoned that if we know the **standard deviation** σ of the population, and therefore the **standard error** $\sigma_{\bar{X}}$ of the **sampling distribution,** we can achieve a confidence level of $1 - \alpha$ by using the range $\bar{X} \pm z_{\alpha/2}(\sigma_{\bar{X}})$.

Now, though, we don't know σ or $\sigma_{\bar{X}}$, so we have to use s and $s_{\bar{X}}$ to estimate these values. This means in turn that we cannot base our confidence intervals on the **normal** distribution and z-scores. Instead, we have to use t distributions and t statistics. Altering the confidence interval formula appropriately, we have:

When the population s.d. is unknown, confidence intervals must use t statistics instead of z-scores.

$$1 - \alpha \text{ confidence interval} = \bar{X} \pm t_{\alpha/2}(s_{\bar{X}}) = \bar{X} \pm t_{\alpha/2}(s / \sqrt{n})$$

Finding the right $t_{\alpha/2}$ to use is a bit more complicated than when we were working with z-scores, because its value will depend on the **degrees of freedom** our $s_{\bar{X}}$ has. As we saw in **Section 6.2.1, distributions** of t statistics get closer in shape to the normal distribution as the **sample size** n (and the degrees of freedom, equal to $n - 1$ for single-sample t statistics) increases. With small n's, the t distribution is relatively flat, meaning that a greater proportion of its

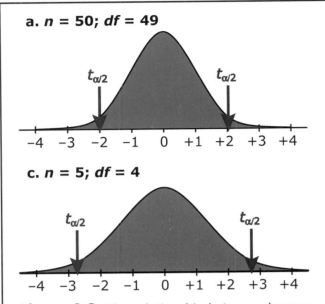

a. n = 50; df = 49

$t_{\alpha/2}$ $t_{\alpha/2}$

−4 −3 −2 −1 0 +1 +2 +3 +4

c. n = 5; df = 4

$t_{\alpha/2}$ $t_{\alpha/2}$

−4 −3 −2 −1 0 +1 +2 +3 +4

Figure 9.2 The relationship between degrees of freedom and $t_{\alpha/2}$ for confidence intervals. This figure is interactive, and includes part (b) on the *Interactive Statistics* website and CD.

area falls in the "tails." This means that the more degrees of freedom we have, the smaller our value for $t_{\alpha/2}$ will be, as illustrated in Figure 9.2.

The larger the sample, the smaller the confidence interval.

Thus a larger sample will result in a smaller *t* value in the confidence interval formula, which will in turn lead to a narrower confidence interval. Note also that raising *n* also lowers the standard error, further slimming down the confidence interval. If you think about it, this makes sense: You should be more confident in your estimate of μ if you collect a large amount of data than if you collect a small amount of data.

Interactive Page 219

You can find the exact value of $t_{\alpha/2}$ needed for any **confidence interval** by using the *t* **Distribution Area** Calculation Tool:

- First, make sure the lower-right radio button—the one labeled "Beyond *t* (two-tailed)"—is clicked.

- Next, enter the **alpha level** for the amount of confidence you want in the "Area under Curve" text box. For example, for a 95% confidence interval, α = area = .05; for a 99% confidence interval, α = .01.

- Enter the **degrees of freedom** for your analysis in the *df* box. (Remember that for a single-sample design, *df* = *n* − 1.)

- *Make sure the "t" text box is empty* and click the "calculate" button (or press the "enter" key on your keyboard). The appropriate value for $t_{\alpha/2}$ will be placed in the "t" text box.

Alternatively, if you don't have the Calculation Tool available, you can use a statistical table to look up the critical value of **t**. Instructions on using this table are provided in the **Box** on statistical tables on p. 89.

Box: Statistical Tables

To make sure you understand the confidence interval procedure, find the value of $t_{\alpha/2}$ we need to construct a 99% confidence interval for Steve's study, the facts of which are reproduced below. Then calculate the lower and upper boundaries of the confidence interval, using the formula derived on the previous page:

$$1 - \alpha \text{ confidence interval} = \bar{X} \pm t_{\alpha/2} (s_{\bar{X}})$$

Steve, the head of marketing for Kidstuff, is asked to determine how many stuffed animals the typical 4-year-old American child has. He asks nine Kidstuff employees who have 4-year-olds to count and report the number of stuffed animals owned by their children. They provide the following set of data:

18 9 16 13 15 19 21 19 23

The mean of this sample is 17.0, with a standard deviation of 4.27.

Interactive Page 220

Figure 9.3 illustrates how to find $t_{\alpha/2}$ for a 99% confidence interval with 9 – 1 = 8 *df*:

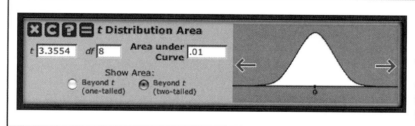

Figure 9.3 Using the *t* **Distribution Area** Calculation Tool to find $t_{\alpha/2}$. Enter an area and *df and leave the "t" text box blank,* then click the "calculate" button to find the *t* value that corresponds to the given *p*-value.

Given this value for $t_{\alpha/2}$ (3.36), our **confidence interval** is:

$$\bar{X} \pm t_{\alpha/2} (s / \sqrt{n}) = 17 \pm 3.36(4.27 / 3) = 12.23 \text{ to } 21.77$$

Steve could thus report to his superiors that he is 99% sure that the number of stuffed animals owned by the average American 4-year-old is somewhere between 12.23 and 21.77.

To make this statement, Steve must assume that the nine children in his dataset represent an unbiased, random sample of the **population** of all American 4-year-olds. Confidence intervals calculated using t also require the other **assumption** underlying hypothesis tests with t: The **sampling distribution** must be **normal.**

Assumptions for confidence intervals

Unfortunately, Steve may very well be in violation of both these assumptions. His sample was clearly not randomly drawn, and children of Kidstuff employees may be more affluent on average than the general population. Furthermore, the population **distribution** is probably positively skewed, since a kid can't have any fewer than 0 stuffed animals but could potentially have hundreds. A sample size of nine is not large enough for the **Central Limit Theorem** to guarantee a normal sampling distribution from such a skewed population. Steve would be better off assessing a larger and more **representative** sample before he comes to any firm conclusions.

Try **Exercise 9.1** a couple of times to test your understanding of the single-sample confidence interval procedure before going on to the next section, which discusses the (relatively slight) changes in the procedure necessary to calculate confidence intervals for two-sample experimental designs.

Interactive Page 221

9.2.2 Confidence Intervals for Two Samples

The procedure for constructing **confidence intervals** for two-sample experimental designs is virtually identical to that for single-sample designs:

1. Compute the average difference between the two samples—$\overline{D}$ for a **related-samples** design or ($\overline{X}_2 - \overline{X}_1$) for an **independent-samples** design. This value serves as the **point estimate** for the size of the experimental effect and also as the center of the confidence interval.

Steps for constructing confidence intervals with two-sample experimental designs

2. Compute the relevant **estimated standard error** for the **sampling distribution** of the **effect size**—$s_{\overline{D}}$ or $s_{\overline{X}_1 - \overline{X}_2}$.

3. Find $t_{\alpha/2}$ for the confidence level you want (e.g., 95%) using the procedure covered in the previous section. Remember that for a related-samples design, $df = n - 1$ (that is, the number of pairs of observations you have, minus one), whereas for an independent-samples design, $df = n_1 + n_2 - 2$ (the number of observations in both samples, minus two).

4. The confidence interval is then equal to the sample mean effect size plus or minus the critical t value times the estimated standard error. The formulas are:

$$\overline{D} \pm t_{\alpha/2}(s_{\overline{D}})$$

for related-samples designs and

$$(\overline{X}_1 - \overline{X}_2) \pm t_{\alpha/2}(s_{\overline{X}_1 - \overline{X}_2})$$

for independent-samples designs.

To test your understanding, see if you can construct the 95% confidence interval for the following scenario:

> Julie is a writer for a national fitness magazine. She decides to do a story investigating whether runners who participate in multiple marathons train harder for their first marathon or for their second marathon. To start her investigation, she conducts a survey of 15 such runners, asking them to report the average number of miles they ran per week during the two months leading up to their first marathon and during the two months leading up to their second marathon. Her data are as follows:

Avg. miles/week for 1st marathon	Avg. miles/week for 2nd marathon
46	58
64	56
31	26
49	67
39	44
34	45
52	69
59	62
27	30
58	55
50	79
63	66
49	52
51	44
60	60

Note that Julie is employing a **related-samples** design: she is collecting two related pieces of information from each respondent. Therefore, she should construct a **confidence interval** for the **difference scores,** which are calculated by subtracting the miles/week for each subject's 1st marathon from the miles/week for the subject's 2nd marathon. Here are Julie's data again, with the D scores included:

Avg. miles/week for 1st marathon	Avg. miles/week for 2nd marathon	D score
46	58	12
64	56	–8
31	26	–5
49	67	18
39	44	5
34	45	11
52	69	17
59	62	3
27	30	3
58	55	–3
50	79	29
63	66	3
49	52	3
51	44	–7
60	60	0

And here are the calculations we need to construct a 95% confidence interval for the D scores:

- Sample mean difference (miles/week for 2nd marathon – miles/week for 1st marathon) = 5.40

- Estimated standard error = $10.32 / \sqrt{15}$ = 2.66

- t for 95% confidence interval = 2.14

- Lower bound of 95% confidence interval = 5.40 – 2.14(2.66) = –.29

- Upper bound of 95% confidence interval = 5.40 + 2.14(2.66) = 11.09

Thus Julie can be 95% sure that the average difference μ_D between training for second marathons compared to first marathons is within the range –.29 to 11.09 miles/week. This confidence interval is illustrated in Figure 9.4. Positive D scores indicate that runners are training more for their second marathons than for their first, while negative D scores indicate the opposite. The range in which we can be 95% confident that the true **population mean** lies is shown by the red bracket in the figure. If we had to pick a single **point estimate,** we would be best off choosing the **sample mean,** 5.40, the location of which is indicated by the blue line in the figure.

Note that Julie's confidence interval includes 0. This means that we cannot say with 95% confidence that the true population mean is greater than (or less than) 0. In other words, we cannot be com-

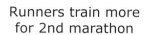

Runners train more
for 2nd marathon

Runners train more
for 1st marathon

Figure 9.4 Illustration
of the confidence interval
for Julie's data.

pletely sure that runners train harder for second marathons than for first marathons.

Any time a 95% confidence interval for a two-sample experimental design includes 0, we also know that a **hypothesis test** on the data would not reject H_0 at the α < .05 level. You can confirm this fact for the present situation by going ahead and running the hypothesis test. You should find that $t(14) = 2.03$ and $p = .062$.

This relationship between hypothesis tests and confidence intervals should not surprise you, given the relationship between confidence levels and α levels noted in **Section 9.2.**

Exercise 9.2 and **Exercise 9.3** walk you through the confidence interval procedure for related-samples designs and independent-samples designs.

If a 95% confidence interval includes 0, a hypothesis test on the data would not be significant at the $p = .05$ level.

Interactive Page 223

9.3 Interpreting Confidence Intervals

Now that we know how to construct **confidence intervals,** let's spend some time discussing how to interpret them. First of all, it is important to understand exactly what we are, and are not, claiming when we report a 95% (or 90% or 80% or any other %) confidence interval:

• We are **not** claiming that 95% of the scores in the population fall within the interval. Our claim is about the value of μ, not about the values of individual scores.

• We are **not** claiming that 95% of μ's fall within the interval. This statement would imply that μ is changeable, but this is wrong: There is only one μ.

• We are **not** claiming that there is a 95% chance that μ is equal to the $\overline{X}$ we observed. As pointed out in the first section of this chapter, the likelihood that any sample mean is *exactly* equal to μ is essentially 0.

• What we **are** claiming is that if we ran our study 100 times, using 100 different samples of subjects, 95 of the confidence intervals we constructed would include μ. Therefore, *on the basis of the* *one sample we have collected, we can be 95% sure that μ is within the interval constructed.*

Confidence intervals can be a useful tool for researchers to use when describing the results of scientific studies. They are especially valuable when the variable assessed in the study is easily interpretable, and when it is clear that certain ranges of values are either desirable or undesirable. Take the following example:

> William, a medical researcher, is interested in the effect of a long-established but controversial treatment for a certain type of cancer. It is widely accepted that the treatment extends patients' lives, but there are a number of fairly severe side effects that make the treatment difficult to undergo.
>
> In a retrospective study, William obtains the histories of 100 patients diagnosed with the cancer, half of whom received the new treatment and half of whom did not. He finds that on average, the patients who received the treatment lived 13.5 years after their diagnoses, whereas the patients who did not receive the treatment survived for an average of 5.2 years after their diagnoses. Thus the patients who were treated lived an average of 8.3 years longer than those who did not receive the treatment. William goes on to calculate and report a 99% confidence interval for the difference of 6.9 to 9.7 years.

Imagine you've just been diagnosed with this type of cancer, and you're considering whether or not to submit to the treatment William studied. William's confidence interval tells you that the average increase in life expectancy is almost certainly substantial—even if the true mean falls near the low end of the interval, patients who undergo the treatment would still be living about 7 years longer than those who aren't treated.

However, we must remember, as pointed out on the previous page, that a confidence interval assesses the range in which the mean value might fall, not the range of individual values. In fact, you can be sure that the range of individual variation is much larger than the range of the confidence interval. William's result in no way guarantees that any one individual will live at least 6.9 years longer if they undergo the treatment.

Another important thing to consider when interpreting confidence intervals is that the population mean is more likely to fall in some

Confidence intervals do *not* indicate the range of individual variation in a population.

portions of the interval than in others. Suppose, for example, that you read a study reporting a 95% confidence interval of 12.0 to 16.0. To the statistically unsophisticated, this might sound like the mean could with equal **probability** be anywhere within this range. In fact, however, it is more than twice as likely that the mean is near the center of the interval (between 13 and 15) than near either boundary (between 12 and 13 or between 15 and 16). Always remember that the **sample mean** itself (which is always right at the center of the confidence interval) is the best guess for the true location of the **population mean.**

> The population mean is more likely to be close to the center than close to the boundary of a confidence interval.

9.4 Chapter Summary/Review

This Chapter Summary/Review is interactive on the Interactive Statistics website and CD. Also be sure to go through all the **Review Exercises** for this chapter.

- In addition to or instead of testing **hypotheses,** experimenters sometimes seek to estimate the values of important **parameters** for experimental populations. The best point estimate for μ in a single-sample experiment is the **sample mean** $\overline{X}$, the best **point estimate** for μ_D in a related-samples experiment is the sample mean of the **difference scores** $\overline{D}$, and the best point estimate for $\mu_1 - \mu_2$ in an independent-samples experiment is the difference between sample means $\overline{X}_2 - \overline{X}_1$ (**Section 9.1**).

- However, it is unlikely that any **sample** statistic will perfectly estimate the **population** parameter in question. An interval estimate allows us to have more confidence in our guess (**Section 9.1**). **Section 9.2** goes through the logic of constructing **confidence intervals** in the special case where σ is known. An interval providing a confidence level of $1 - \alpha$ runs from $\overline{X} - z_{\alpha/2}(\sigma_{\overline{X}})$ to $\overline{X} + z_{\alpha/2}(\sigma_{\overline{X}})$, where $z_{\alpha/2}$ is the z-score for which the proportion of the standard **normal** curve that falls beyond z is equal to α. For example, the 68-95-99.7 rule tells us that for a 95% confidence interval (where $\alpha = .05$), $z_{\alpha/2}$ should be about 2.0 (**Section 9.2**).

- To generalize to cases in which σ is unknown, we substitute $t_{\alpha/2}$ for $z_{\alpha/2}$ and the relevent estimated standard error for $\sigma_{\overline{X}}$ in the confidence interval formula (**Section 9.2.1**). The formulas are:

Single-sample design: $\overline{X} \pm t_{\alpha/2} (s_{\overline{X}})$

Related-samples design: $\overline{D} \pm t_{\alpha/2} (s_{\overline{D}})$

Independent-samples design: $(\overline{X}_1 - \overline{X}_2) \pm t_{\alpha/2} (s_{\overline{X}_1 - \overline{X}_2})$

- To find the correct $t_{\alpha/2}$ to use, you must first calculate the **degrees of freedom** for your sample, equal to $n - 1$ for single-sample and related-samples designs and $n_1 + n_2 - 2$ for an independent-samples design. You can then use the **t Distribution Area** Calculation Tool to look up $t_{\alpha/2}$ (**Section 9.2.1**).

- Several important principles about confidence intervals were noted in the chapter. In general, the larger the sample size you have, the more precise (i.e., the more narrow) your confidence interval will be (**Section 9.2.1**). The same **assumptions** hold for confidence intervals as for hypothesis tests (**Section 9.2.1**). If a $1 - \alpha$ confidence interval for a two-sample experiment includes zero (that is, if the bottom of the interval is negative and the top of the interval is positive), a hypothesis test for the effect will always fail to be **significant** at the α level (**Section 9.2.2**).

- **Section 9.3** makes some important points to keep in mind when interpreting confidence intervals. The range of individual variation in a population will always be larger than the range of values for μ specified by a confidence interval. Also, the true population mean is more likely to be near the center of a confidence interval than near either of the boundaries of the interval.

Chapter 10
Inference for Three or More Means: Analysis of Variance (ANOVA)

10.1 Introduction to ANOVA

You may have encountered an **ANOVA** in readings for another class, and you may know that it stands for ANalysis Of VAriance. But you may be surprised to learn that ANOVA is not usually used

ANOVA is used to test hypotheses involving two or more means.

to test **hypotheses** about variances. Instead, it is an approach to analyzing data that uses **variance** statistics to test hypotheses about the relationships between two or more **means.** For example:

> Karla prepares a survey designed to measure teenagers' attitudes toward schooling. The higher the composite score from the survey, the more positive the student's attitude is about school. She gives the survey to high school students of five ethnic backgrounds: African-American, Hispanic, Asian-American, Middle Eastern, and European-American. The mean scores for each group of students are shown in Figure 10.1.

Do the results of Karla's study indicate that there are **significant** differences between the different ethnic groups in their attitudes toward school? We could address this question by conducting **independent-samples *t* tests** on each pair of experimental conditions. However, this strategy would require 10 separate hypothesis tests:

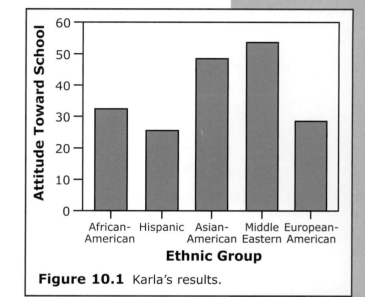

Figure 10.1 Karla's results.

- African-American vs. Hispanic
- African-American vs. Asian-American
- African-American vs. Middle Eastern
- African-American vs. European-American
- Hispanic vs. Asian-American
- Hispanic vs. Middle Eastern
- Hispanic vs. European-American
- Asian-American vs. Middle Eastern
- Asian-American vs. European-American
- Middle Eastern vs. European-American

This multitude of tests would cause at least two problems. First, it would be extremely difficult to digest the results of so many significance tests at one time. Imagine, for example, the paragraph we would need to describe the results:

For the comparison of African-American vs. Hispanic teenagers, the difference was not significant, $t(45) = 1.23$, $p = .23$. For the comparison of African-American vs. Asian-American teenagers, the difference was significant, $t(34) = 2.56$, $p = .015$. For the comparison of African-American vs. Middle Eastern teenagers, the difference was significant, $t(29) = 3.53$, $p = .0014$. For the comparison of African-American vs. European-American teenagers, the difference was not significant, $t(42) = 1.04$, $p = .30$. For the comparison of Hispanic vs. Asian-American teenagers, . . .

The second and more serious problem with analyzing a single experiment using 10 t tests has to do with the **Type I error rate**—the chances of incorrectly rejecting a true **null hypothesis** (see Section 7.1.1). For the first hypothesis test you do in an experiment, the **probability** of a Type I error is set at the **alpha** rate you adopt, usually .05 in behavioral science experiments. The scientific community accepts this amount of risk (a 1/20 chance of claiming a significant effect when in fact none exists).

What do you think the chances are of getting at least one Type I error when you conduct 10 different hypothesis tests on a single experiment?

Interactive Page 228

As noted on the previous page, the probability of making a **Type I error** when conducting a single **hypothesis test** is .05 (assuming this is the value you adopt for α).

Now consider what happens when you do more hypothesis tests in the same experiment. There is a .95 chance that you won't make a Type I error in that first test you did. But if you get through that test unscathed, there is a .05 chance that you'll make a Type I error in the second test. Thus the likelihood of a Type I error in one of the first two tests is $.05 + .95(.05) = .0975$ (we use **The Multiplication Rule for Probabilities** to calculate the second term in this equation; see the Box on p. 82).

Box: The Multiplication Rule for Probabilities

The chances of making it through these first two tests without an error are $1 - .0975 = .9025$. But you're not out of the woods yet: There's still a .05 chance that you'll make a Type I error in the third test. So now, the overall likelihood of an error jumps to $.05 + .95(.05) + .9025(.05) = .142625$.

By now, you should get the idea. The probability of at least one Type I error is about 1 minus $(1 - \alpha)$ to the Cth power (where C is equal to the number of hypothesis tests being conducted), which works out to .40 for $C = 10$ comparisons. In other words, there is a 4/10 chance that at least one of the comparisons in Karla's experiment

Comparing three or more means using t-tests inflates the Type I error rate in an experiment.

would come out to be statistically significant, even if all the populations Karla studied actually had identical means!

Interactive
Page 229

By beginning the analysis of her experimental results (see Figure 10.1) with an **ANOVA,** Karla will be able to describe the results much more succinctly and will largely avoid the inflation of **Type I error** rates described on the previous page.

The basic idea behind ANOVA is as follows. Suppose all the conditions in Karla's experiment are really identical. That is, suppose all five ethnic groups really have the same attitudes toward schooling. In this case, the five groups of subjects in Karla's experiment are really just five **samples** from the same **population** (at least with respect to their school attitudes). Because of sampling error, we wouldn't expect the **sample means** to be identical, but we wouldn't expect them to vary too much from each other.

Now, turn these statements around: If the sample means *do* vary more than we would expect due to sampling error, we can make a reasonable argument that the samples are *not* drawn from the same population. So in an ANOVA, we calculate the **variance** between the sample means and compare it to the amount of variance we would expect due to sampling error. If the former variance exceeds the latter variance by a large enough margin, we conclude that the conditions in the experiment were not all identical.

If the variance between sample means is greater than we would expect from sampling error, we can conclude that the samples were drawn from different populations.

We'll expand upon this brief introduction in **Section 10.2,** then show how to calculate *F*, the test statistic in ANOVA. In **Section 10.3,** we'll describe the process of testing hypotheses using *F* when our data come from multiple independent samples of subjects. In **Section 10.4,** we'll adapt the ANOVA procedure to deal with data from related samples. **Section 10.5** will go through procedures to use when further analyzing experimental results after the initial *F* test, and **Section 10.6** will compare ANOVA to the independent-samples *t* test. Finally, we will end this long chapter with a relatively brief section (**Section 10.7**) covering ANOVAs for experimental designs that include multiple **independent variables.**

Interactive
Page 230

10.1.1 ANOVA Terminology

Before we dive into the logic of ANOVA, we need to introduce some new terminology, and review some old terms, that we'll need in this chapter.

If you've taken a research methods course, you should already be familiar with the terms **independent variable** and **dependent variable.** The dependent variable in an experiment or research project is the set of numbers the researcher collects—for example, the

Glossary Terms: independent variable, dependent variable

survey responses Karla collected in the previous section. The independent variable can be defined as the variable on which the dependent variable depends. In Karla's project, the independent variable is ethnic group, which can take on one of five values (African-American, Hispanic, etc.). Generally speaking, the independent variable is what the researcher is controlling, and the dependent variable is what the researcher measures to assess the effects of the independent variable.

Until we got to Chapter 8, we were working almost exclusively with dependent variables, because we had a single set of measured scores in each research project. Although we didn't use the terminology at the time, the experimental designs covered in Chapter 8 all had an independent variable with two possible values (e.g., listening to NPR or watching CNN in Kimberly and John's experiment; see **Section 8.3**). Now, in this chapter, we're seeing independent variables with more than two possible values.

In ANOVA terminology, an independent variable is often called a **factor,** and the values that a factor can assume are called **levels.** We will only cover in detail analyses of a single factor, even though ANOVA can be used to analyze multifactor experimental designs.

Glossary Terms: factor, level

The letter k will be used to represent the number of levels in a factor (i.e., the number of conditions in a single-factor experiment; $k = 5$ for the factor in in Karla's study). Subscripts will be used to represent the numbers of subjects (n_1, n_2, n_3, . . . n_k), the means ($\bar{X}_1$, $\bar{X}_2$, $\bar{X}_3$, . . . $\bar{X}_k$), and so on, for each level.

We will sometimes need to refer to the total number of subjects in the experiment (that is, $n_1 + n_2 + n_3 + ... + n_k$), a number we will designate as N. The mean of all N subjects will be designated $\bar{X}_T$ (for "$\bar{X}$ Total").

Finally, we should review the distinctions between the terms variability, variance, and standard deviation:

- **Variability** (a.k.a. "spread," **Section 2.4**) is a generic term referring to the extent to which the scores in a distribution are spread out along the X axis of the distribution's histogram.

- The **standard deviation** (s for a sample and σ for a population) is the statistic we've used most often so far to measure variability.

- In this chapter, we'll more often use the **variance** statistic (s^2 and σ^2), which is equal to the standard deviation squared.

- The variance of a set of scores is calculated from another quantity called the **sum of squares** (SS; see **Section 3.3**). SS is calculated by taking each score, subtracting the mean of the scores, squaring this deviation, and summing all the squares. The variance of a sample is equal to SS divided by $n - 1$. Here are the formulas for SS and the sample variance (also viewable in the **Formula Reference** Activity):

$$SS = \Sigma(X - \bar{X})^2 \qquad s^2 = \frac{SS}{n-1}$$

- Finally, in ANOVA terminology, we refer to the sample variance as the "**mean squared deviation**," abbreviated MS.

If you feel you need to review the sum of squares calculations, go through the **Standard Deviation Helper** Activity and/or **Exercise 3.6.**

10.2 The Logic of ANOVA

Let's face it: **Analysis of variance** is a pretty complicated business. We'll attempt to get through the job of explaining it all by breaking it up into three stages. First, in this section we'll explain the logic behind ANOVA using two simplifying assumptions—that we have only three groups of subjects in the experiment (i.e., the factor we're analyzing has only three levels), and that the **sample sizes** in the three groups are equal. Then in the next section we'll work through a specific example where these simplifying assumptions hold. Then we'll eliminate the simplifications in subsequent sections.

In this portion of the chapter (this section through **Section 10.3.2**), we will limit ourselves to experimental designs in which the **independent variable** is manipulated between subjects. That is, the version of ANOVA we're talking about here is used when we have **independent samples** of subjects, each sample contributing data to one condition of the experiment. In **Section 10.4**, we'll show how the procedure can be modified to analyze data from **related samples.**

Ready or not, here we go!

As in other **hypothesis** testing situations, we will start with a **null hypothesis** stating that nothing is different, then try to prove the null wrong. If we can do this, we've indirectly shown that something happened in the experiment. For an ANOVA on an experiment with three conditions, H_0 states that:

$$\mu_1 = \mu_2 = \mu_3$$

In other words, the null hypothesis asserts that it is as if the three samples of subjects were all drawn from the same population with a common mean μ.

Suppose the **distribution** of this **population** looked like the one shown in Figure 10.2a, with the locations of the three means indicated by the arrows in the figure. Drawn like this, the null hypothesis doesn't

Glossary Term: mean squared deviation (MS)

The null hypothesis in an ANOVA predicts that all population means are equal.

a. If the underlying population were this variable, it would seem unlikely to draw three sample means as variable as pictured here . . .

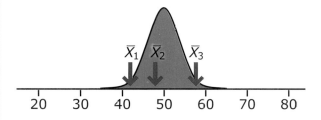

b. . . . But if the underlying population were this variable, it would not be surprising to see three sample means this far apart.

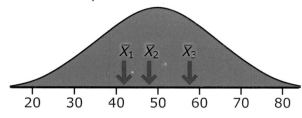

Figure 10.2 A theoretical population distribution, with the locations of three sample means indicated by red arrows.

appear very likely: the condition means appear to be too spread out for the subjects to have all been drawn from this distribution.

But what if the common population distribution were actually as pictured in Figure 10.2b? Now the null hypothesis isn't looking so bad. If there were this much overall **variability** in the scores, it would seem perfectly reasonable for the sample means to be as variable as they turned out to be.

What these two figures make clear is that the amount of variance we should expect in the condition means depends on the amount of "background" variance in the scores. In physical science research, this background variability is usually due to measurement errors on the part of experimenters. For this reason, we call the background variablility "error" variability, or **error variance.**

Glossary Term: error variance

To evaluate the null hypothesis in an **ANOVA,** we want to compare the **variability** that we would expect if scores only differed due to chance factors—the error **variance**—to the variability in scores that came about due to the different conditions we included in the experiment—the variance between conditions. We will present one way to calculate these two variances here, then present an alternate calculation process a little later on.

To estimate the **error variance** in an **independent-samples** ANOVA, we can do the same thing we did when calculating the **independent-samples *t* statistic**: take the weighted average of the variances in each of the individual samples in our dataset. In *t* tests, we called this weighted average the **pooled variance.** For ANOVA, we call it the MS_E—the "Mean of the Squared Error deviations." When we have three **samples,** all of the same size (as we're assuming in this section), the calculation of MS_E is very simple:

$$MS_E = \frac{s_1^2 + s_2^2 + s_3^2}{3}$$

FOR THREE EQUALLY SIZED SAMPLES

When all samples are the same size, we can estimate the error variance by calculating the average of the variances in each sample.

Now for the **between-condition variance,** designated MS_B in ANOVA terminology. What we want here is a statistic that reflects the amount of variability attributable to the fact that we included three different conditions in our experiment. To calculate this value, we start with the variance of the three condition means, which is calculated like so:

$$s_{\overline{X}}^2 = \frac{(\overline{X}_1 - \overline{X}_T)^2 + (\overline{X}_2 - \overline{X}_T)^2 + (\overline{X}_3 - \overline{X}_T)^2}{2}$$

FOR THREE EQUALLY SIZED SAMPLES

Glossary term: between-condition variance

(Remember from **Section 10.1.1** that $\overline{X}_T$ is the overall mean of all the scores in the dataset and k is the number of conditions.)

We cannot, however, compare $s_{\overline{X}}^2$ to the error variance directly, because MS_E reflects the variability between individual *scores,* whereas $s_{\overline{X}}^2$ reflects the variability between sample *means* of scores. To make our measure of between-condition variance comparable to our measure of error variance, we multiply $s_{\overline{X}}^2$ by n (the number of subjects in each condition)*:

$$MS_B = n(s_{\overline{X}}^2)$$

FOR THREE EQUALLY SIZED SAMPLES

When all samples are the same size, we can estimate the variance between conditions by calculating the variance between condition means times the sample size.

---

*Why do we multiply by n here? You could just take it on faith, but if that's not good enough, here's a short explanation. According to H_0, the samples from all three of our conditions are really just three samples from the same population. Therefore, if H_0 is correct, $s_{\overline{X}}^2$ is an estimate of the variance of the **distribution of sample means** for this population. We know from the **Central Limit Theorem** that $s_{\overline{X}}^2 = s^2 / n$. So to get back to a statistic that represents the variance for individual *subjects,* rather than variance for *means,* we rearrange this formula: $MS_B = s^2 = n \times s_{\overline{X}}^2$.

Note that, like the formula given above for MS_E, this formula for MS_B is only valid when all the n's are equal. We'll give more general formulas, which will hold true with unequal sample sizes, in **Section 10.2.3.**

Interactive
Page 233

Once we have values for MS_B (the **between-condition variance**) and MS_E (the error variance), we're ready to evaluate the **null hypothesis.** As in the case of a **t** test, we do so by calculating a **test statistic.** For **ANOVA,** the test statistic we use goes by the letter **F** (for Sir Ronald Fisher, the inventor of the statistic), and is calculated by dividing MS_B by MS_E:

$$F = MS_B / MS_E$$

If $\mu_1 = \mu_2 = \mu_3$, as the null hypothesis proposes, the three condition means should be equivalent to three sample means drawn from the same **population.** And if this is true, then MS_B should be nothing more than an estimate of the **variance** of this population. Likewise, MS_E should also be an estimate of the population variance. Therefore, if H_0 is true, MS_B and MS_E *should be two estimates of the same variance,* and when we divide MS_B by MS_E, we should get a value close to 1.

The *F* statistic is calculated by dividing the variance between conditions by the error variance.

Alternatively, if the underlying population means of the three conditions are different, then the variability between conditions (MS_B) should be greater than the error variability (MS_E). The greater the experimental effects, the larger the discrepancy between MS_B and MS_E, and the bigger our F ratio will be.

The larger our *F* statistic, the less likely it is that the subjects in each condition were all drawn from populations having exactly the same mean.

Interactive
Page 234

If you're still a little confused at this point, don't worry. In the next section we'll go through a specific example, then in the section after that we'll go through the **ANOVA** calculations in a different way. Here's a brief summary of what we've covered so far:

- To analyze whether or not there are significant differences between three condition **means,** we compare the **variance** between the means (the variance between conditions) with the amount of variance we would expect to see by chance alone (the variance due to error).

- If all our conditions include **independent samples** with identical **sample sizes,** we can calculate the **error variance,** MS_E, by averaging the variances within each condition.

- We can calculate the **between-condition variance,** MS_B, by computing the variance between the sample means and multiplying by n (again, assuming that all n's are equal).

Summary of ANOVA logic

- Once we have MS_B and MS_E, we divide the former by the latter to get an **F statistic.**

- If there are no differences between the conditions, MS_B should be about equal to MS_E, so F should be somewhere around 1.0. If at least one of the conditions does differ from the others, MS_B should be greater than MS_E, so F should be greater than 1.

We'll put off until **Section 10.2.2** discussion about exactly how big F has to be for an effect to be considered "significant." Try **Exercise 10.1** for a quick check on your basic understanding of error variance and between-condition variance, then go on to the the next section to see our first specific example of using ANOVA to analyze experimental results.

10.2.1 An Example

Yonnick, a cognitive psychologist, is studying ways to help people attach names to faces. He runs an experiment in which subjects meet seven confederates and are then asked to recall the confederates' names one week later. Five subjects are tested in each of three conditions:

- **Nametag condition**: The confederates wear nametags, but no one actually introduces them to the subjects.

- **Other-Introduction condition**: The experimenter introduces each of the confederates to each subject, but the confederates themselves do not speak.

- **Self-Introduction condition**: Each of the confederates introduces himself or herself to the subjects.

The independent variable in the experiment is type of introduction (Nametag, Other-Introduction, or Self-Introduction). The dependent variable is the number of confederates' names each subject recalls. The data come out as follows:

Nametag condition	Other-Intro condition	Self-Intro condition
0	3	6
1	4	3
1	1	7
3	5	6
0	2	3

The formulas for MS_B, MS_E, and F that we're working with for now are:

$$MS_B = ns_{\bar{X}}^2 = (n)\frac{(\bar{X}_1 - \bar{X}_T)^2 + (\bar{X}_2 - \bar{X}_T)^2 + (\bar{X}_3 - \bar{X}_T)^2}{2}$$

$$MS_E = \frac{s_1^2 + s_2^2 + s_3^2}{3}$$

Remember that these formulas are quite specific—they only apply to an **independent-samples** ANOVA involving a single **factor** with three **levels** (i.e., three conditions) where all three conditions have the same **sample size**.

The formulas above are only valid if all n's are equal.

ANOVA calculations are tedious, but you have to learn how to work through them in order to really understand what **analysis of variance** is all about. Try the calculations yourself for the data above before going on.

Interactive
Page 236

ANOVA calculations for Yonnick's dataset:

n for each condition = 5

$\bar{X}_1$ (Nametag condition) = 1.0

$\bar{X}_2$ (Other-Intro condition) = 3.0

$\bar{X}_3$ (Self-Intro condition) = 5.0

$\bar{X}_T$ = 3.0

MS_B = 20.0

s_1^2 (Nametag condition) = 1.5

s_2^2 (Other-Intro condition) = 2.5

s_3^2 (Self-Intro condition) = 3.5

MS_E = 2.5

F = 8.0

As shown in the answers above, $F = MS_B / MS_E = 20.0 / 2.5 =$ **8.0** for Yonnick's data. Since **F** is much larger than 1, it appears that the **variance** between conditions is greater than we would expect to see by chance alone. In a few pages we'll learn how to attach a **p-value** to this F statistic.

If you got all the calculations correct, give yourself a pat on the back. If not, don't worry too much about it, and in either case don't bother trying to memorize the formulas for MS_B and MS_E from the previous page. In the next section, we'll learn more general formulas that you will need to commit to memory.

10.2.2 Partitioning Variance

The formulas introduced in **Section 10.2** are useful for defining the concepts behind **ANOVA,** but as noted several times already, they only work when **n** is the same in all conditions. Now let's look at analysis of variance in another way. This second look will serve two purposes. First, it gives you a second perspective on the logic of the ANOVA calculations, and second, it will lead us to a set of formulas for calculating MS_B and MS_E that generalizes to situations in which conditions have different sample sizes.

Figure 10.3b shows the name-recall scores of Yonnick's 15 subjects from the previous section, without differentiating between the three experimental conditions. The overall mean for the 15 scores is $\bar{X}_T =$ 3 (represented by the red arrow in the figure). Given this mean, we can calculate the total **sum of squared deviations,** SS_T, as follows:

$$SS_T = \Sigma(X - \bar{X}_T)^2$$

$$= (0 - 3)^2 + (0 - 3)^2 + (1 - 3)^2 + \ldots + (7 - 3)^2 = 70$$

Figure 10.3b illustrates the **deviations** that go into the calculation of SS_T (Figure 10.3 parts a and c, which are not shown, are interactive and animated on the *Interactive Statistics* website and CD).

In Figure 10.3e, the scores are color-coded to show which condition each one comes from. The three condition means, 1, 3, and 5, are also shown. Follow along for a bit now while we work through some formulas that might not seem too relevant at first.

From the formula above, the deviation of each of the 15 scores from the overall mean $\bar{X}_T$ is:

$$X - \bar{X}_T$$

We can also write:

$$X - \bar{X}_T = (\bar{X}_i - \bar{X}_i) + (X - \bar{X}_T)$$

where $\bar{X}_i$ is the condition mean for the score in question. For example, for the first subject in the Self-Introduction condition, X (the subject's score) is 6, $\bar{X}_i$ (the mean score for the Self-Introduction condition) is 5.0, and $\bar{X}_T$ (the overall mean for all 15 subjects) is 3.0. Note that all we did here was to add a number ($\bar{X}_i$) and then subtract it right back out of the expression. Now we can rearrange the expression like so:

$$X - \bar{X}_T = (\bar{X}_i - \bar{X}_T) + (X - \bar{X}_i)$$

The formula we've ended up with can be interpreted as follows:

Calculating the total sum of squared deviations

b. Total variability, as measured by the deviations of each score from the overall mean ($\overline{X}_T$).

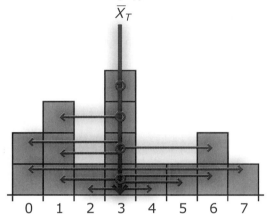

Figure 10.3 Graphical illustration of how variance is partitioned in an independent-samples ANOVA (this figure is interactive and animated, and includes parts (a) and (c) on the *Interactive Statistics* website and CD). Total variability (b) is separated into variability due to condition differences (d) and variability due to within-condition error (e). You will probably find it easier to follow the logic of this section if you work through it while interacting with Figure 10.3 on the *Interactive Statistics* website or CD.

d. Between-condition variability, as measured by the deviations of the $\overline{X}_i$'s from $\overline{X}_T$.

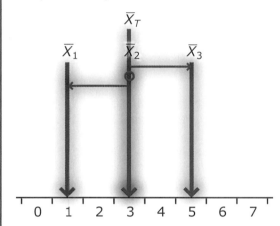

e. Error variability, as measured by the deviations of each score from $\overline{X}_i$ of its condition.

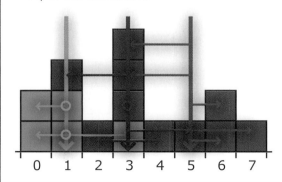

The deviation of each subject's score from the overall mean $(X - \bar{X}_T)$

= the deviation of the mean of the subject's condition from the overall mean $(\bar{X}_i - \bar{X}_T)$

+ the deviation of the subject's individual score from the mean of the subject's condition $(X - \bar{X}_i)$

In other words, the total amount of **variability** in the scores (measured by $X - \bar{X}_T$) can be partitioned into

- The variability caused by differences between conditions (measured by $\bar{X}_i - \bar{X}_T$), at least part of which can be attributed to the effects of Yonnick's experimental manipulation, and

- The variability caused by differences between individual scores within each condition (measured by $X - \bar{X}_i$), which must be due to the factors we've designated as "error."

Parts d and e of Figure 10.3 illustrate this concept graphically. Figure 10.3d shows the deviations of the three condition means from the overall mean, and Figure 10.3e shows the deviations of the scores from their condition means.

Partitioning total variability into variability due to differences between conditions and variability due to error

Interactive Page 238

On the previous page, we established that for each score in our sample dataset, $X - \bar{X}_T = (\bar{X}_i - \bar{X}_T) + (X - \bar{X}_i)$

We'd like to use the three components of this formula, each of which is essentially a **deviation score,** to measure the total variability in the dataset, the variability due to differences between conditions, and the error variability. But as noted in **Section 3.3,** raw deviation scores aren't very useful for measuring variability, because when you add up a set of deviation scores, they always sum to 0! To overcome this problem, we can square all three deviation scores for each subject:

$(X - \bar{X}_T)^2$	$(\bar{X}_i - \bar{X}_T)^2$	$(X - \bar{X}_i)^2$
$(0 - 3)^2$	$(1 - 3)^2$	$(0 - 1)^2$
$(1 - 3)^2$	$(1 - 3)^2$	$(1 - 1)^2$
$(1 - 3)^2$	$(1 - 3)^2$	$(1 - 1)^2$
$(3 - 3)^2$	$(1 - 3)^2$	$(3 - 1)^2$
$(0 - 3)^2$	$(1 - 3)^2$	$(0 - 1)^2$
$(3 - 3)^2$	$(3 - 3)^2$	$(3 - 3)^2$
$(4 - 3)^2$	$(3 - 3)^2$	$(4 - 3)^2$
$(1 - 3)^2$	$(3 - 3)^2$	$(1 - 3)^2$
$(5 - 3)^2$	$(3 - 3)^2$	$(5 - 3)^2$
$(2 - 3)^2$	$(3 - 3)^2$	$(2 - 3)^2$
$(6 - 3)^2$	$(5 - 3)^2$	$(6 - 5)^2$
$(3 - 3)^2$	$(5 - 3)^2$	$(3 - 5)^2$
$(7 - 3)^2$	$(5 - 3)^2$	$(7 - 5)^2$
$(6 - 3)^2$	$(5 - 3)^2$	$(6 - 5)^2$
$(3 - 3)^2$	$(5 - 3)^2$	$(3 - 5)^2$

If we now sum these squared deviations, we get three quantities:

$$\Sigma (X - \overline{X}_T)^2 = 70 = SS_T = \text{total variability}$$
$$\Sigma (\overline{X}_i - \overline{X}_T)^2 = 40 = SS_B = \text{variability between conditions}$$
$$\Sigma (X - \overline{X}_i)^2 = 30 = SS_E = \text{variability within conditions}$$

Note that the SS's add together just like the components of the deviation formula: $SS_T = SS_B + SS_E$.

Now we're ready to relate the ANOVA calculations in this section to the ones we went through in **Section 10.2.** Recall that the F statistic is calculated by dividing the mean squared deviation between conditions (MS_B) by the *mean* squared deviation due to error (MS_E). As we learned in **Section 3.3.2** and in the **Box** on degrees of freedom (see p. 55), to get from an **SS** to an **MS** (a.k.a. a **variance**), we divide by the **degrees of freedom** associated with the sum of squares. The df represents how many numbers going in to the SS calculation are free to vary independently:

- For SS_B, we have 3 condition means, each of which is being compared to 1 overall mean. Two of the $\overline{X}_i$'s are free to vary, but the third $\overline{X}_i$ is restricted by the requirement that the deviations must add up to 0. Therefore, $df_B = 3 - 1 = 2$.

- For SS_E, we are comparing each of the 15 individual scores to its condition mean. Of the 5 scores in the Nametag condition, 4 are free to vary, but the fifth score is restricted by the requirement that the 5 deviation scores must add up to 0. Similarly, 4 of the 5 scores in the Other-Introduction condition and 4 of the 5 scores in the Self-Introduction condition are free to vary. Therefore, $df_E = (5 - 1) + (5 - 1) + (5 - 1) = 15 - 3 = 12$.

- We can also calculate the df associated with SS_T. Here, we are comparing each of the 15 individual scores to the overall mean ($\overline{X}_T$). Fourteen of the scores are free to vary, but the fifteenth is restricted by the requirement that the deviations must add to 0. Therefore, $df_T = 15 - 1 = 14$.

Satisfyingly, the df's add together just like the SS's: $df_T = df_B + df_E$.

**Interactive
Page 239**

We can now recalculate our **F statistic** using the **SS's** we calculated on the previous page:

Source	SS / df = MS	F
Between Conditions	40 / 2 = MS_B = 20.0	MS_B / MS_E = **8.0**
Error	30 / 12 = MS_E = 2.5	
Total	70 / 14	

Sidebar:

$MS = SS / df$

**Box:
Degrees of freedom**

Degrees of freedom between conditions, error df, and total df

This table, minus the arithmetic symbols, is called an "ANOVA summary table," and it is a great way to present the results of an **analysis of variance.** One benefit of presenting the information this way is that it allows you to easily check to make sure that $SS_T = SS_B + SS_E$ and that $df_T = df_B + df_E$. (Note, however, that MS_T, which would be 70 / 14 = 5.0, does *not* equal $MS_B + MS_E$. For this reason, and because it is not actually used in the calculation of F, MS_T is not included in the ANOVA summary table.)

Interactive Page 240

10.2.3 Final Formulas

In this section we'll go through a set of formulas, based on the expressions we developed in the previous section, that you can use to compute an independent-samples *F* value for any number of conditions and with **sample sizes** that may be unequal. Before you start using these formulas, you'll have to calculate some preliminary statistics:

• k: the number of conditions in your dataset

• N: the total number of subjects in all conditions

• $\bar{X}_T$: the **mean** score for all subjects in all conditions

• n_i, $\bar{X}_i$: the number of subjects and mean in each condition i (e.g., for an experiment with three conditions, you will need n_1, n_2, n_3, $\bar{X}_1$, $\bar{X}_2$, and $\bar{X}_3$)

Once these preliminaries are out of the way, you can dive into the **ANOVA** calculations:

$$SS_B = \Sigma n_i(\bar{X}_i - \bar{X}_T)^2$$

$$SS_E = \Sigma(X - \bar{X}_i)^2$$

$$df_B = k - 1$$

$$df_E = N - k$$

$$MS_B = SS_B / df_B$$

$$MS_E = SS_E / df_E$$

$$F = MS_B / MS_E$$

Note that for SS_B you are summing k terms (one for each condition), whereas for SS_E you are summing N terms (one for each subject in the experiment). Also note that if you have the SS_i's for each condition separately, you can calculate SS_E as ΣSS_i (e.g., for an experiment with three conditions, $SS_E = SS_1 + SS_2 + SS_3$).

The ANOVA summary table

Preliminary calculations necessary for ANOVA

General formulas for calculating F

Although not strictly necessary, you should also calculate:

$$SS_T = \Sigma(X - \overline{X}_T)^2$$

$$df_T = N - 1$$

To check your work, make sure that $SS_T = SS_B + SS_E$ and that $df_T = df_B + df_E$.

Interactive
Page 241

Now let's use the formulas given on the previous page to recalculate **F** for Yonnick's experiment. Here are the data again:

Nametag condition	Other-Intro condition	Self-Intro condition
0	3	6
1	4	3
1	1	7
3	5	6
0	2	3

The **Descriptive Statistics** Calculation Tool can greatly ease the **ANOVA** calculations. Here's a set of step-by-step instructions for calculating F with the help of this Tool:

Calculation Tool: Descriptive Statistics

1. First enter *all* the scores in the dataset into the Tool and click the "calculate" button. This will give you the total number of scores ($N = 15$), the overall mean ($\overline{X}_T = 3$), and the total sum of squares ($SS_T = 70$).

2. Next use the Tool to calculate n_i, $\overline{X}_i$, and the **sum of squared deviations** (SS_i) for each condition. Add the SS_i's together to get SS_E. For Yonnick's data, $SS_E = SS_1 + SS_2 + SS_3 = 6 + 10 + 14 = 30$.

3. Now use your n_i's, $\overline{X}_i$'s, and $\overline{X}_T$ to calculate SS_B:

$$SS_B = n_1(\overline{X}_1 - \overline{X}_T)^2 + n_1(\overline{X}_2 - \overline{X}_T)^2 \ldots n_k(\overline{X}_k - \overline{X}_T)^2$$
$$= 5(1 - 3)^2 + 5(3 - 3)^2 + 5(5 - 3)^2$$
$$= 20 + 0 + 20 = 40$$

Calculation Tool: Expression Calculator

5. Note that you can use the **Expression Calculator** Tool to calculate each of the terms in this formula. For example, enter "5(1 – 3)^2" and click the "calculate" button in the **Expression Calculator** to get the value for the first condition in Yonnick's experiment. (The caret symbol "^" means "raise to the power of.")

6. Now that you have all your sums of squares, make sure that $SS_T = SS_B + SS_E$.

7. Then calculate your **degrees of freedom**: df_B = the number of conditions minus 1 (3 − 1 = 2 for Yonnick's experiment), df_E = the total number of scores in the dataset minus the number of conditions (15 − 3 = 12), and df_T = the total number of scores in the dataset minus 1 (15 − 1 = 14). Make sure that $df_T = df_B + df_E$. (You shouldn't need any kind of calculator for the df calculations!)

8. Finally, use the Desktop Calculator, the **Expression Calculator** Tool, or your own long division skills to calculate $MS_B = SS_B / df_B$ (40 / 2 = 20), $MS_E = SS_E / df_E$ (30 / 12 = 2.5), and $F = MS_B / MS_E$ (20 / 2.5 = 8).

The **ANOVA Helper** Activity can help you organize, check, and practice these calculations. To use the Helper, first enter the scores for each condition in the three large text boxes in the upper left corner of the window (the Helper will only Help with ANOVAs including three or fewer conditions). Then start performing the calculations described above, and enter your intermediate calculations in the blanks in the main portion of the window. Note that you can press the "tab" key on your keyboard to move from field to field in the window.

Click the "Check Answers" button at any time—or simply hit the "enter" or "return" key after typing any value into a text box—to check your calculations. Any incorrect answers will be colored red; correct calculations will remain in black. Roll your mouse over the bullet (small circle) to the left of each text box to see a hint for what's supposed to go there. If you give up, click the bullet to fill in the correct value (it will appear in blue so you can keep track of which fields you needed help with).

Activity: ANOVA Helper

10.3 Testing Hypotheses with Independent-Samples ANOVA

Now that you know how to calculate an **F** ratio, you're ready to test **hypotheses** using this **inferential statistic.** We will use the same four-step procedure we've been using with t statistics:

Step 1: Establish the hypotheses. As noted earlier in the chapter, the initial **null hypothesis** in an analysis of variance states that all the condition means are identical:

$$H_0: \mu_1 = \mu_2 = \mu_3 \ldots = \mu_k$$

In a two-sample t test, there are three possible **alternative hypotheses** that a researcher can choose: $\mu_1 \neq \mu_2$, $\mu_1 > \mu_2$, and $\mu_1 < \mu_2$. When we move up to more than two samples and **ANOVA,** the number of alternatives skyrockets. For example, in a one-factor ANOVA with four levels, H_0 would be wrong if $\mu_1 = \mu_2 = \mu_3$, but μ_4 was different from the other three, or if $\mu_1 = \mu_3$ and $\mu_2 = \mu_4$, but $\mu_1 \neq$

There are many possible alternatives to the null hypothesis in an ANOVA.

μ_2 and $\mu_3 \neq \mu_4$, or if $\mu_1 = \mu_2$, but $\mu_2 \neq \mu_3$ and $\mu_3 \neq \mu_4$, etc. The possibilities are not endless, but they are quite numerous. So for the initial analysis of variance on a dataset, we simply state:

H_a: at least one population mean differs from the others

Figure 10.4 illustrates these null and alternative hypotheses graphically, for an ANOVA with three conditions. Compare this figure to the ones showing possible alternatives in single-sample experiments (**Section 5.3.1**) and two-sample experiments (**Section 8.3**).

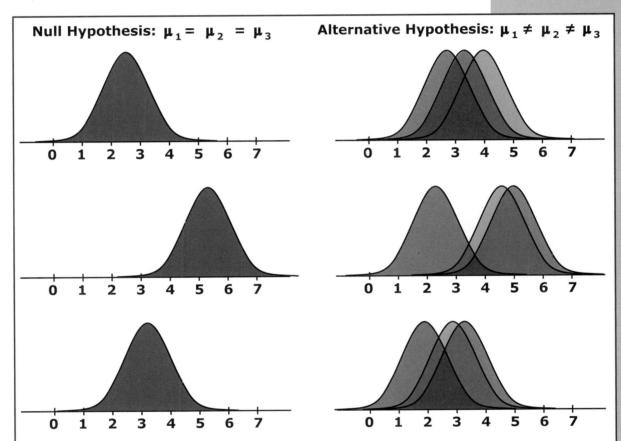

Null Hypothesis: $\mu_1 = \mu_2 = \mu_3$ **Alternative Hypothesis: $\mu_1 \neq \mu_2 \neq \mu_3$**

Figure 10.4 Graphical illustration of the null and alternative hypotheses for an independent-samples ANOVA with three conditions (this figure is animated on the *Interactive Statistics* website and CD). The null hypothesis states that the means of all three samples are exactly equal (since the distributions overlap precisely, they appear to be a single distribution in the figure). The alternative hypothesis states that there is some difference between the means. Neither hypothesis makes any claim about the absolute magnitude of any of the μ's, and H_a makes no claim about the magnitude or direction of any effects.

Step 2: Collect and describe the sample data. As in the case of the ***t* statistic,** calculation of an F statistic requires measures of the **sample size, central tendency,** and **variability** of each group of scores in the dataset. In fact, it is possible to calculate F using exactly the same **descriptive statistics** (n, $\overline{X}$, and s) we used in calculating t. However, it is both conceptually clearer and computationally easier to use the F formulas involving **SS's** instead of s's (these formulas are given in **Exercise 197** and in the **Formula Reference** Activity).

Step 3: Compute the test statistic. We've spent quite a lot of time on this already. The F statistic is a ratio of the **variance** in the data that is attributable to differences between conditions, divided by the variance that is attributable to individual differences ("error") within conditions.

Step 4: Evaluate the null hypothesis. To do this, we need a **distribution** of all possible values for the test statistic. Just as there were multiple t distributions, there are multiple F distributions, depending on the degrees of freedom for the experimental design.

We'll discuss these distributions in the next section. But to jump ahead a bit, once we determine the right distribution we will assess where our particular test statistic is located in the distribution, then calculate the **probability** of arriving at such a test statistic if all that was at work in the experiment were random factors.

In other words, the final product of our F test will be the same as the final product of a **z test,** a single-sample t test, or a two-sample t test: an objective measure of the probability that we would observe the sample means we observed, if the null hypothesis were correct. If we judge this p-value to be low enough (traditionally, if $p < .05$), we can conclude that the null hypothesis is probably *incorrect,* and declare that a **significant** effect occurred in the experiment.

In the next section we'll see what the F distributions look like and discuss how to use these distributions to translate F ratios into p-values. Then we'll go through the four steps of hypothesis testing using a new example.

10.3.1 *F* Distributions

Before we see our first F **distribution,** let's review some general points about distributions of **test statistics** that we covered in earlier chapters. When we were first introduced to **hypothesis testing** in **Chapter 5,** we learned to compare **z-score** test statistics to the standard **normal** distribution, shown in Figure 10.5. The rationale we developed for comparing test statistics to the normal distribution was as follows:

Activity: Formula Reference

The end result of an ANOVA is, like the end result of a t test, a p-value.

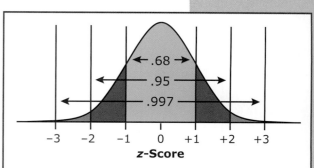

Figure 10.5 The standard normal distribution, with the 68-95-99.7 rule superimposed.

- Suppose we took a million different **samples** from the **population** studied in an experiment, calculated a z-score test statistic for each sample, and plotted all 1,000,000 z-scores in a **histogram.**

- The **Central Limit Theorem** guarantees that this histogram would be shaped exactly like the normal curve.

- Furthermore, if the **null hypothesis** were true, the distribution of z-score test statistics would be centered on 0 and would have a **standard deviation** of 1.

- Therefore, we can use our detailed knowledge of the shape of the normal curve, together with the **probability rules** we developed in **Chapter 4,** to determine the probability of randomly drawing from a normal distribution a z-score as large as the one calculated from our **sample mean.** For example, the 68-95-99.7 rule (see **Section 4.2.2** and Figure 10.5) tells us that about 95% of the z-scores in a normal distribution will be between –2.0 and +2.0. This means that if the null hypothesis is true, the probability of getting a z-score test statistic as extreme as 2.0 is $1 - .95 = .05$. So if the test statistic we observe is greater than 2.0, the likelihood of the null hypothesis being correct is less than $p = .05$.

Review: finding the p-value associated with a z-score test statistic

In Chapter 6, we learned that the distribution of a million t **statistics** would look similar to a normal curve but would be a little flatter, depending on the **degrees of freedom** involved in calculating t. Again, we can use our detailed knowledge of the shape of t distributions to calculate p-values associated with the one value of t found in an experiment.

The same logic holds for F **statistics**: Mathematical definitions exist that precisely describe the shape of F distributions, allowing us to calculate p-values in the same way we did with z-scores and t statistics.

The p-values for F tests are found with the aid of F distributions, just like p-values for z and t tests.

However, the F distributions look quite different from the z-score or t distributions. Although you may never have seen an F distribution before, the information you've learned already about how F statistics are calculated should tell you something about what these distributions will look like. Recalling that F is calculated by dividing the **between-condition variance** by the **error variance,** and remembering a few definitions from **Section 2.4,** see if you can answer the following questions:

- Will any F statistics fall below 0? That is, will we ever see a negative F statistic?

- The z-score and t distributions are symmetrical. Will the F distributions be symmetrical as well, or will they be skewed?

Interactive Page 244 Figure 10.6 shows a number of **F** distributions, revealing answers to the queries posed on the previous page:

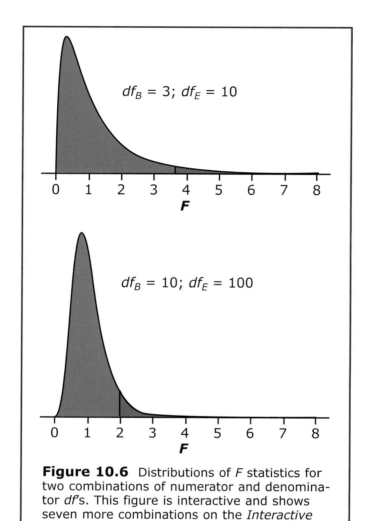

Figure 10.6 Distributions of F statistics for two combinations of numerator and denominator *df*'s. This figure is interactive and shows seven more combinations on the *Interactive Statistics* website and CD.

- *F* statistics can **never** fall below 0. *F* is calculated as a ratio of two **variances.** Variances can never be negative, so both the numerator and the denominator in an *F* calculation will always be positive. The result will always be a positive *F* statistic.

- While *F* scores cannot be lower than 0, they can in theory be infinitely large. A large *F* ratio is produced whenever there is very little **error variance** (meaning a small denominator in the *F* formula) and/or a great deal of **between-condition variance** (resulting in a large numerator). A distribution that is limited on the left side but can spread out infinitely far on the right is bound to be positively **skewed.**

F distributions have a left boundary at 0 and are positively skewed.

As with distributions of **t statistics,** F distributions vary depending on the **degrees of freedom** involved in calculating the test statistic. But remember that when calculating F we are dealing with two degrees of freedom—one for the numerator ($df_B = k - 1$) and one for the denominator ($df_E = N - k$). This means that there is one **distribution** for F statistics with $df = (1, 1)$ (one degree of freedom in the numerator and one in the denominator), another distribution for $df = (1, 2)$, another for $df = (1, 3)$, another for $df = (2, 2)$, and so forth

There is a different F distribution for each combination of df's in the numerator and denominator.

Figure 10.6 illustrates the variability of F distributions. In addition to showing the general shape of the F distribution for each df combination, Figure 10.6 also highlights in pink the portion of the **tail** of the distribution containing the most extreme 5% of all possible F values. If the F statistic for a hypothesis test falls in this area, **p** will be less than .05, and the null hypothesis will be rejected (assuming an alpha setting of .05). (For the distributions with $df_E = 3$, you need an F value greater than 8 to be able to reject H_0, so none of the area in these graphs is colored pink.)

If you play around a bit with the interactive Figure 10.6 on the *Interactive Statistics* website and CD, you should notice two trends:

The more subjects in an experiment, the more likely it is that the null hypothesis will be rejected.

1. If you keep the numerator df (df_B) constant and increase the denominator df (df_E), the "rejection region" of the F distribution will shift to the left. This means that a smaller F statistic is needed to reject H_0. Remember that df_E is equal to $N - k$. So what we learn here is that by running more subjects in an experiment (increasing N), the likelihood of rejecting H_0 increases. This relationship also holds, you may recall, for experiments whose results are assessed with t tests (see **Section 7.4.3**).

2. If you keep the denominator df constant and increase the numerator df, the rejection region of the F distribution will also shift to the right. This trend might appear to indicate that adding extra conditions to your experiment will, like adding subjects, give you more power to reject H_0. But this is an *invalid* conclusion to draw, because adding conditions will also tend to reduce the value of the F statistic you calculate in the experiment (for reasons we won't go into here).

To find the **p-value** associated with an F statistic, use the **F Distribution Area** Calculation Tool, which works just like the **t Distribution Area** and **Normal Distribution Area** Tools. The only difference is that you have to enter two df values in the F Tool. For example, you should find that for Yonnick's F of 8.0, with $df_B = 2$ and $df_E = 12$, the p-value is .0062. Test the Tool out yourself by looking up the p-value associated with the same F but with the df's reversed.

Calculation Tool: F Distribution Area

(If you have to find the *p*-value for an *F* ratio and don't have access to a computer, you may need to use a statistical table. The procedure for looking up *p*-values in an *F* Distribution Table is described in the **Box** on statistical tables on p. 89; see Figure 4 of the Box.)

Box: Statistical Tables

10.3.2 Another Example

We're ready now to work through the entire **hypothesis** testing procedure with a new example:

> For her doctoral dissertation, Freida decides to study changes in people's level of anxiety across the lifespan. She searches the literature and finds a test that provides a quantitative measure of anxiousness, ranging from 1 (least anxious) to 9 (most anxious). She then administers this test to groups of 15-year-olds, 20-year-olds, 30-year-olds, and 50-year-olds. The anxiety scores are:

Age	Scores
15	2, 4, 1, 4, 4, 1, 5, 5, 3, 2, 2
20	9, 2, 7, 8, 7, 8, 7, 8, 8, 6
30	4, 5, 6, 6, 4, 3, 4, 8, 5
50	7, 4, 3, 3, 7, 4, 6, 8, 5, 5

> Can Freida conclude that people at different ages have significantly different levels of anxiety?

Step 1. The **null hypothesis** states that the mean anxiety scores are equal for the **populations** of 15-year-olds, 20-year-olds, 30-year-olds, and 50-year-olds: $\mu_1 = \mu_2 = \mu_3 = \mu_4$. The **alternative hypothesis** states that at least one of the populations differs from the others.

Step 2. The numbers of subjects, the means, and the sums of squared deviations for each condition are:

Age	*n*	$\bar{X}$	*SS*
15	11	3.0	22.0
20	10	7.0	34.0
30	9	5.0	18.0
50	10	5.2	27.6

Overall, there are $N = 40$ subjects whose mean score is $\bar{X}_T = 5.0$. The total **sum of squared deviations** is $SS_T = 186.0$.

Step 3. The F calculations are:

$$SS_B = \Sigma n_i(\overline{X}_i - \overline{X}_T)^2 = 11(3.0 - 5.0)^2 + 10(7.0 - 5.0)^2 + 9(5.0 - 5.0)^2 + 10(5.2 - 5.0)^2 = 84.4$$

$$SS_E = \Sigma(X - \overline{X}_i)^2 = SS_1 + SS_2 + SS_3 + SS_4 = 22.0 + 34.0 + 18.0 + 27.6 = 101.6$$

$$df_B = k - 1 = 4 - 1 = 3$$

$$df_E = N - k = 40 - 4 = 36$$

$$MS_B = SS_B / df_B = 84.4 / 3 = 28.13$$

$$MS_E = SS_E / df_E = 101.6 / 36 = 2.82$$

$$F = MS_B / MS_E = 28.13 / 2.82 = \textbf{9.97}$$

Checking our work, $SS_B + SS_E = 84.4 + 101.6 = 186.0$, equal to the calculated value of SS_T. Also, $df_T = N - 1 = 40 - 1 = 39$, which is equal to $df_B + df_E$ (3 + 36 = 39).

Step 4. Entering our F value (9.97) and **degrees of freedom** (3 and 36) into the **F Distribution Area** Calculation Tool, we find that $p = .0001$. Freida's dissertation is looking good!

F statistics are usually reported along with their numerator and denominator degrees of freedom, the MS_E for the analysis, and the p-value. So Freida will write "$F(3, 36) = 9.97$, $MS_E = 2.82$, $p = .0001$."

Reporting the results of an F test

Now is the time to try your hand at doing some independent-samples ANOVA calculations yourself. Do **Exercise 10.2, Exercise 10.3,** and **Exercise 10.4** at least one time each before moving on to the next section, because the concepts behind and calculations for related-samples ANOVA, to which we turn next, are based on those for independent-samples ANOVA. (You can also fine-tune your ANOVA calculation skills in the **ANOVA Helper** Activity, which will generate random sets of data for you to practice with.)

**Activity:
ANOVA Helper**

Interactive
Page 246

10.4 ANOVA for Related Samples

Now that we've covered the basic tenets of ANOVA with **independent-samples designs,** let's look at how the analysis changes when we have **related samples** of data. Here's an example to work with:

> Harry works for the marketing department of a long-distance telephone company. The company wants to run a promotion in which customers' calls to their mothers are free for one month. They want to run

the promotion in May, November, or December (around Mother's Day, Thanksgiving, or Christmas), and they ask Harry to find out in which month people call their mothers most often.

To answer this question, Harry conducts a small survey in which he asks five subjects to estimate how many mom-calls they made in each month (a real marketing study would use many more subjects, but we'll keep things small for our example). He collects the following responses:

Subject	May	Nov.	Dec.	Subject Means
Subject A	3	4	5	4.0
Subject B	5	6	7	6.0
Subject C	5	4	6	5.0
Subject D	6	7	8	7.0
Subject E	2	2	5	3.0
Condition Means	4.2	4.6	6.2	5.0

Figure 10.7a shows a graph of Harry's raw data (this type of display is sometimes called a dotplot). It appears that mothers generally got more calls in December than in the other two months. Indeed, the condition means in the data table indicate that the subjects made roughly two more calls in December. However, there is a lot of **variability** in the data. To assess whether or not we should trust the observed differences between months, we could conduct an **ANOVA** on the data using the procedures introduced in earlier sections. The summary table for this ANOVA would look like this:

Source	SS	df	MS	F	p
Between Conditions	11.20	2	5.60	2.05	.17
Error	32.80	12	2.73		
Total	44.00	14			

Given the relatively large **p-value** of .17, this analysis would appear to indicate that the difference between the three months is unreliable. That is, there is too much background variability in the data for the two-call December increase to be declared significant.

But look now at the subject means in the data table above (the far-right column of the table). These means vary from a low of 3.0 (subject E) to a high of 7.0 (subject D). If we draw lines connecting the datapoints for each subject, as shown in Figure 10.7b, it becomes obvious that most of the variability in the data stems from these individual differences between subjects.

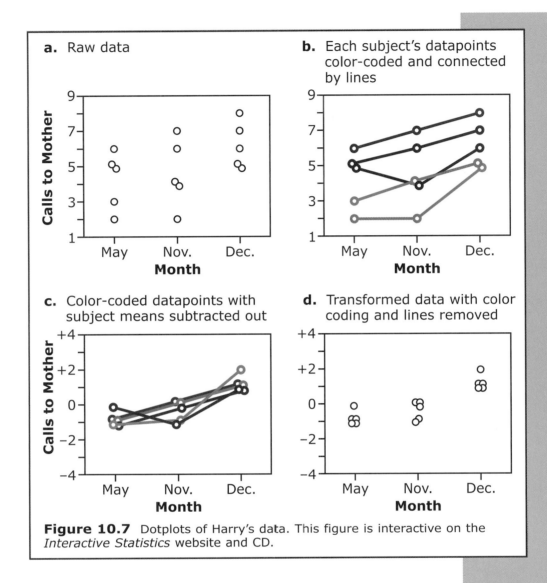

Figure 10.7 Dotplots of Harry's data. This figure is interactive on the *Interactive Statistics* website and CD.

Interactive Page 247

The fact that all five subjects called their mothers more times in December than in either of the other two months indicates that regardless of one's baseline level of maternal phone contact, one tends to call more in December than in May or November. This is exactly the kind of conclusion Harry would like to draw from his study. The problem is that when the subject **variability** is added into the error term of the **ANOVA,** it drowns out the month-to-month variation.

If we could take this subject variability out of the error term in the *F* ratio, we would have a much better chance of detecting a **significant** effect. One way to remove the variability between subjects is to subtract each subject's means from all of his or her scores. This will create a new variable, which we might call "relative calling fre-

Individual differences between subjects can overwhelm differences between conditions.

quency." After applying an adjustment to the denominator *df*, explained below, an ANOVA on this transformed dataset will reveal whether or not the individual differences are spoiling an otherwise significant effect.

Subject A called his or her mother 3, 4, and 5 times in May, November, and December, for an average of 4.0 calls over the three months. Subtracting 4.0 from each of the raw datapoints, we get relative calling frequencies of −1, 0, and +1. The **transformed** data for all five subjects are shown in the second table below and in Figure 10.7c.

Absolute Calling Frequencies (raw data):

Subject	May	Nov.	Dec.	Subject Means
Subject A	3	4	5	4.0
Subject B	5	6	7	6.0
Subject C	5	4	6	5.0
Subject D	6	7	8	7.0
Subject E	2	2	5	3.0
Condition Means	4.2	4.6	6.2	5.0

Relative Calling Frequencies (transformed data):

Subject	May	November	December	Subject Means
Subject A	−1	0	+1	0
Subject B	−1	0	+1	0
Subject C	0	−1	+1	0
Subject D	−1	0	+1	0
Subject E	−1	−1	+2	0
Condition Means	−0.8	-0.4	+1.2	0

SS_B for our transformed data is:

$$SS_B = \Sigma n_i (\overline{X}_i - \overline{X}_T)^2 = 5(-0.8 - 0)^2 + 5(-0.4 - 0)^2 +$$
$$5(1.2 - 0)^2 = 11.20$$

Note that this is exactly the same as the SS_B we calculated on the previous page for the raw data. This is not a coincidence. Subtracting the subject means from each datapoint does not change the variation between conditions at all, so SS_B is unaffected by this transformation.

SS_E, however, *will* change. With the raw data, SS_E was 32.8. For the transformed data, we have:

Subtracting each subject's mean from his or her datapoints leaves the *SS* between conditions unchanged, but reduces the error *SS*.

$$SS_E = SS_1 + SS_2 + SS_3 = 0.8 + 1.2 + 0.8 = 2.80$$

The reason that SS_E is so much smaller now than before should be pretty clear if you inspect the two data tables. The SS_E is equal to the sum of the **SS's** for each condition. In the May condition, the raw data vary from a low of 2 calls to a high of 6 calls. But the relative calling frequencies for this month are all either 0 or –1. As a result, the SS for this condition is 10.8 for the raw data, but only 0.8 for the transformed data. The contrast is easiest to see by comparing Figure 10.7d (the transformed data with subject lines and colors removed) with Figure 10.7a (the original dotplot of the raw data).

To get from SS's to an **F** ratio, we first have to divide SS_B and SS_E by df_B and df_E. The df_B is the same for our transformed data as it was for the raw data: $k - 1 = 2$. However, subtracting the subject means from each datapoint causes us to lose $n - 1 = 4$ **degrees of freedom** in the calculation of SS_E. Since df_E was 12 before, in our new analysis $df_E = 12 - 4 = 8$.

In the process of subtracting out the subject means, we lose $n - 1$ df.

We can now calculate the **MS's** and the **F** ratio for our transformed data:

$$MS_B = SS_B / df_B = 11.20 / 2 = 5.60$$

$$MS_E = SS_E / df_E = 2.80 / 8 = 0.35$$

$$F = MS_B / MS_E = 5.60 / 0.35 = 16.00$$

We don't even have to consult the **F Distribution Area** Calculation Tool to know that an F ratio this big (16.0) is **significant** (if F is greater than 10, you will always have a significant effect as long as df_E is at least 4). To be exact, the **p-value** for $F_{(2, 8)} = 16.0$ is .0016. Compare this to the p-value of .17 for the raw data ANOVA, and you should get a sense of the **power** of **repeated-measures** experimental designs!

In the next section, we'll see how to calculate the proper SS_E for a **related-samples design** (that is, SS_E with subject variability removed) more directly, without going through the extra step of subtracting the subject means from every datapoint.

10.4.1 Partitioning Variance and Testing Hypotheses

In **Section 10.2.2,** we showed how we can partition the **sum of squared deviations** of each datapoint from the overall mean, SS_T, into the **variability** between conditions and the variability within conditions. For an **independent-samples** ANOVA, we called the variability between conditions SS_B and the variability within conditions SS_E, since the latter served as the error term in our **ANOVA.** For a **related-samples** ANOVA, we rename the latter variability SS_W (for "Sum of Squares Within conditions").

SS_E for independent-samples ANOVA is renamed SS_W for related-samples ANOVA.

To calculate SS_T, SS_B, and SS_W in a related-samples ANOVA, we can use the same formulas as before (see **Section 10.2.3**):

$$SS_T = \Sigma(X - \overline{X}_T)^2$$

$$SS_B = \Sigma n_i(\overline{X}_i - \overline{X}_T)^2$$

$$SS_W = \Sigma(X - \overline{X}_i)^2$$

As we saw in the previous section, with a related-samples dataset we can break SS_W down even further into the variability due to subject differences, SS_S, and the within-condition variability left over after SS_S is taken out. This leftover variability is the error term that we need for the F denominator in a related-samples ANOVA, so we'll call it SS_E. Thus we have:

$$SS_W = SS_S + SS_E$$

SS_S is calculated as the sum of squared deviations of subject means ($\overline{X}_S$'s) from the overall mean, multiplied by the number of conditions*:

$$SS_S = \Sigma k(\overline{X}_S - \overline{X}_T)^2$$

Since k is the same for every subject, we can simplify this formula by moving k outside the summation symbol. Similarly, we can simplify the SS_B formula in the related-samples ANOVA calculations by moving n outside the summation symbol (since exactly the same subjects participate in each condition in a related-samples design, we know that **n** will be the same for all conditions).

Once we know SS_W and SS_S, we can find SS_E (the within-condition error variability that is *not* due to individual variations between subjects) by subtraction. So now we have a complete set of SS's for a related-samples ANOVA:

$$SS_T = \Sigma(X - \overline{X}_T)^2$$

$$SS_B = n\Sigma(\overline{X}_i - \overline{X}_T)^2$$

$$SS_W = \Sigma(X - \overline{X}_i)^2 = \Sigma SS_i$$

$$SS_S = k\Sigma(\overline{X}_S - \overline{X}_T)^2$$

$$SS_E = SS_W - SS_S$$

---

*Why do we multiply by k in this formula? For the same reason we multiply by n when calculating SS_B: SS_S is really a sum of the squared deviations of subject means from the overall mean *for every score in the dataset*. But since $\overline{X}_S$ is the same for subject A's May, November, and December datapoints, we can simply take the squared deviation of subject A's $\overline{X}_S$ from $\overline{X}_T$ once, then multiply by the number of conditions k.

SS_T, SS_B, and SS_W are calculated exactly the same in related-samples ANOVA as in independent-samples ANOVA.

To find SS_E in related-samples ANOVA, subtract SS_S from SS_W.

Let's run through these calculations for our sample data, repeated below:

Subject	May	Nov.	Dec.	Subject Means
Subject A	3	4	5	$\bar{X}_{S1} = 4.0$
Subject B	5	6	7	$\bar{X}_{S2} = 6.0$
Subject C	5	4	6	$\bar{X}_{S3} = 5.0$
Subject D	6	7	8	$\bar{X}_{S4} = 7.0$
Subject E	2	2	5	$\bar{X}_{S5} = 3.0$
Condition Means	$\bar{X}_1 = 4.2$	$\bar{X}_2 = 4.6$	$\bar{X}_3 = 6.2$	$\bar{X}_T = 5.0$
Condition SS's	$SS_1 = 10.80$	$SS_2 = 15.20$	$SS_3 = 6.80$	

$$SS_T = (3 - 5.0)^2 + (5 - 5.0)^2 + (5 - 5.0)^2 + \ldots = 44.00$$

$$SS_B = 5((4.2 - 5.0)^2 + (4.6 - 5.0)^2 + (6.2 - 5.0)^2) = 11.2$$

$$SS_W = 10.80 + 15.20 + 6.80 = 32.80$$

$$SS_S = 3((4.0 - 5.0)^2 + (6.0 - 5.0)^2 + (5.0 - 5.0)^2 +$$
$$(7.0 - 5.0)^2 + (3.0 - 5.0)^2) = 30.00$$

$$SS_E = SS_W - SS_S = 32.80 - 30.00 = 2.80$$

Note that the value we get for SS_E using the **variance** partitioning procedures in this section, 2.80, is exactly the same as the value we got in the previous section by subtracting the subject means from every datapoint. This is as it should be: Either method removes the variation due to individual subject differences from the within-subjects error variation, so both methods should always produce the same results.

Note that in a **related-samples** ANOVA, you can check your work by confirming that:

$$SS_T = SS_B + SS_S + SS_E$$

We also saw in **Section 10.2.2** how to partition the total **degrees of freedom** in an **ANOVA** dataset into the df between conditions and the df within conditions:

$$df_T = df_B + df_W$$

We divide df_W into the degrees of freedom for subjects and the left-over error in the same way we divided SS_W up:

$$df_W = df_S + df_E$$

Degrees of freedom are partitioned in the same way as sums of squares.

You're probably savvy enough by now to know how to calculate df_S. Given that we have 5 subjects in our sample experiment, what do you think df_S is going to be?

**Interactive
Page 250**

We have exactly the same number of **degrees of freedom** for subjects (df_S) in a **related-samples** ANOVA as we do in a **related-samples t test:** the number of subjects, minus one (= 4 for our sample experiment).

The total degrees of freedom in a related-samples ANOVA (df_T) is equal to the total number of datapoints N minus one. Since each of the n subjects contributes k scores to the dataset, $N = kn$, so:

$$df_T = kn - 1$$

Degrees of freedom between conditions (df_B) and within conditions (df_W) are calculated exactly the same as in an **independent-samples ANOVA,** and df_E is calculated by subtracting df_S from df_W, just as $SS_E = SS_W - SS_S$. Thus we have:

Final formulas for df's in related-samples ANOVA

$$df_B = k - 1$$

$$df_W = N - k = kn - k = k(n - 1)$$

$$df_S = n - 1$$

$$df_E = df_W - df_S = k(n - 1) - (n - 1) = (k - 1)(n - 1)$$

You can check your df calculations by making sure that:

$$df_T = df_B + df_S + df_E$$

For our sample experiment:

$$df_T = kn - 1 = (5)(3) - 1 = \mathbf{14}$$

$$df_B = k - 1 = 3 - 1 = \mathbf{2}$$

$$df_W = k(n - 1) = 3(5 - 1) = \mathbf{12}$$

$$df_S = n - 1 = 5 - 1 = \mathbf{4}$$

$$df_E = (k - 1)(n - 1) = (3 - 1)(5 - 1) = \mathbf{8}$$

**Interactive
Page 251**

Now that we finally have all our **SS's** and **df's,** we can organize them into an **ANOVA** summary table, calculate our **MS's** and **F,** and find the **p-value** for our sample experiment. Values for MS_B, MS_E, and F are calculated exactly as in **independent-samples** ANOVA:

$$MS_B = SS_B / df_B = 11.20 / 2 = 5.60$$

$$MS_E = SS_E / df_E = 2.80 / 8 = 0.35$$

$$F = MS_B / MS_E = 5.60 / 0.36 = 16.00$$

Again, note that we get the same F value here as we did in the previous section when we subtracted the subject means from the raw datapoints.

We also look up the p-value for a **related-samples** F statistic exactly the same way as we did in an independent-samples ANOVA. Thus we have:

We assess F statistics exactly the same way in related-samples and in independent-samples ANOVAs.

Source	SS	df	MS	F	p
Between	11.20	2	5.60	16.00	.0016
Within	32.80	12			
Subjects	30.00	4			
Error	2.80	8	0.35		
Total	44.00	14			

And that's all there is to a related-samples ANOVA! The **ANOVA Helper** Activity introduced in **Section 10.2.3** for independent-samples analyses also works for related-samples designs. Just click the radio buttons to switch back and forth between the two types of ANOVAs.

Activity: ANOVA Helper

Interactive Page 252

10.4.2 Comparing Independent-Samples and Related-Samples ANOVA

Figure 10.8 illustrates how **variability** is partitioned in independent-samples vs. related-samples ANOVAs. With an **independent-samples design** (Figure 10.8a), total variability (SS_T) is divided into between-conditions variability (SS_B) and error variability (SS_E). The relative amounts of each type of variability vary from experiment to experiment, as shown in the figure, but SS_B and SS_E together always add up to SS_T.

In the first stage of variability partitioning in a **related-samples** ANOVA (Figure 10.8b), SS_T is divided into SS_B and variability within conditions, which we now call SS_W. Up to this point, the two types of ANOVAs are identical. But in related-samples ANOVA, we go on to divide SS_W into two parts: variability due to differences between subjects (SS_S) and leftover error variability (SS_E).

Related-samples ANOVA can be viewed as a two-stage process in which the first stage is identical to independent-samples ANOVA.

In an independent-samples design, we know that subject variability is leading to increased error within each condition (and is thus

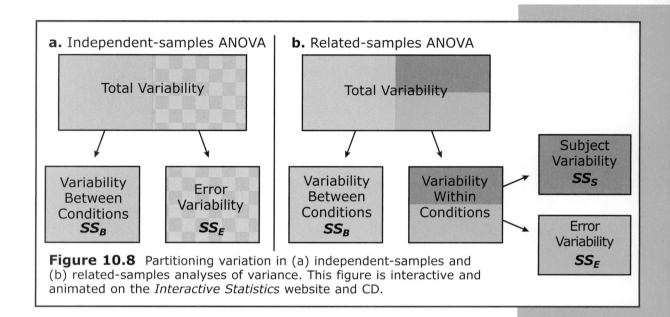

a. Independent-samples ANOVA

Total Variability

Variability Between Conditions
SS$_B$

Error Variability
SS$_E$

b. Related-samples ANOVA

Total Variability

Variability Between Conditions
SS$_B$

Variability Within Conditions

Subject Variability
SS$_S$

Error Variability
SS$_E$

Figure 10.8 Partitioning variation in (a) independent-samples and (b) related-samples analyses of variance. This figure is interactive and animated on the *Interactive Statistics* website and CD.

decreasing our F ratio). But we're powerless to do anything about it, since subject variability and the rest of the error variability are jumbled up and thus indistinguishable from each other. This mixture of subject variability and error variability is represented in Figure 10.8a by the checkerboard pattern of SS_E.

In a related-samples design, we are able to determine exactly how much of the within-condition variability is due to subject error by examining the subject means. Thus the subject variability (blue) is clearly separated from the error variability (green) in Figure 10.8b.

As was the case with two-sample *t* tests, the ability to account for subject variability imparts a tremendous **power** advantage. Practically speaking, subtracting out the subject variability reduces the denominator of the F ratio, thus increasing the magnitude of F and lowering the final *p*-value for the experiment.

Note, however, that a related-samples design does not necessarily increase the *magnitude* of an experimental effect. In fact, if there are practice effects or fatigue effects (see **Section 8.5**), the **effect size** might actually go down in a related-samples design as compared to an independent-samples design. However, the likelihood of being able to detect that some effect occurred is generally much greater in a related-samples experiment. For this reason, researchers often try to conduct two experiments, one with each type of design, and present the results as a package.

Exercise 10.5, Exercise 10.6, and **Exercise 10.7** cover the basic calculations for related-samples ANOVA. If you can, try these exercises now before moving on to the next section.

Related-samples research designs are almost always more powerful than independent-samples designs.

10.5 Further Analysis of Multicondition Datasets

The **null hypothesis** in a two-sample **t** test states that $\mu_1 = \mu_2$. If we reject this H_0, the conclusion drawn from the experiment is pretty simple: If μ_1 is not the same as μ_2, then $\mu_1 \neq \mu_2$, as stated in the **alternative hypothesis** for an independent-samples **t** test.

In the **F** test we performed on Freida's data from **Section 10.3.2,** H_0 stated that $\mu_1 = \mu_2 = \mu_3 = \mu_4$. We soundly rejected this null hypothesis. But, as noted in **Section 10.3,** there are many possible alternatives to this H_0, and the analysis we have conducted so far does not distinguish between them.

Therefore, the **omnibus** **F** test we've been using up to now is really only the starting point for a complete analysis of the data. A significant omnibus **F** tells us that *something* happened in the experiment, but really tells nothing at all about what that something is. For Freida's experiment, which we worked through in the previous section, were 15-year-olds significantly less anxious than subjects of other ages? Were 20-year-olds significantly more anxious than everyone else? Was the difference between 50-year-olds and 30-year-olds significant?

An omnibus ANOVA does not reveal which conditions were significantly different from which.

To answer these questions, we first need to look at the data. And the best way to look at the data is to graph them. (Actually, looking at your data should be the first thing you do when you've collected it, but many researchers don't put much stock in their assessment of an experiment's results until an omnibus **ANOVA** confirms the fact that some kind of significant effect was observed.)

A good graph is essential for interpreting the results of an experiment with three or more conditions.

With a good graph, one can usually see which conditions are and aren't different from each other in an experiment. Comparisons of two conditions in a multicondition experiment are called **pairwise comparisons.**

But the scientific community generally demands more than eyeball assessments to declare an effect significant. The most common way to clinch the case for significant pairwise comparisons is through the use of **post hoc** (a Latin phrase meaning "after the fact") tests. The name given to these tests reflects the fact that they are only valid when done after obtaining a significant result with an omnibus **F** test. We'll delve into the wonderful world of post hocs after we cover some basic graphing techniques.

Glossary Term: post hoc tests

10.5.1 Graphing Results

The two types of graphs most commonly used to present the results of a multicondition experiment are bar graphs and line graphs.

To construct either one, start by drawing an X and a Y axis (or a box, whose left side can serve as the Y axis and right side the X axis), as shown in Figure 10.9. On the X axis, draw evenly spaced tick marks for each condition in your experiment (i.e., if you have four conditions, draw four tick marks). Then label each tick mark with the name of one of your conditions (Figure 10.9; parts a, b, c, and e, not shown here, are on the *Interactive Statistics* website and CD).

If there is a natural order to put the conditions in, use it (for example, if your experiment includes high school freshmen, sophomores, juniors, and seniors, use this order; putting seniors to the left of freshmen will be confusing). If there is no natural order, you are free to arrange the conditions in whatever way you think looks best.

Next, divide the Y axis into equally spaced intervals and label them in a way that makes sense. "Making sense" is obviously a fairly subjective guideline, and researchers often misrepresent their data (intentionally or unintentionally) by choosing a misleading Y-axis scale. Figure 10.9 shows a good set of labels for graphing the results of Freida's anxiety data; common sense, previous research, and a mentor should guide you when graphing your own data.

Once your axes are set up, you're ready to plot the data. For a bar graph, draw a bar starting on the X axis above each tick mark, and rising just enough so that the top of the bar is even with the point on the Y axis representing the mean for that condition (Figure 10.9d).

For a line graph, draw a marker (typically a small circle or square) centered on each X-axis tick mark and at the height representing the mean for the condition. Then connect each marker with a straight line (Figure 10.9f).

Many graphs include error bars, which give some indication of the amount of **variability** present in the data. There

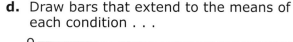

Constructing a bar or line graph

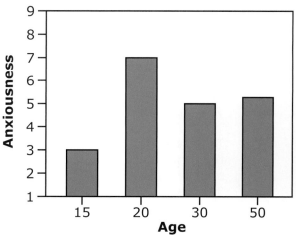

d. Draw bars that extend to the means of each condition . . .

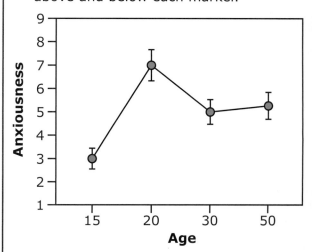

f. Add error bars if possible. Here, the bars stretch one standard error of the mean above and below each marker.

Figure 10.9 Drawing a bar or line graph. This figure is interactive and has additional parts on the *Interactive Statistics* website and CD.

are different schools of thought as to exactly how big the error bars in a graph should be. In Figure 10.9f, we show the line graph with error bars added to show the **estimated standard error** of each condition mean ($s / \sqrt{n}$). Other researchers prefer to show **standard deviations, confidence intervals,** or other values. Any of these choices is fine, as long as you make sure to specify in the caption of your graph what the error bars represent.

Interactive Page 255

Now that you know how to construct two types of graphs, when should you choose to use one or the other?

As a general rule, bar graphs are most appropriate when the **independent variable** in an experiment is qualitative in nature, whereas line graphs should only be used when the levels of the independent variable differ quantitatively. Here are some examples:

In general, use bar graphs for qualitative IV's and line graphs for quantitative IV's.

- A developmental psychologist observes first, third, and fifth graders during recess periods and records the percentage of time the children spend playing with classmates of the opposite sex. Here, the **levels** of the independent variable are quantitatively related to each other—third graders are two years ahead of first graders and two years behind fifth graders. So a line graph is clearly called for in this case (Figure 10.10).

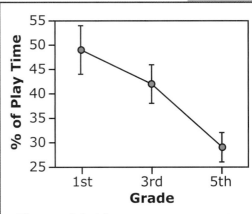

Figure 10.10 Play time by grade.

- A theology graduate student polls groups of Catholics, Protestants, Jews, and atheists. Each respondent answers a series of questions that together provide a general rating of the person's political leanings, from 1 (most liberal) to 9 (most conservative). The independent variable here (religious affiliation) is clearly qualitative in nature, so a bar graph would be most appropriate (Figure 10.11).

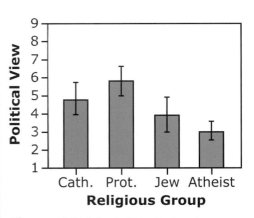

Figure 10.11 Political view by religious group.

- A cognitive psychologist records subjects' times as they read aloud, as fast as possible, three lists of 10 sentences each. In the first list all the sentences are perfectly well formed (e.g., "The wise doctor told the man to exercise more"). In the second list, the sentences are grammatical but nonsensical

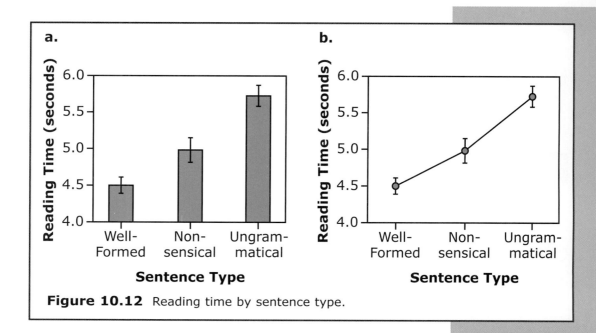

Figure 10.12 Reading time by sentence type.

("Loving rulers ask for flying green walls"). In the third list, words are scrambled completely haphazardly ("Dig books key wooden it drawings run").

The results of this last experiment could be plotted with either type of graph. The researcher could argue that the three types of sentences form a continuum of comprehensibility, and that the reading times should therefore be represented by a line graph. On the other hand, it is impossible to say exactly how much harder it is to comprehend grammatical–nonsensical sentences than it is to comprehend well-formed sentences—indeed, assessing differences in comprehensibility would presumably be the whole point of the experiment. In this sense, a bar graph would seem to be called for. Figures 10.12a and b show the data plotted both ways.

One last note about graphing results: In practice, very few 21st century researchers actually draw their own graphs by hand. Instead, most use computer software to generate graphs for them. However, the decisions discussed in this section (what type of graph to plot, what scale to use for the Y axis, whether or not to include error bars, etc.) have to be made regardless of whether a graph is being plotted by a computer or by hand. Software packages will generally try to make these decisions for you, and if you don't know what the relevant decisions are, your computer may produce a misleading graph without you even knowing it! So don't dismiss this section as irrelevant just because your professor requires you to use a computer to do your graphing.

Interactive
Page 256

10.5.2 Post Hoc Tests

From the graph of Freida's experiment we constructed two pages ago (Figure 10.9f, reproduced at right), it appears:

- That 15-year-olds are substantially less anxious than the other three ages.

- That 30- and 50-year-olds are about equally anxious.

- That 20-year-olds are more anxious than any of the other ages.

One way to confirm these conclusions **inferentially** is to conduct a series of **post hoc tests.** Each post hoc test will assess the **significance** of the difference between one pair of conditions.

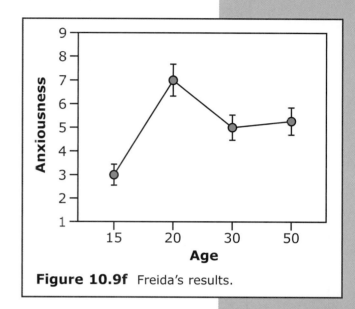

Figure 10.9f Freida's results.

The simplest way to do post hoc tests is called Fisher's least significant differences method, usually abbreviated LSD (this is the same Fisher for whom the **F** statistic is named). The LSD procedure starts with doing the **omnibus ANOVA** that we've spent most of this chapter covering. If the omnibus F test is not significant (that is, if the **p-value** is not judged to be low enough to reject H_0), we stop right there, and don't even try to test any of the pairwise differences between conditions.

LSD tests should only be done if the omnibus F test is significant.

If, however, the omnibus F is significant, the LSD procedure states that we can test each **pairwise comparison** using the standard formula for a two-sample **t test,** with the exception that we substitute MS_E for the **sample variance** in the denominator of the t formula:

An LSD test is exactly like an independent-samples or related-samples t test, except that MS_E is used to calculate t.

- For an **independent-samples design,** MS_E replaces the pooled variance term, s_p^2 (recall from **Section 10.2** that MS_E is, like s_p^2, simply the weighted average of the variances of each condition in the experiment).

- For a **related-samples design,** MS_E replaces the variance of the **difference scores,** s_D^2.

Thus the t statistic for an LSD test

For independent samples: $t = \dfrac{(\bar{X}_1 - \bar{X}_2)}{\sqrt{MS_E}\sqrt{\dfrac{1}{n_1} + \dfrac{1}{n_2}}}$

For related samples: $t = \dfrac{\bar{D}}{\sqrt{MS_E}\,/\,\sqrt{n}}$

For both designs, the t statistic has df_E **degrees of freedom**. The substitution of MS_E for the variance in the denominator of the t formula gives us two advantages:

1. Since we are assuming that the variances of all the experimental populations are equal, MS_E is, at least in theory, a better estimate than s_p^2 or s_D^2 of the underlying variance in the sample means. In other words, we're better off averaging the variation in all the conditions in the experiment (as is done in calculating MS_E) than averaging the variances of just the two conditions we're comparing (as is done in calculating s_p^2 or s_D^2 for a "normal" t test).

2. MS_E has $N - k$ degrees of freedom in an independent-samples design, whereas s_p^2 has only $n_1 + n_2 - 2$ degrees of freedom. For Freida's experiment, this means we get 36 df for each LSD test, whereas we would only have between 17–19 df if we were testing pairwise comparisons using independent-samples t tests. Remember that the greater the df, the lower t has to be in order to be able to reject H_0 (for example, for a t value of 2.22, $p = .051$ with 10 df, but $p = .038$ with 20 df). For a related-samples design, there is a similar df boost when conducting post-hoc tests compared to individual t tests.

For Freida's experiment, MS_E is 2.82. So, what value will we use for the **estimated standard error** term (the denominator in the t formula) in our LSD test of the difference between 20-year-olds and 50-year-olds (for both of these conditions, $n = 10$)?

Interactive Page 257

The MS_E in Freida's omnibus **F** test was 2.82. Therefore, the **estimated standard error** for the LSD test of the 20-year-old vs. 50-year-old conditions, both of which included 10 subjects, is:

$$s_{\bar{X}_1 - \bar{X}_2} = \sqrt{MS_E}\sqrt{\dfrac{1}{n_1} + \dfrac{1}{n_2}} = \sqrt{2.82}\sqrt{\dfrac{1}{10} + \dfrac{1}{10}} = 0.75$$

Since the difference between the sample means of these two conditions was 1.8 (= 7.0 − 5.2), the **t statistic** for this comparison is

1.8 / .73 = 2.40. With *df* = 36 (remember that we use the *t* distribution with df_E **degrees of freedom** for LSD tests), the **p-value** comes out to .022, and we can declare the difference between these two conditions **significant.**

The following table shows the LSD tests for all the pairwise comparisons in Freida's experiment:

Condition 1 (n, $\bar{X}$)	Condition 2 (n, $\bar{X}$)	Mean Difference	$s_{\bar{X}_1 - \bar{X}_2}$	t	p
15 (11, 3.0)	20 (10, 7.0)	−4.0	.73	−5.45	< .0001
15 (11, 3.0)	30 (9, 5.0)	−2.0	.76	−2.65	.012
15 (11, 3.0)	50 (10, 5.2)	−2.2	.73	−3.00	.005
20 (10, 7.0)	30 (9, 5.0)	−2.0	.77	2.59	.014
20 (10, 7.0)	50 (10, 5.2)	1.8	.75	2.40	.022
30 (9, 5.0)	50 (10, 5.2)	−0.2	.77	−0.26	.80

Happily, the LSD procedure confirms what we can see with our eyes in the graph of the results: All **pairwise comparisons** are significant except for the one between 30- and 50-year-olds.

By the way, you may be wondering where the name "least significant differences" comes from. It refers to an alternate way of conducting LSD tests, which was more practical to use before the advent of computers in statistical analysis. If you find the **critical value** for *t* in an LSD test (that is, the value of *t* for which *p* will be exactly .05 given the *df* in the test), it is possible to work backward from the formula for *t* and find the smallest possible difference between two means that will produce a significant pairwise comparison. Any pairwise comparison for which $\bar{X}_1 - \bar{X}_2$ is greater than this "least significant difference" can be declared statistically significant.

This method of conducting LSD tests disposes with the need to calculate a separate *p*-value for each comparison, which was a great advantage when such calculations had to be done by laboriously poring through a long table of numbers. But now that computers make it trivially easy to find the exact *p*-value associated with any *t* statistic, it usually makes more sense to evaluate the significance of each pairwise comparison separately, as we did above.

Interactive Page 258

Although the LSD method is the most conceptually clear way to conduct **post hoc pairwise comparisons,** it is fairly rare to see an LSD test reported in a scientific journal. The reason for the disfavor of Fisher's procedure goes back to a point we made at the very start of this chapter: If you conduct 6 inferential tests in an experiment

Finding the "least significant difference" in an ANOVA

The LSD test is unpopular because it carries a high risk of Type I errors.

and set the **alpha level** for each at .05, your chance of making a **Type I error** is .05 for each test, but there is a .26 chance of making at least one Type I error over the course of the whole experiment.

The LSD procedure "protects" us somewhat from this inflation of the Type I error rate by demanding that the **omnibus** ANOVA must be found to be significant before we even attempt any pairwise comparisons. But this amount of protection isn't good enough for most researchers, many of whom feel that a 1/4 chance of claiming a **significant** effect when there isn't one is unacceptable.

In response to this concern, statisticians have developed alternate procedures for conducting post hoc pairwise comparisons. In fact, one could argue that the statisticians have gone a bit overboard here: The commercial statistics package SPSS offers to do 18 different types of post hoc tests! Here are brief summaries of some of the popular alternatives to LSD:

- In the Bonferroni procedure, **t** and **p** are calculated exactly as in LSD tests, but when deciding whether or not to declare a comparison significant, we divide alpha by the total number of comparisons being done. So, for example, if we're doing 6 comparisons and we want an overall alpha rate of .05, p must be less than .05 / 6 = .008 in order for a comparison to be considered significant by a Bonferroni test. Obviously, this makes it much more difficult to be able to claim significant differences between conditions.

- In Tukey's HSD procedure, an "honestly significant" difference is computed for the experiment. The HSD is based on the MS_E, the total number of conditions in the experiment, the number of subjects in each condition, and a modified distribution of t statistics called the "Studentized range." Any pair of means whose difference is greater than the HSD is declared significant. The modification to t essentially "raises the bar" for declaring a comparison significant, in a similar way to the Bonferroni procedure.

- The Newman–Keuls procedure is considerably more complicated than the others listed here. Here's how the procedure might work for an experiment with four conditions, where $\overline{X}_1 < \overline{X}_2 < \overline{X}_3 < \overline{X}_4$ (that is, $\overline{X}_1$ is the smallest mean and $\overline{X}_4$ is the largest). First, we would compare $\overline{X}_1$ to $\overline{X}_4$. Suppose this largest difference is significant. In this case, we would declare that $\overline{X}_1 \neq \overline{X}_4$, and go on to compare $\overline{X}_1$ to $\overline{X}_3$ and $\overline{X}_2$ to $\overline{X}_4$, this time using a slightly more liberal criterion for rejecting H_0 (that is, the chances of declaring the differences significant would be slightly greater here than in the first comparison). Suppose the former comparison is significant but the latter is not. We would then declare that $\overline{X}_2 = \overline{X}_3 = \overline{X}_4$. But, we would allow that $\overline{X}_1 \neq \overline{X}_3$, and we would go on to test whether or not $\overline{X}_1 = \overline{X}_2$, using an even more liberal criterion for this final comparison. As you can imagine, this procedure was not

Alternative post hoc tests that reduce the chances of making a Type I error

all that popular before researchers had computers to keep track of all this information.

- The Scheffé test is based on the *F* statistic, rather than the *t* statistic. It is the safest (least likely to incorrectly declare two means significantly different) and least **powerful** (most likely to incorrectly declare two means statistically equivalent) of all the post hoc tests.

Interactive Page 259

Each of the different **post hoc** procedures described on the previous page (as well as others that you might encounter, including the Duncan, Dunnet, Dunnet-C, Dunnet-T3, Games–Howell, and Ryan-Einot-Gabriel-Welsch tests) trades off safety for **power** in some way or another. Unfortunately, there is no cut-and-dried way to choose between the many options, which highlights the underappreciated fact that statistics are not nearly as objective as most people believe them to be.

There is no one standard way to do a post hoc test.

For this text, we'll just stick to the simplest procedure, the LSD test. If you ever have to analyze data of your own and need to conduct post hoc tests, we offer the following two pieces of advice:

1. Choose one type of test and stick with it. Suppose you conduct two experiments, the first of which happens to produce huge effects and the second much smaller effects. It is not acceptable to use Scheffé tests in Experiment 1 and then revert to LSD tests in Experiment 2. For this reason, many researchers prefer a middle-of-the-road test such as Tukey's HSD (which is more conservative than LSD but less so than Scheffé).

2. No matter which test you choose, always remember that rejecting the null hypothesis for a pairwise comparison provides proof that the conditions are different, but failing to reject H_0 does *not* prove that the conditions are identical. Drawing the latter conclusion when doing post hoc tests can lead to serious logical fallacies. For example, in an experiment with three conditions, it is quite common to reject H_0 for the μ_1–μ_3 comparison, but fail to reject H_0 for the μ_1–μ_2 and μ_2–μ_3 comparisons. If μ_1 is identical to μ_2 and μ_2 is identical to μ_3, then logically, μ_1 must also be identical to μ_3, a conclusion directly contradicted by the first post hoc test!

A nonsignificant post hoc test does not mean that the two conditions are identical.

To avoid such logical loops, never make statements like "post hoc tests showed that there was no difference between condition A and condition B." Instead, say something wimpier, such as that the conditions were "statistically indistinguishable" or "approximately equal."

Exercise **10.8** and Exercise **10.9** walk you through the process of conducting LSD post hoc tests with independent-samples and related-samples ANOVA designs.

10.6 *F* vs. *t*

In the examples we've considered so far in this chapter, there have always been three or more experimental conditions. However, it's perfectly possible to conduct an **ANOVA** on an experimental design with only two conditions. For example, suppose we're analyzing an experiment in which Condition 1 included 20 subjects whose average score was 10.5, with a standard deviation of 2.3, and Condition 2 included 20 subjects whose average score was 8.4, with a standard deviation of 2.6.

With a little arithmetic (which is simplified by the fact that the sample sizes in the two conditions are equal), the paragraph above provides everything we need to calculate an *F* statistic:

$k = 2$

$N = 40$

$\overline{X}_T = (10.5 + 8.4) / 2 = 9.45$

$n_1 = 20$

$\overline{X}_1 = 10.5$

$SS_1 = (n_1 - 1)s_1^2 = (19)(2.3)^2 = 100.51$

$n_2 = 20$

$\overline{X}_2 = 8.4$

$SS_2 = (n_2 - 1)s_2^2 = (19)(2.6)^2 = 128.44$

$SS_B = \Sigma n_i (\overline{X}_i - \overline{X}_T)^2 = 20(10.5 - 9.45)^2 + 20(8.4 - 9.45)^2 = 22.05 + 22.05 = 44.10$

$SS_E = SS_1 + SS_2 = 228.95$

$df_B = k - 1 = 2 - 1 = 1$

$df_E = N - k = 40 - 2 = 38$

$MS_B = SS_B / df_B = 44.10 / 1 = 44.10$

$MS_E = SS_E / df_E = 228.95 / 38 = 6.03$

$F = MS_B / MS_E = 44.10 / 6.03 = 7.31$

p (from the **F Distribution Area** Tool) = .01

"But why," you may be thinking at this point, "would you analyze these data with an *F* test? Didn't we learn in Chapter 8 that the **independent-samples *t* statistic** is specifically designed to test hypotheses in research designs with two conditions and different samples of subjects? So shouldn't we use the *t* test to analyze the data?" Good point. These data certainly can be analyzed with an independent-samples *t* test. Go ahead and calculate the *t* statistic for this data, along with the associated ***p*-value** before going on.

Interactive Page 261

On the previous page, we calculated ***F*** and asked you to calculate ***t*** for two conditions where $n_1 = 20$, $\overline{X}_1 = 10.5$, $s_1 = 2.3$, and $n_2 = 20$, $\overline{X}_2 = 8.4$, and $s_2 = 2.6$. The results of the *t* test look like this:

> *t* statistic   2.71
>
> *df* associated with this *t* statistic   38
>
> *p*-value   .01

Recall from the previous page that an **ANOVA** revealed that $F(1, 38) = 7.31$, and that ***p*** was also equal to .01 for that analysis. The fact that the *p*-values for the *F* test and *t* test are identical is not a coincidence: *The independent- and related-samples* t *tests can be considered special cases of the independent- and related-samples* F *tests with 1 degree of freedom in the numerator.* Therefore, the *p*-values will always be identical for an *F* and a *t* test on the same set of data. The test statistics themselves are related by the formula:

$$F = t^2$$

A *t* test is identical to an *F* test with $df_B = 1$.

Another indication of the close relationship between the *F* and *t* statistics is the fact that the same **assumptions (Section 5.5)** must be met for results of the two tests to be valid:

Assumptions for independent sample *F* tests

1. The subjects in the experiment must be drawn at random from the population being studied (**Section 6.4.1**).

2. The **distributions of sample means** for all conditions in the experiment must be **normal (Section 6.4.2)**. This assumption will be met as long as a) the underlying population distributions are themselves approximately normal, and/or b) there are at least 20 or so scores in each condition.

3. The **population variances** for all the conditions in the experiment must be equal. As in the case of the **independent-samples *t* test (Section 8.4)**, ANOVA is resistant to relatively minor violations of this assumption, but if there are big differences in the variances of your conditions, you shouldn't be analyzing them together in the same ANOVA.

To check this last assumption, look at the sample variances of each of the groups of scores in your dataset. Divide the largest of the s^2's by the smallest of the s^2's. If the resulting ratio is less than 2 (or less than 4, if the sample sizes are smaller than 10), it's probably OK to proceed with an ANOVA.

If one or more assumptions for ANOVA are violated, you can often use a **nonparametric statistic** as an alternative. For example, the Kruskal–Wallis *H* test (an extension of the Mann–Whitney *U* test; see **Section 8.4.1**) can be used to test for differences between the medians of three or more conditions in an independent-samples design. The median test, another nonparametric alternative to independent-samples ANOVA, will be discussed in detail in Section **12.4.1**.

10.7 Multiple-Factor ANOVA

See if you can figure out what is different about the following research example compared to ones we've considered previously in this chapter:

> Laura, a nutrition researcher, wants to find out whether people are most likely to eat foods high in saturated fat at breakfast, lunch, or dinner. She suspects that there might be sex differences in these patterns of fatty-food consumption. So, she designs an experiment in which 20 men and 20 women are asked to carefully record exactly what they eat for breakfast, lunch, and dinner every day for two weeks. Laura then examines each subject's food inventories and records how many total servings of fatty foods were eaten at each meal.

How many independent variables (see **Section 10.1.1**) does Laura's experiment involve?

The research examples we looked at earlier in this chapter all involved one **independent variable** (e.g., Months in Harry's long-distance phone call research or Age in Freida's anxiety study; researchers usually capitalize the titles they give to independent variables when writing about them).

Laura's experiment, however, includes two independent variables. In **ANOVA** terminology, we say that the research design includes two **factors,** Sex and Meal. In this particular design, one factor, Sex, is manipulated between subjects and has two **levels** (that is, we have two **independent samples** for this factor, one sample of men and one of women), while the other factor, Meal, is manipulated within subjects and has three levels (there are three **related samples** for this factor: one for breakfasts, one for lunches, and one for dinners).

More succinctly, we can say that Laura's experiment uses a 2 × 3 **mixed-factors design** ("mixed" refers to the mixture of between- and within-subjects factors).

Research designs with multiple independent variables (i.e., two or more factors)

Glossary term: mixed-factors design

Just as with single-factor designs, we can analyze the results of multiple-factor designs by computing the total **variance** in the data, partitioning the variance by source, and computing **F** ratios. Note, however, that the last word of that last sentence was plural: Multiple-factor ANOVAs produce multiple F ratios. More specifically, with the present two-factor design, we will get three F's:

- One will assess the significance of the **main effect** of Sex

- One will assess the significance of the main effect of Meal

- One will assess the significance of the **interaction** between Sex and Meal

A main effect is simply the effect of one factor when the other factor(s) are ignored. For example, there would be a main effect of Sex in Laura's study if men ate significantly more or less fatty-food servings overall than women.

An interaction occurs when the effect of one factor depends in some way on the level of the other factor. We would have an interaction if, for example, men ate more fatty foods at breakfast than at lunch, while women ate more fatty foods at lunch than at breakfast. This pattern of results is illustrated in Figure 10.13a.

Sometimes, though, main effects don't interact, as illustrated in Figure 10.13b. In this theoretical set of results, the pattern of fatty-food consumption across the three meals is identical for men and women—the effect of Meal does not depend on the Sex effect in any way.

Note that there are main effects of Sex and Meal in both sets of theoretical results shown in Figure 10.13. Averaged across all levels of Meal, both graphs show men eating more fatty-food servings than

Glossary Term:
main effect

Glossary Term:
interaction

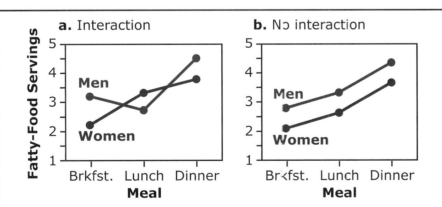

Figure 10.13 Two possible patterns of results in Laura's experiment. In the pattern shown in part (a), the two factors of Meal (breakfast, lunch, or dinner) and Sex (men or women) interact. In the pattern shown in part (b), the two factors do not interact. Both factors show main effects in both patterns of results.

women (that is, the line for men is above the line for women in Figures 10.13a and b). And averaged across men and women, both graphs indicate that more fatty-food servings are consumed at dinner than at either of the other two meals.

The computations involved in a multiple-factor **ANOVA** are just extensions of the concepts introduced in this chapter. But obviously, these computations get more complicated as the number of F's increases. And researchers don't always stop with two factors—some experiments involve three, four, or even more factors. Analysis of a four-factor design involves 14 F ratios (four main effects, six one-way interactions, three two-way interactions, and one four-way interaction)!

Furthermore, the computations are different when designs include only between-subjects factors, only within-subjects factors, or some of each. Complications also arise when there are different numbers of subjects in different conditions, when one or two datapoints are missing from a design that includes a within-subjects factor, and for all kinds of other reasons.

What we're trying to say here is that a full description of how to compute the F's in all the different kinds of multiple-factor ANOVAs is beyond the scope of an introductory statistics class, so we won't attempt it. What we will do, however, is explore how the end results of a multiple-factor ANOVA are interpreted.

Suppose Laura ran her experiment and obtained the results shown in Figure 10.14a (note that these results are not the same as either of the two theoretical patterns shown on the previous page). The summary table for the **ANOVA** on these data might look like this:

Source	SS	df	MS	F	p
Sex	15.89	1	15.89	0.70	.41
Subjects	856.91	38	22.55		
Meal	98.65	2	49.33	4.65	.0120
Sex × Meal	45.87	2	22.93	9.99	.0001
Error	375.13	76	4.94		
Total	1392.50	119			

Let's look more closely at the main effects and interaction tested in this ANOVA:

Sex: The first row of the table shows the **SS, df, MS, F,** and **p-value** for the **main effect** of Sex. This F ratio is calculated as the MS for Sex divided by the MS for Subjects, shown in the second row of the summary table ($F = 15.89 / 22.55 = 0.70$). The fact that the *p*-value for this main effect is so high (well above .05) indicates that, averaged across all meals, both sexes actually eat about the same amount of fatty-food servings.

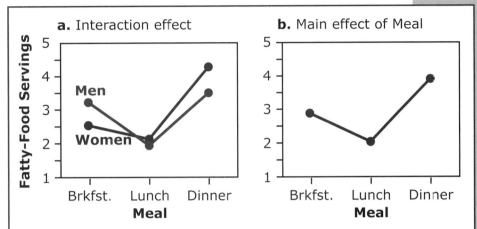

Figure 10.14 Actual results of Laura's experiment: (a) means for each Meal, broken down by the two Sexes; (b) means for each Meal averaged across the two Sexes.

Meal: The third row of the summary table shows the ANOVA statistics for the main effect of Meal. Since this is a within-subjects factor, we use the Error *SS* (which has the within-subjects error taken out, as described in **Section 10.4.1**) to calculate the denominator of the *F*-ratio for this main effect (*F* = 49.33 / 4.94 = 9.99). Here, the *p*-value is less than .05, indicating that averaged across men and women, there are **significant** differences in the number of servings of fatty foods eaten at each meal.

Note, however, that this *F* ratio does *not* indicate which meal includes the most servings of fatty foods. To find that out, we have to look at the **descriptive statistics** (i.e., the **means** for each meal), which are graphed in Figure 10.14b. This graph shows that for both sexes put together, lunch was the least fatty meal, dinner was the most fatty, and breakfast was somewhere in between. We could use any of the **post hoc tests** discussed in **Section 10.5.2** to help determine which of the **pairwise comparisons** here were significant.

Graphs, not *F* statistics, tell us what actually happened in an experiment.

Sex × Meal: The fourth row of the table shows that the **interaction** of the two factors was also significant. The *F*-ratio for the interaction in this design uses the same error term as the within-subjects main effect (*F* = 22.93 / 4.94 = 3.65).

Again, we have to look at the means (or a graph of the means) to find out exactly what the effect is. Examining Figure 10.14a, we see that the fat content of lunches was about the same for both sexes, but men ate fattier breakfasts than women, while women ate fattier dinners than men.

This example highlights the value of including multiple factors in the same experiment. If Laura had conducted an experiment including the Sex factor only, she probably would have observed no differences, and might have concluded that men and women do not differ at all in their fatty-food consumption. In fact, though, the two sexes *do* differ substantially, in their *pattern* of fatty-food consumption across meals. Interaction effects often reveal such discrepancies in response patterns for different groups of subjects.

10.8 Chapter Summary/Review

This Chapter Summary/Review is interactive on the *Interactive Statistics* website and CD. Also be sure to go through all the **Review Exercises** for this chapter.

- When the results of a research study include three or more means, **t** statistics alone should not be used to analyze the data, because doing so would inflate the Type I error rate for the study. Instead, such results should be assessed using an **analysis of variance** (ANOVA), in which the observed variation between sample means is compared to the variation we would expect given the **variability** of the individual scores (**Section 10.1**).

- The test statistic for ANOVA goes by the one-letter symbol F, and is calculated as the ratio of the **between-condition variance** to the **error variance** in the data. For an **independent-samples design** in which all n's are equal, MS_E, the denominator of the F ratio, is simply the mean of the variances for each sample. MS_B, the numerator of the F ratio, is equal to n times the variance of the sample means (**Section 10.2**).

- A more general method for calculating F in an independent-samples design is to partition the total sum of squared deviations SS_T, which is equal to $\Sigma(X - \bar{X}_T)^2$, into SS_B (equal to $\Sigma n_i(\bar{X}_i - \bar{X}_T)^2$) and SS_E (equal to ΣSS_i). The total **degrees of freedom** for the analysis, $df_T = N - 1$, can be similarly partitioned into $df_B = k - 1$ and $df_E = N - k$ (N is the total number of scores in the dataset and k is the number of conditions in the design). Dividing each SS by its corresponding df gives the appropriate MS, and dividing MS_B by MS_E gives the F ratio for an independent-samples ANOVA. The ANOVA summary table presents the intermediate calculations for calculating F in a well-organized and standardized way (**Section 10.2.2**).

- A theoretical **distribution** of all possible F statistics can be mathematically defined for each combination of df_B and df_E. All F distributions are positively skewed and fall entirely to the right of 0 (i.e., negative F values are impossible). The **F Distribution Area**

Calculation Tool will calculate the p-value associated with any given F / df_B / df_E combination (**Section 10.3.1**).

- In a **related-samples** ANOVA, SS_T and SS_B are calculated exactly the same as in an independent-samples ANOVA. However, SS_E is now calculated by taking SS_W (the error term in independent-samples ANOVA) and subtracting out SS_S, which measures variability between individual subjects. SS_S is calculated as $k\Sigma(\overline{X}_S - \overline{X}_T)^2$, and SS_E is found by subtraction: $SS_E = SS_W - SS_S$. The df_E is equal to $(k - 1)(n - 1)$. Values for MS_B, MS_E, and F are calculated exactly as in independent-samples ANOVA, and the p-value is found from the appropriate F distribution exactly as before (**Section 10.4.1**). As was the case for two-sample t tests, related-samples ANOVAs are considerably more powerful (i.e., more likely to reject H_0 when an effect actually exists) than independent-samples ANOVAs (**Section 10.4.2**).

- Graphs are essential for interpreting the results of studies that produce three or more means, because it is very difficult to grasp the pattern of effects when looking at raw numbers. Bar graphs should generally be used to picture the results when the **independent variable** in an experiment is categorical, whereas line graphs can be used when the **levels** of a **factor** are quantitatively related to each other (**Section 10.5.1**).

- A significant omnibus ANOVA indicates that the condition means in a dataset vary in some way from each other, but says nothing about which means are actually different from which. **Post hoc tests** can be used to assess the statistical significance of **pairwise comparisons** between means. In the simplest post hoc procedure, called the LSD test, the MS_E from the omnibus ANOVA is used in place of s_p^2 or s_D^2 in the standard formulas for t, and the resulting **test statistic** is assessed exactly as in a two-sample t test. The LSD test is relatively unpopular, however, because it is viewed by many statisticians as being too prone to Type I errors. A host of alternative post hoc procedures, such as the Scheffé test and Tukey's HSD test, use various means to protect researchers from making such errors (**Section 10.5.2**).

- A **bidirectional,** two-sample t test is mathematically equivalent to a single-factor F test with $df_B = 1$. The two test statistics are related by the formula $F = t^2$. In accord with this tight relationship, the assumptions underlying F tests are identical to those underlying t tests (**Section 10.6**).

- When an experimental design includes more than one independent variable, the results of the experiment can be analyzed with a multiple-factor ANOVA. This analysis tests the significance of the **main effects** of each factor as well as the **interactions** between each factor. An interaction occurs if the effect of one factor depends in some way on the effect of another factor (**Section 10.7**).

Chapter 11
Describing Relationships:
Correlation and Regression

11.1 Covariability

Imagine you interviewed 15 male members of the Springfield State College marching band, asking each of the interviewees to report their heights (in inches) and weights (in pounds). You might get the following two sets of scores, whose **distributions** are shown in Figure 11.1.

Subject	Height	Weight
1	70	185
2	73	202
3	72	209
4	64	169
5	70	205
6	72	165
7	61	162
8	66	204
9	63	165
10	69	150
11	64	150
12	72	203
13	61	141
14	69	180
15	60	154

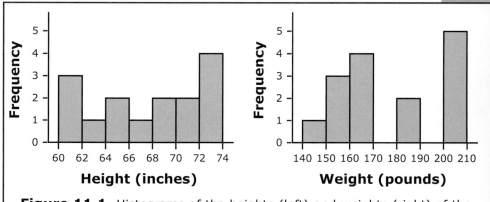

Figure 11.1 Histograms of the heights (left) and weights (right) of the 15 band members whose data appear in the table above.

As is the case for most datasets, the heights and weights you collected vary. As we've seen, measuring the amount of **variability** in an individual **variable** is crucial to understanding the meaning of the variable.

But the dataset we have here features a new kind of variability that we haven't discussed before. In addition to varying separately, the band members' heights and weights also vary together, a phenomenon called **covariability.** That is, the shorter you are, the less you tend to weigh; or, looking at it the other way, the taller you are, the more you tend to weigh.

We can visualize this **relationship** with a type of graph called a **scatterplot,** shown in Figure 11.2. To create a scatterplot, we draw scales on the X and Y axes to match the ranges of the two variables (here, height is represented on the X axis and weight on the Y axis).

Glossary Term: covariability

Then for each individual in the dataset, we move over to the point on the X axis that corresponds to his height (e.g., for the first band member in the table above, 70 inches), then move up to the point on the Y axis that corresponds to his weight (185 pounds), and plot a point at that location. Then we do the same thing for each of the other individuals in the dataset.

Now suppose that you gathered two additional pieces of information from your 15 band members: their grade point average (GPA) at Springfield State and the inseam length of their band uniform pants. The scatterplots on the left and right sides of Figure 11.3 plot the relationships between height vs. GPA and height vs. pants length; the height–weight scatterplot is reproduced in the middle of Figure 11.3.

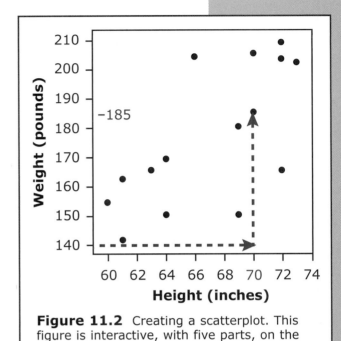

Figure 11.2 Creating a scatterplot. This figure is interactive, with five parts, on the *Interactive Statistics* website and CD.

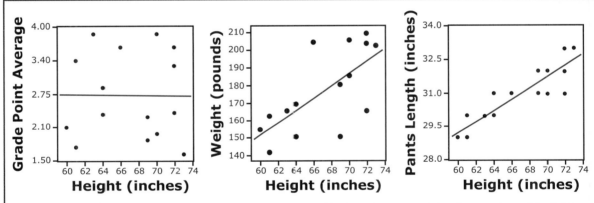

Figure 11.3 Scatterplots of height vs. GPA (left), height vs. weight (center) and height vs. pants length (right). Which graph indicates the strongest relationship?

Which of the three variable pairs in Figure 11.3 appear most closely related?

Interactive Page 268

As the **scatterplots** in Figure 11.3 illustrate, different pairs of variables have different amounts of **covariability.** There appears to be little if any **relationship** between height and GPA, while the relationship between height and pants length appears even stronger than that between height and weight.

The covariability between a pair of variables ranges from no relationship to a very strong relationship.

This chapter is all about covariability. First we'll learn in **Section 11.3** about the Pearson product-moment **correlation** (a term almost always shortened to "correlation"), the most common measure of the extent to which two variables covary. Correlation values range from .00 (no covariability at all) to ±1.00 (perfect correspondence between the two variables). The correlations for the variable pairs shown in Figure 11.3 are −.01, .69, and .88, from left to right.

A good way to summarize a relationship between two variables is to draw a line through their scatterplot such that the distances between the datapoints and the line are minimized (see the middle graph in Figure 11.3). This is called a **regression line,** and it is useful for predicting new values on one variable given a value on the other variable. For example, if you interview a 16th band member and find that he is 67 inches tall, you could use the regression lines to predict that he weighs 176.1 pounds and has a pants length of 30.9 inches. (Since the correlation between height and GPA is so small, it wouldn't make sense to try to predict your new subject's GPA.) **Section 11.4** explains how to construct and interpret regression lines.

In addition to prediction, correlation and regression statistics can be used for other research purposes, as described in **Section 11.5** and **Section 11.5.2.** Researchers using correlational techniques often want to generalize from samples of subjects to larger populations. This goal is accomplished using **inferential statistics** and **hypothesis testing,** as described in **Section 11.5.1.**

Section 11.6 introduces a statistical technique called **multiple regression/correlation,** with which we can examine the relationships between three or more variables all at once. Whole books are written and courses taught about multiple regression, so the details presented here will be necessarily sketchy. Finally, **Section 11.6.1** will briefly describe how all the inferential statistics we've covered so far (*t*-tests, ANOVAs, and correlations), as well as many additional analyses, can all be thought of as variations on the same theme, a "Grand Theory" of statistics called the General Linear Model.

11.2 Characteristics of Relationships

Interactive
Page 269

The **covariability** between any two **variables** can be generally characterized by three factors: the form of the **relationship,** the direction of the relationship, and the degree of relationship. We'll go through each of these factors in this section.

The **form** of a relationship refers to the general shape taken on by the datapoints in a **scatterplot.** The only form of relationship that we'll be learning how to deal with is the **linear** form. In a linear relationship, all the datapoints in the scatterplot fall more or less on a single, straight line. The relationship between height and weight, shown again in Figure 11.4a, would be considered linear (as would the relationship between height and pants length).

Relationship form

Although we won't learn how to assess them in this book, you should be aware of some of the other forms of relationships, which can all be lumped together under the general term **curvilinear** (because if a shape isn't a straight line it must have some sort of curve in it).

For example, the relationship between the amount of practice one puts in and one's tuba-playing ability is shown in Figure 11.4b. Ability increases rapidly with practice at first, but once you achieve a certain amount of competence, extra practice doesn't help as much as it used to.

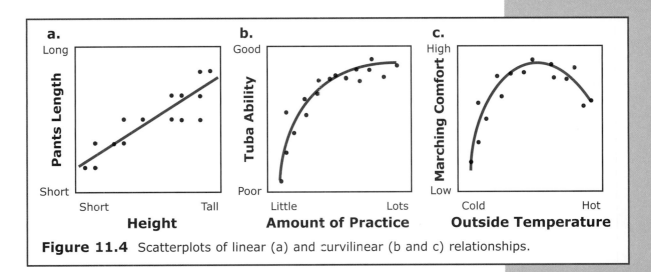

Figure 11.4 Scatterplots of linear (a) and curvilinear (b and c) relationships.

An even more complex relationship exists between the comfort level of band members and outside temperature, as shown in Figure 11.4c. Marching is uncomfortable both when it is very cold and when it is very hot. The comfort level is highest at in-between temperatures.

The second general characteristic of a relationship is its **direction.** In positive relationships, increases on one variable are associated with increases on the other variable, whereas in negative relationships, increases on one variable are associated with decreases on the other variable.

In a scatterplot of a negative relationship, the datapoints tend to move down on the Y axis as they move right on the X axis. But in a scatterplot of a positive relationship, Y values get higher (more positive) as X values get higher.

The **form** and **direction** of a relationship are qualitative characteristics—a relationship is either **linear** or **curvilinear,** and either positive or negative. In contrast, the third characteristic, **degree** of rela-

Glossary Term: curvilinear relationship

Relationship direction

Interactive Page **270**

Degree of relationship

tionship, is quantitative, meaning that it varies continuously. We've already seen examples of variations in degree of relationship in Figure 11.3 (**Section 11.1**).

Figure 11.5 shows the full range of degrees of relationships. Figure 11.5a shows a **perfect relationship,** in which the values of the *Y* variable are perfectly predictable from the values of the *X* variable (and vice versa).

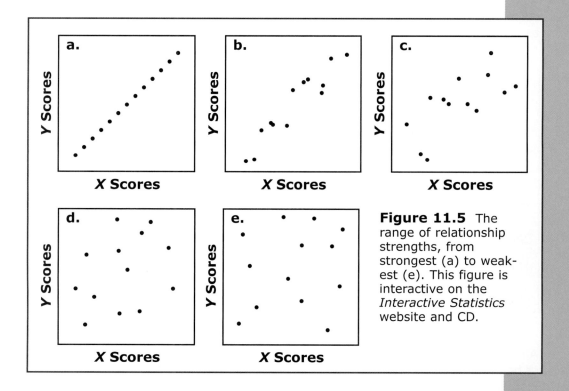

Figure 11.5 The range of relationship strengths, from strongest (a) to weakest (e). This figure is interactive on the *Interactive Statistics* website and CD.

In contrast, Figure 11.5e shows a pair of variables that is not related at all. Here, knowing the value of the *X* variable tells you nothing at all about the value of the *Y* variable. Figures 11.5b, c, and d illustrate strong (but not perfect), intermediate, and weak relationships, respectively.

In the next section, we'll learn how to measure the degree of linear relationship between two variables with the Pearson **correlation** statistic.

11.3 The Pearson Correlation

Our goal in this section is to develop a statistic that measures the **degree** to which two variables are **related.** That is, we want the statistic to have the smallest possible value when the variables are not related at all, the highest possible value when the variables are perfectly related, and intermediate values for intermediate degrees of relationship. Figure 11.6a shows **scatterplots** of a pair of strongly related **variables** (left) and a pair of weakly related variables (right) to use in the upcoming discussion.

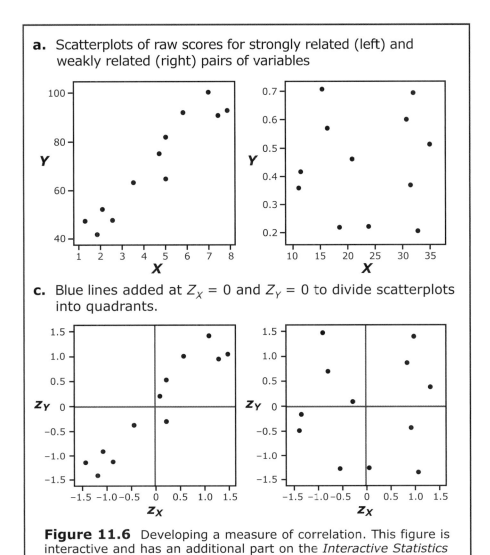

a. Scatterplots of raw scores for strongly related (left) and weakly related (right) pairs of variables

c. Blue lines added at $Z_X = 0$ and $Z_Y = 0$ to divide scatterplots into quadrants.

Figure 11.6 Developing a measure of correlation. This figure is interactive and has an additional part on the *Interactive Statistics* website and CD.

We'd like our measure of relationship to be unit-free—that is, we'd like every pair of perfectly related variables to have the same correlation value regardless of the scales of measurement for the variables. As we learned in **Section 3.6,** we can get rid of measurement scales by transforming raw data into **z-scores.** So this is just what we've done in Figure 11.6b for the raw scores from Figure 11.6a. To refresh your memory, the z-score formula is:

$$z = \frac{X - \mu}{\sigma}$$

In Figure 11.6c, we've divided each scatterplot into quadrants by placing vertical and horizontal lines at the locations of $z = 0$ on the X and Y axes. Note that for the strongly related pair of variables (left scatterplot), 11 datapoints fall in either the upper right or lower left quadrant, whereas only one datapoint falls in either of the other two quadrants. But for the weakly related variable pair (right scatterplot), the datapoints are equally divided between all four quadrants.

Now consider what would happen if we took each datapoint in the two scatterplots and multiplied the z-score for the X variable by the z-score for the Y variable:

- For datapoints in the upper right quadrant, the product of the z-scores would be positive, since a positive number times a positive number will always produce a positive number.

- The same is true for datapoints in the lower left quadrant: A negative number times a negative number will also produce a positive number.

- However, for datapoints in the upper left and lower right quadrants, one z-score will be positive and the other negative, so the product of the two z-scores will always be negative.

The interactive figure on the *Interactive Statistics* website and CD allows you to roll your mouse over the dots in the scatterplots of Figure 11.6c to see the products of the z-scores for each datapoint.

Now suppose we add all these products together for each variable pair. For the strongly related pair, all but one of the products will be positive, so we will end up with a pretty large sum. But for the weakly related pair, the positive products will be more or less cancelled out by the negative products, so we'll end up with something close to 0. To be exact, the sum of the products of the z-scores is 11.35 for the strong relationship and 0.37 for the weak relationship.

Interactive Page 272

On the previous page, we established that by multiplying the **z-scores** for the X and Y **variables** and taking the sum of these products, we get a number that is larger for stronger **relationships** and smaller for weaker relationships. In this sense, the sum of the z-score products fulfills the basic requirement for a correlation **statistic.**

The only problem is that the sum will also increase with the number of individuals in the dataset, but we would like a perfect relationship for 20 datapoints to produce the same correlation as a perfect relationship for 500 datapoints.

To solve this problem, we divide the sum by the number of subjects n in the dataset (that is, the number of pairs—12 in the case of the

Datapoints in strong relationships will tend to fall almost exclusively in two quadrants of the scatterplot.

For strong positive relationships, almost all the z-score products will be positive.

Average the z-score products to calculate r.

datasets plotted in Figure 11.6). Put it all together, and we get the formula for the **Pearson correlation,** which goes by the symbol r:

$$r = \frac{\Sigma z_X z_Y}{n}$$

The table below shows the steps in calculating r using this formula for the strong relationship (left scatterplot) in Figure 11.6: Thus r for the strong **relationship** in Figure 11.6 is .95. For the weak relationship in Figure 11.6, $r = .03$.

X	z_X	Y	z_Y	$z_X z_Y$
1.1	−1.49	46.0	−1.18	1.76
1.7	−1.22	40.3	−1.46	1.78
1.9	−1.13	50.7	−0.95	1.07
2.5	−0.90	46.2	−1.17	1.05
3.5	−0.46	62.2	−0.39	0.18
4.7	0.09	74.7	0.23	0.02
5.1	0.23	63.9	−0.30	−0.07
5.1	0.23	81.7	0.57	0.13
5.9	0.60	91.8	1.06	0.64
7.2	1.15	100.4	1.48	1.70
7.7	1.35	90.7	1.01	1.36
8.1	1.55	92.7	1.11	1.72
$\mu_X = 4.54$		$\mu_Y = 70.11$		$\Sigma z_X z_Y = 11.34$
$\sigma_X = 2.32$		$\sigma_Y = 20.41$		$n = 12$
r = .95				

Note that the z-score formula requires the **population standard deviation** σ, not the **sample standard deviation** s. So even though we usually calculate correlations for **samples** of scores, we must treat the dataset as a **population** when transforming it into z-scores to calculate r. The difference between the population and sample standard deviations is that in calculating σ, we divide **SS** by the total number of scores (n), whereas in calculating s, we divide SS by the number of scores minus one ($n − 1$; see **Section 3.3.2**).

It's not obvious from the formula, but for positive relationships, r will always be between .00 and 1.00. The latter value will be obtained only for an absolutely perfect positive relationship.

Can you guess what the value of r will be for a perfect *negative* relationship?

Interactive
Page 273

As we saw a couple of pages ago, when two **variables** are positively correlated, most of the datapoints will fall in the upper right or lower left quadrants of their scatterplot (Figure 11.7a).

Use the population standard deviation when calculating z-scores.

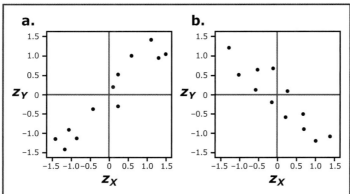

Figure 11.7 For a positive relationship (a), most datapoints will fall in the upper right or bottom left quadrants of a scatterplot. For a negative relationship (b), most datapoints will fall in the other two quadrants—the upper left and bottom right of the scatterplot.

In contrast, when the variables are negatively correlated, most of the datapoints will fall in the opposite two quadrants—the upper left and bottom right (Figure 11.7b). When we multiply the **z-scores** in Figure 11.7b together, negative z-scores will usually be paired with positive z-scores, so the products will usually be negative. Thus r will end up being a negative number. For a perfect negative correlation, r will equal −1.0.

For strong negative relationships, most of the products of the z-scores will be negative

So, to sum up the basic qualities of the Pearson **correlation,** r allows us to assess two of the characteristics of **relationships** outlined in the previous section:

Basic qualities of the Pearson correlation r

1. The sign of r tells us whether the direction of the relationship is positive (r greater than 0) or negative (r less than 0).

2. The magnitude (absolute value) of r tells us the degree of relationship, from no relationship at all (r = .00) to a perfect relationship (r = 1.0 or −1.0).

What about the third characteristic, the **form** of the relationship? As we've already noted, r measures the degree of **linear** relationship. That is, it only makes sense to calculate r in the first place if we assume that the relationship is linear.

The Pearson correlation r can only be used to assess linear relationships.

To decide whether this assumption is reasonable, you have to inspect the **scatterplot.** For example, the relationship shown in Figure 11.8 is quite strong, but is not linear. If you blindly calculated r without inspecting the scatterplot of these data, you would find that r = .09 and incorrectly conclude that the two variables were very weakly related.

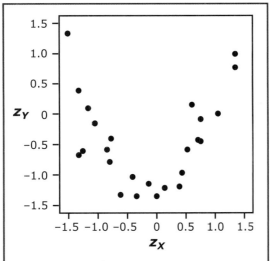

Figure 11.8 A strong, but nonlinear, relationship; *r* should not be used to assess the strength of such relationships.

Interactive
Page 274

11.3.1 The Correlation Helper Activity

To calculate a **correlation** by hand, it helps to draw out a table like the one presented earlier and reprinted here:

X	z_X	Y	z_Y	$z_X z_Y$
1.1	−1.49	46.0	−1.18	1.76
1.7	−1.22	40.3	−1.46	1.78
1.9	−1.13	50.7	−0.95	1.07
2.5	−0.90	46.2	−1.17	1.05
3.5	−0.46	62.2	−0.39	0.18
4.7	0.09	74.7	0.23	0.02
5.1	0.23	63.9	−0.30	−0.07
5.1	0.23	81.7	0.57	0.13
5.9	0.60	91.8	1.06	0.64
7.2	1.15	100.4	1.48	1.70
7.7	1.35	90.7	1.01	1.36
8.1	1.55	92.7	1.11	1.72
$\mu_X = 4.54$		$\mu_Y = 70.11$		$\Sigma z_X z_Y = 11.34$
$\sigma_X = 2.32$		$\sigma_Y = 20.41$		$n = 12$
$r = .95$				

The **Correlation Helper** Activity will aid you in performing these calculations. To use the Helper, you first need to enter the scores you want to analyze in the two text boxes in the upper left corner of the

Activity: Correlation Helper

Helper window. Make sure to enter the X and Y scores in the same order in the two boxes. You can use spaces, commas, carriage returns, and/or parentheses to separate scores. For example, if your dataset is:

X	Y
9	25
34	12
26	2
15	33
14	31
9	25
2	23

You should enter "9 34 26 15 14 9 2" in the X text box and "25 12 2 33 31 25 23" in the Y text box.

You can also click on the "generate" link in the Helper to generate a random set of data to practice on.

Once you have your data entered in the upper left text box, transfer the scores to the X and Y columns. Then calculate the **means** and **standard deviations** of the X and Y variables and enter these statistics at the bottom of the X and Y columns. You can use the Descriptive Statistics Calculation Tool to get these means and standard deviations. (Remember to calculate the **population standard deviations.**)

Next you need to calculate **z-scores** for each of the X and Y raw scores. Then multiply the z-scores together and enter the products in the far-right column of the Helper. Add the products together and enter the sum, along with the number of scores, in the designated text fields near the bottom of this column. Then divide the sum of the products by **n** to get r and enter this value in the bottom-right text field.

Note that you can press the "tab" key on your keyboard to move from field to field in the table. Click the "Check Answers" button at any time, or simply hit the enter or return key after typing any value into a text box, to check your calculations. Any incorrect answers will be colored red; correct calculations will remain in black.

Roll your mouse over the bullet (small circle) to the left of each text box to see a hint for what's supposed to go there. If you give up, click the bullet to fill in the correct value (it will appear in blue so you can keep track of which fields you needed help with).

Another Activity, the **Scatterplot Helper,** allows you to generate a **scatterplot** for a set of data, which, as we've already seen, is an essential part of analyzing any **relationship.** This Activity includes detailed instructions on its use.

Activity: Scatterplot Helper

Interactive Page 275

11.3.2 Understanding *r*

Now that we've been introduced to the correlation statistic *r*, let's delve further into what this statistic tells us.

First of all, you need some idea of what different values of *r* indicate. We've already seen that *r* = .00 indicates that the two variables in question are not related at all. And we know that *r* = 1.00 and *r* = −1.00 are indicative of perfect **relationships,** where scores on the *X* variable perfectly predict scores on the *Y* variable (and vice versa).

r = .00: no relationship

r = 1.0 or *r* = −1.0: perfect relationship

But what about values between 0 and 1 (or between 0 and −1— unless noted otherwise, assume for the rest of this chapter that everything we say about positive *r* values holds also for negative *r* values)? How high does *r* have to be before we can say that two **variables** are meaningfully related?

Well, the wishy-washy response to this question is that it depends on a) the research context in which you're working and b) what you mean by meaningful. A correlation of *r* = .20 might be considered very important in some contexts, while an *r* of .40 might be considered relatively low in other contexts.

But many behavioral scientists accept as a general convention that *r* values of .10, .30, and .50 should be considered small, medium, and large effects, respectively. Correlations greater than .60 will be considered fairly strong by nearly any researcher in nearly any context, and correlations less than .05 are almost always considered meaningless (that is, two variables whose correlation is less than .05 are usually taken to be practically unrelated).

r = .10: small effect

r = .30: medium effect

r = .50: large effect

Interactive Page 276

Now that we have some general guidelines for interpreting *r* values, let's look at a more complex issue. Consider the following:

> Ms. Tyler, the cross-country running coach at a large high school, is trying to bolster the size of her team by recruiting students who play other sports to also run cross country. She performs a study in which she looks at the records of all the boys and girls in the past 10 years who ran on the school cross-country team and also played on the school basketball team. It turns out that there is a high correlation, *r* = .55,

between the athletes' best cross-country times and the average number of points they scored per game in their basketball careers. Can Coach Tyler use these results to argue that running cross country improves basketball-playing ability?

As tempting as Coach Tyler's proposed conclusion may be, we cannot jump to it. As an old statistical adage says, **correlation** *does not imply causation.*

Correlation ≠ causation

Figure 11.9 diagrams four potential scenarios that can lead to a strong correlation between variables X (in our example, cross-country ability, as measured by athletes' best times) and Y (basketball ability, as measured by average points per game). In this type of figure, solid lines represent causative links, with the arrow indicating the direction of causation. Dashed lines represent **relationships** in which neither variable is causing the other.

Scenario (a) is the one that represents Coach Tyler's hypothesis, that X causes Y. For example, the leg strength and cardiovascular endurance built up during the cross-country season may help athletes to become better basketball players.

The simplest possible alternative is scenario (b): Y may in fact cause X. Perhaps the competitive skills learned on the basketball court make athletes better racers, causing them to be able to achieve better cross-country times. This explanation may seem less compelling than Coach Tyler's X causes Y hypothesis, but statistically speaking, there is no way to determine which is more likely.

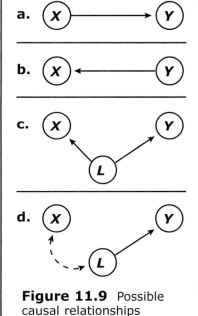

Figure 11.9 Possible causal relationships between two related variables X and Y.

Scenarios (c) and (d) in Figure 11.9 show how a third variable L, called a **lurking variable** (it is unaccounted for in the bivariate correlation, so it is "lurking in the shadows" of our analysis), may be at least partially responsible for the correlation between X and Y. In scenario (c), L causes both X and Y. For example, it may very well be that the amount of raw athletic ability one has determines both how fast one can run and how many points one is able to score.

Glossary Term: lurking variable

In scenario (d), L is a confounding variable that has a hand in causing Y and is tangentially associated with X. Perhaps the hand–eye coordination that is an essential element to being a high scorer in basketball happens to co-occur with the leg strength needed to be a fast cross-country runner. Hand–eye coordination (the L variable) is the real cause of Y (basketball scoring), but because of L's association with X (running speed), it appears that X causes Y.

The differences between scenarios (c) and (d) are subtle and are not as important as the general point that a lurking variable can, and very often does, lead to an inflation of the correlation between two other variables X and Y. It is also possible that all four of the scenarios in Figure 11.9 are partly correct: X may partially cause Y, Y may partially cause X, and there may be any number of L's that partially cause X, Y, or both, all at the same time. And there are even more possible causal paths that we haven't diagrammed here. Life is complicated.

All this is not to say that correlations cannot be used as support for an X causes Y argument. For example, much of the evidence that smoking causes lung cancer is correlational in nature. But there are thousands of studies using a variety of subject populations and methodologies that all converge on a causal link between smoking (X) and lung cancer (Y).

> Strong correlations can be supportive, but not definitive, evidence for a causal link.

Before moving on to the subject of **regression** in the next section, make sure you understand how to calculate *r* by going through **Exercise 11.1** and **Exercise 11.2** at least once.

11.4 Linear Regression

Springfield State College has a very simple system for tuition and fees. Each student must pay a flat $250 registration fee every semester, then pay an additional $100 per credit hour for the courses he or she takes. The table and scatterplot below (Figure 11.10) give the number of credit hours and the total charges for a sample of 8 students who took classes at S.S.C. last semester:

Student	Credit Hours	Charges
1	10	1250
2	4	650
3	12	1450
4	15	1750
5	13	1550
6	8	1050
7	6	850
8	16	1850

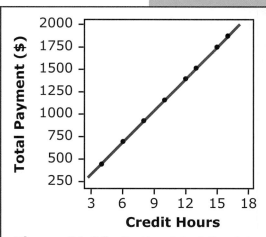

Figure 11.10 Scatterplot of the data at left, with regression line in red.

There is a perfect **linear** relationship between credit hours and charges for Springfield State students. You could confirm this fact by entering the data into the Correlation Helper Activity and calculating *r*, which will come out to 1.00.

But we can also *see* the perfect linear relationship in the **scatterplot,** because we can draw a straight line that runs right through all the points in the plot (shown in red in Figure 11.10). Such a line is called a **regression line.**

Glossary term: regression line

The regression line (like any other straight line) can be compactly described with two parameters: the Y intercept and the slope. The Y intercept, which goes by the symbol *a*, is defined as the height at which the line passes through the Y axis. The slope, which goes by *b*, is the rate at which the line rises and falls from this point—the amount of increase in Y for each unit increase in X. Put the two parameters together and you get the following formula for relating Y and X:

A regression line is specified by its slope (*b*) and its Y intercept (*a*).

$$Y = a + bX$$

Can you specify the values of *a* and *b* for the regression line in Figure 11.10, relating credit hours to S.S.C. tuition and fees?

Interactive Page 278

On the previous page, you were asked to determine the Y intercept and slope for the **regression line** relating credit hours (X) to total payments (Y) for Springfield State College students (see Figure 11.10, reprinted at right).

If you were to register at S.S.C. but not sign up for any classes, you would pay only the registration fee, $250. So $250 is the Y intercept (i.e., 250 is the value of Y when X is 0). And for each credit hour that you take, you have to pay $100, so the slope is $100 (each time X increases by 1, Y increases by 100). Thus the formula for the regression line is:

$$Y = 250 + 100X$$

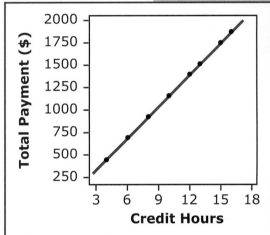

Figure 11.10 Scatterplot of the S.S.S. credit hours–total payments data, with regression line in red.

Now, say we asked the students whose credit hours and charges are plotted in Figure 11.10 to report the average number of hours of homework they did per week during the semester. We should expect amount of homework to also be related to credit hours, but we should not expect this **relationship** to be perfect. In fact, as shown by the **scatterplot** in Figure 11.11a, this relationship is strong but not perfect. The **correlation** *r* for this data is .68.

Although we cannot draw a single straight line that passes right through every datapoint in Figure 11.11a, it is still possible to find the regression line that provides the best possible *fit* to these data. That is, we can find the line that has the smallest possible average distance between the datapoints and the line.

Regression lines minimize the average distance in a scatterplot between datapoints and the line.

We could eyeball the regression line by anchoring a ruler at the point in the scatterplot that corresponds to the mean X value and the mean Y value, then swiveling the ruler back and forth around this point until we found the slope that appears to minimize the discrepancies between the datapoints and the line.

But some fairly complicated calculus, which we won't subject you to here, can do this job faster and more accurately: It turns out that we can calculate precisely the values for a and b that will result in the best-fitting regression line. There are a number of ways of expressing the formulas for a and b, but we'll use these:

$$b = r\frac{\sigma_Y}{\sigma_X} = r\frac{s_Y}{s_X}$$

$$a = \bar{Y} - b\bar{X}$$

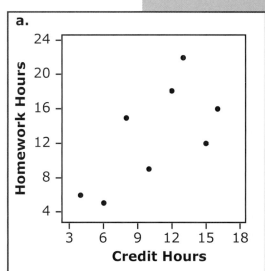

a.

Figure 11.11 Scatterplot of the relationship between credit hours and homework hours. This figure is interactive and animated, and has an additional part, on the *Interactive Statistics* website and CD.

Note that the two formulas for b produce the same results—you will get the same b value regardless of whether you use population or sample standard deviations (as long as you use the same statistic for both the X and Y variables; that is, don't use the **population standard deviation** of X along with the **sample standard deviation** of Y).

The **mean** and sample standard deviation for the number of credit hours in Figure 11.11 are 10.5 and 4.3, respectively, and $\bar{Y}$ and s_Y for hours of homework are 12.9 and 6.0. We already established that r for the **relationship** between these two variables is .68. So b and a for these data are:

$$b = .68(6.0 / 4.3) = 0.95$$
$$a = 12.9 - 0.95(10.5) = 2.93$$

Put these two parameters together, and we get a regression line equation of:

Homework hours = 2.93 + 0.95(Credit hours)

Interactive Page 279

11.4.1 Using Regression for Prediction

On the previous page, we determined that the total amount paid in tuition and fees (Y) as a function of credit hours (X) for Springfield State College students is:

$$Y = 250 + 100X$$

We developed this formula to describe the data for the 8 students whose credit hours and costs we already knew (shown in the **scatterplot** on the left of Figure 11.12). But one nice thing about

Regression equations can be used to predict new values of Y given new values of X.

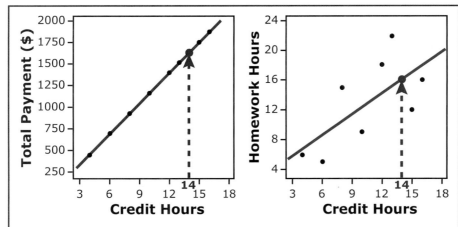

Figure 11.12 Scatterplots of credit hours vs. total payments (left) and credit hours vs. homework hours (right), with regression lines shown in red. Blue dots show predicted values for payments (left) and homework hours (right) given a new value (14) for credit hours.

regression lines is that they can also be used to *predict* new values of Y given new values of X. For example, if a prospective S.S.C. student plans to take 14 credit hours next semester, he can expect to pay (Figure 11.12, left):

$$250 + 100(14) = 1650 \text{ dollars}$$

We can also use regression lines for prediction if the **relationship** between X and Y is less than perfect, as in the case of credit hours (X) vs. hours of homework (Y). On the basis of our 8 original students' data (replotted in the right-hand scatterplot of Figure 11.12), we computed the regression equation for this relationship as:

$$\text{Homework hours} = 2.93 + 0.95 (\text{Credit hours})$$

Our new student planning to take 14 credit hours can expect (right of Figure 11.12):

$$2.93 + 0.95(14) = 16.23 \text{ hours of homework per week}$$

Now, because the relationship between credit hours and homework hours is not perfect, predictions based on the above regression equation cannot be expected to be foolproof. This observation leads to a new question: Exactly how accurate should we expect our predictions for new datapoints to be?

To begin answering this question, consider a somewhat simpler question. Say we have two pairs of variables. The **correlation** for the first pair is $r = -.77$, and the correlation for the second pair is $r = .33$. We calculate regression equations for the two pairs of variables, then use these equations to predict new Y values. Which regression analysis will provide the most accurate predictions?

**Interactive
Page 280**

11.4.2 Accuracy of Regression Predictions

The closer the **relationship** between two variables (that is, the closer the *r* value is to −1.00 or +1.00), the better we will be able to predict new values of *Y* from new values of *X*. Thus a **regression line** based on an *r* of −.77 will provide better predictions (on average) than a regression line based on an *r* of .33.

To get an estimate of exactly how accurate we should expect our predictions for new datapoints to be, we can assess how well the regression line we've constructed fits the data we constructed it from. In other words, we translate our original question about new datapoints to a hypothetical question about our old datapoints:

> "Suppose we only knew the regression equation and the *X* values for our original group of subjects. Now suppose we calculated predicted values of *Y* for these subjects. How well, on average, would these predictions match the actual *Y* values we observed?"

Figure 11.13 shows the credit hours vs. homework hours **scatterplot** that we've been working with. The figure also shows the regression line we calculated from our data:

$$\hat{Y} = 2.93 + 0.95X$$

The symbol $\hat{Y}$ stands for "the estimated value of *Y*." The first subject in our dataset took *X* = 10 credit hours, so $\hat{Y}$ for this subject, the circled red dot in the figure, is:

$$\hat{Y} = 2.93 + 0.95(10) = 12.43$$

Note that this value falls right on the regression line, as it should. Now we can compare $\hat{Y}$ to the actual value of *Y* for this student (*Y* = 9; this is the circled green dot in the figure). The difference between the two scores (Figure 11.13) is 12.43 − 9 = 3.43. We can make the same comparisons for the other students we have data for, as illustrated in the figure.

Below we've constructed a tabular version of the comparisons shown in Figure 11.13. For each value of *X*, we list the actual value of *Y*, the predicted value $\hat{Y}$, and the difference between the actual and predicted values.

The closer the relationship between two variables, the more accurate regression predictions will tend to be.

To assess the probable accuracy of future regression predictions, we compare predicted values of *Y* to the actual *Y* values for the data we've collected.

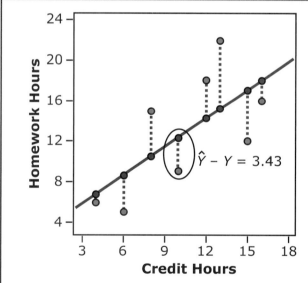

Figure 11.13 Assessing the accuracy of regression predictions. Predicted values for *Y* are shown in red and actual values for *Y* in green. This figure is interactive, with five parts, on the *Interactive Statistics* website and CD.

X	Y	$\hat{Y}$	$\hat{Y} - Y$
10	9	12.43	3.43
4	6	6.73	0.73
12	18	14.33	−3.67
15	12	17.18	5.18
13	22	15.28	−6.72
8	15	10.53	−4.47
6	5	8.63	3.63
16	16	18.13	2.13

We want to find out how accurate, on average, the predictions are. If the differences between Y and $\hat{Y}$ are all close to 0, the predictions are pretty good, whereas if the differences tend to be large, the predictions are pretty bad.

However, calculating the average of the differences won't do us any good, because the differences will always sum to 0 (confirm this using the Desk Calculator and the data above; because of rounding errors, you won't get exactly 0, but it will be very close). You may remember that we faced a similar difficulty way back in **Section 3.3,** when we were developing the formula for the **standard deviation.**

To overcome this difficulty here, we do the same thing we did there: square the **difference scores.** The extent of the differences between predicted and actual scores indicates the amount of error in our predictions, so we call the sum of the squared difference scores SS_E, for "sum of squared error." In case you're wondering, this SS_E is mathematically equivalent to the SS_E we used for ANOVA calculations in **Section 10.2.2.** We'll see how we use SS_E to evaluate regression accuracy on the next page.

On the previous page, we established a procedure for computing SS_E, the sum of squared error, for a **regression line**:

- Find the values $\hat{Y}$ predicted by the regression equation for each X value in the dataset.

- Subtract $\hat{Y}$ from Y for each observation (see Figure 11.14).

- Square these difference scores (multiply each difference score by itself).

- Sum the squared difference scores.

Here's the full table for computing SS_E in the example we've been working with:

Procedure for calculating sum of squared error

$$\hat{Y} = 2.93 + 0.95X$$

X	Y	$\hat{Y}$	$\hat{Y} - Y$	$(\hat{Y} - Y)^2$
10	9	12.43	3.43	11.77
4	6	6.73	0.73	0.53
12	18	14.33	−3.67	13.47
15	12	17.18	5.18	26.83
13	22	15.28	−6.72	45.16
8	15	10.53	−4.47	19.98
6	5	8.63	3.63	13.18
16	16	18.13	2.13	4.54

Thus SS_E for our example is 135.46. Now, how should we evaluate this value?

We want to compare the amount of error given the values of $\hat{Y}$ predicted by the regression equation to the amount of error we would expect if we didn't have a regression equation at all.

If we didn't know anything about the **relationship** between credit hours and amount of homework, the best we could do in predicting a student's amount of homework would be to guess that he or she would have an average amount. That is, in the absence of any better predictor, the best estimate for the amount of homework any one student would have is the **mean** amount of homework everyone else has.

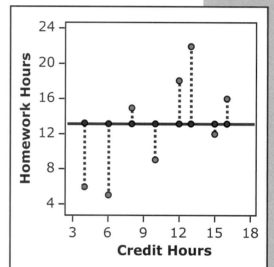

Figure 11.14 The error variance for Y values predicted by the mean of Y. This figure is interactive and has two parts on the *Interactive Statistics* website and CD.

So to get comparison value for SS_E, we add up the squared deviations of the observed values of Y from the mean value of Y ($\bar{Y}$; Figure 11.14):

Y	$\bar{Y}$	$\bar{Y} - Y$	$(\bar{Y} - Y)^2$
9	12.9	3.9	15.21
6	12.9	6.9	47.61
18	12.9	−5.1	26.01
12	12.9	0.9	0.81
22	12.9	−9.1	82.81
15	12.9	−2.1	4.41
5	12.9	7.9	62.41
16	12.9	−3.1	9.61

Without a regression equation, the best predicted Y value for any new subject will be the mean of Y.

We call the sum of these squared deviations SS_T, because it represents the Total amount of variation in Y—the maximum amount of error we could possibly find for any regression equation. Here, SS_T = 248.88. (Notice that SS_T is exactly the same as the SS used to calculate **variance** and **standard deviation.** It's also identical to the SS_T we used in ANOVA.)

If we divide SS_E by SS_T, we get the proportion of the total variance in Y that is left over after using our regression equation to help predict the value of Y. In other words, SS_E / SS_T = 135.46 / 248.88 = .54 of the variance in amount of homework remains unaccounted for after we take into account the number of credit hours students are taking.

If we subtract this proportion from 1, we get the "glass half full" version of our prediction success, rather than the "glass half empty" version. That is, to find the proportion of variance we've accounted for with our regression equation, we use the following equation:

The proportion of Y variance explained by applying the regression equation

Proportion of variance accounted for

$$= 1 - \frac{SS_E}{SS_T} = 1 - \frac{\Sigma(\hat{Y} - Y)^2}{\Sigma(\bar{Y} - Y)^2}$$

In our running example, we can say that predictions based on credit hours account for 1 − .54 = .46 of the **variability** in amount of homework for our 8 Springfield State students. Therefore, we can expect that predictions of future students' homework time based on credit hours will be similarly accurate.

To make sure you understand the SS_E and SS_T calculations, let's work through the procedure for finding the proportion of variance in Y accounted for by X for the following small set of scores:

X	Y
6	4
5	8
9	3

First, we need the **regression** equation for predicting Y from X. When you do this type of problem in the **Review Exercises,** you'll be asked to find this yourself, but here we'll just give it to you:

$$\hat{Y} = 11.92 + (-1.04)(X)$$

Now we calculate the predicted $\hat{Y}$ values and calculate SS_E (use the text boxes below to help organize your calculations):

X	$\hat{Y}$	Y	$\hat{Y} - Y$	$(\hat{Y} - Y)^2$
6	5.68	4	1.68	2.82
5	6.72	8	−1.28	1.64
9	2.56	3	−0.44	0.19
			Total (SS_E):	4.65

Now we calculate the mean of Y, $(4 + 8 + 3) / 3 = 5.0$, and use it to calculate SS_T:

$\bar{Y}$	Y	$\bar{Y} - Y$	$(\bar{Y} - Y)^2$
5	4	1	1
5	8	−3	9
5	3	2	4
		Total (SS_T):	14

Finally, we divide SS_E by SS_T and subtract the result from 1 to find the proportion of Y variance accounted for by X.

Proportion of Y variance accounted for by $X = 1 - (SS_E / SS_T) = 1 - (4.65 / 14) = .67$.

Interactive Page 283

Thanks to the Magic of Math, it turns out that we can determine the proportion of **variance** in Y accounted for by a **regression** equation without going through the hassle of calculating SS_E and SS_T, because:

$$1 - \frac{SS_E}{SS_T} = r^2$$

For example, the correlation for the three pairs of scores above is $r = -.82$, so $r^2 = .67$, the same value we calculated using the $1 - SS_E / SS_T$ method.

r-squared gives the proportion of variance in Y accounted for by X.

The value r^2 is often illustrated with something called a Venn diagram, three of which are shown in Figure 11.15. The pink circles in the diagrams represent the total amount of variation in three different Y variables, whereas the light blue circles represent variation in related X variables.

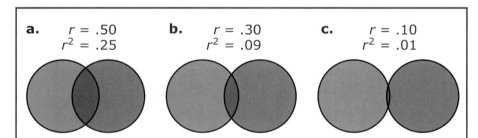

| a. | $r = .50$ | b. | $r = .30$ | c. | $r = .10$ |
| | $r^2 = .25$ | | $r^2 = .09$ | | $r^2 = .01$ |

Figure 11.15 Venn diagrams illustrating the proportion of overlapping variance between two variables when $r =$ a) .50, b) .30, and c) .10.

Venn diagrams do a pretty good job of representing the true meaning behind r^2, which can be tricky to understand. If the correlation between two variables X and Y is $r = .50$, a researcher will often state that "r^2 = .25 of the variation in Y is explained by X," as illustrated in Figure 11.15a.

All this really means is that about 25% of the variation in Y *covaries* with X. In other words, knowing X for an individual (along with the regression equation relating X to Y) gets us about 25% of the way toward predicting the true value of Y. This is represented in the Venn diagram by the fact that 25% of the Y circle overlaps the X circle.

Note, however, that it is also the case that 25% of the X circle overlaps the Y circle, and we could just as accurately say that .25 of the variation in X is explained by Y. This point goes back to the one made in **Section 11.3.2,** that correlation does not prove causation.

Here are a few more things to note about r^2:

- Since r varies between −1.00 and 1.00, r^2 varies between 0 and 1.00.

- When we square a number less than 1, the result is a number smaller than the one we started with. Thus for a correlation of r = .30, which we defined earlier as a medium-strength **relationship,** X is only explaining 9% of the variance in Y (r^2 = .09; see Figure 11.15b). And when the correlation is r = .10, only 1% of the variance in Y is explained by X (Figure 11.15c)! This observation helps explain why we said before that variables with correlations of less than .05 are generally considered to be unrelated.

- If there is absolutely no relationship between two variables, r = .00 and r^2 = .00. Remembering the equations for b ($r \times (s_Y / s_X)$) and a ($\bar{Y} - b\bar{X}$), note that when r is 0, the regression equation will be Y = $\bar{Y}$ + 0X, confirming that when there is no relationship between X and Y, we might as well just use $\bar{Y}$ to predict the value of Y.

Exercise 11.3 and **Exercise 11.4** go through the basic process of calculating regression lines and using them to make predictions, and **Exercise 11.5** lets you work on assessing the accuracy of predictions using regression.

11.4.3 The Effect of Outliers

In **Section 3.4,** we saw how extreme scores, called **outliers,** can have a huge impact on **nonresistant descriptive statistics** such as the **standard deviation.** The **correlation** coefficient r and **regression** slope b are based on similar mathematical principles as the standard deviation, so these statistics are also sensitive to the effects of outliers.

Figure 11.16 shows two pairs of variables that appear at first glance to be about equally related, as indicated by the green regression

Interpreting r^2

The correlation coefficient and slope (r and b) are not resistant to the influence of outliers.

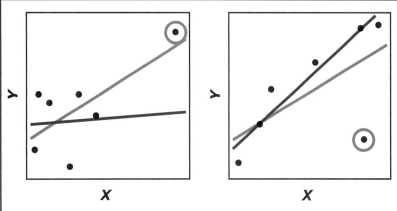

Figure 11.16 Two relationships with outliers. This figure is interactive, with additional parts, on the *Interactive Statistics* website and CD.

lines. The correlation for the left pair is $r = .70$, whereas for the right pair it is slightly smaller, $r = .65$.

However, note that in the left graph one datapoint falls well above and to the right of the other points (this datapoint is circled in the figure). This is the outlier in this dataset, and with this observation removed, r falls to .06, as shown by the red regression line. (Note that the slope of the regression line goes from being steeply sloped to almost flat, as we would expect given the close **relationship** between r and b.)

The outlier in the left graph is relatively easy to spot, since its X and Y values are extreme on their own. That is, the X value of the outlier is more than 2 standard deviations above the mean of X, and the Y value is about 1.7 standard deviations above the mean of Y. The standard methods of identifying outliers in single variables, outlined in **Section 3.4.2,** will also flag this outlier.

In contrast, the outlier in the right graph, also circled in the figure, is more insidious, because its X and Y values are not extreme when taken alone (they are both within one standard deviation of the means). Taken together, though, this datapoint is very different from the other observations, and when it is removed (red regression line), the correlation for the rest of the datapoints rises to $r = .96$ (the regression line also gets correspondingly steeper). Examining the **scatterplot** is the only good way to spot **bivariate outliers** such as this one.

To confirm that a datapoint is an outlier, of either of the two types described above, you can do what we've done here: remove the suspect datapoint, recalculate r and b, and see how much these values change. A large shift—greater than about .2 in r for a dataset with fewer than 10 datapoints, or greater than about .1 in r for a

Glossary Term:
bivariate outlier

larger dataset—will confirm that something funny is going on with that datapoint.

Test your outlier detection skills on the following set of data:

Subject ID	X	Y
A	11	68
B	9	71
C	2	2
D	8	65
E	7	47
F	6	12
G	11	91
H	10	70
I	16	120
J	11	42
K	4	110
L	8	32
M	15	91
N	11	77
O	8	20

Which subject produced an outlying point? (Hint: use the **Scatterplot Helper** Activity to make a scatterplot of the data.)

Interactive Page 285

The **outlier** in the dataset given on the previous page is subject K. This subject's X and Y values are not unusual on their own. But when these two values are taken together, this subject's datapoint clearly falls outside of the fairly well-defined cluster of points formed by the other subjects' data in the scatterplot.

So how should one deal with outliers in correlational data? For datapoints whose X and/or Y values are extreme on their own, the same general strategies used for single variables (**Section 3.4.2**) are also used here:

Strategies for dealing with outliers

- If there are only a few extreme values and they appear to be anomalies, these datapoints can be excluded from the **correlation/regression** calculations. However, the outliers should be noted when the data are reported, and some attempt should be made to explain why the outliers occurred. Sometimes the anomalous datapoints are the most interesting aspect of a study.

- Data **transformations** (see the **Box** on transformations on p. 65) on the X and Y variables can be used to reduce the influence of the outliers on r and b.

Box: Transformations

- There is also an alternative correlation measure, called the Spearman correlation, which is more **resistant** than the Pearson correlation to the influence of outliers. Using the Spearman correlation rather than the Pearson correlation in this case is analogous to reporting the median instead of the mean for a single variable. However, there is no widely used resistant alternative to *b*.

Data transformations and Spearman correlations don't help correct for bivariate outliers such as the one in the scatterplot above, so the only option for dealing with these datapoints is to exclude them from the calculations and try to account for how they came about.

Interactive Page 286

11.4.4 The Danger of Extrapolation

Another important consideration when interpreting **correlation** and regression analyses has to do with the range of *X* values studied. As we've seen, one of the nice things about a **regression line** is that it allows us to predict values of *Y* given new values of *X*.

As we've also seen, the values of *r* and r^2 give us some indication of how much we should trust such predictions. However, even if the correlation for our dataset is very high, we cannot predict what will happen to *Y* outside of the range of *X* values included in our sample.

Figure 11.17 illustrates what can happen if we extrapolate a regression equation in this way. Suppose we are measuring the effect of air temperature (*X*) on perceived comfort level (*Y*). If we only considered temperatures between 10 and 40 degrees Fahrenheit, we would observe a positive **relationship,** as shown in the upward-slanting regression line in the figure: the warmer it is, the more comfortable we feel.

However, if we instead considered only the range of temperatures between 80 to 110 degrees Fahrenheit, we would observe a *negative* relationship, as shown in the down-ward-slanting regression line, with warmer temperatures leading to less comfort.

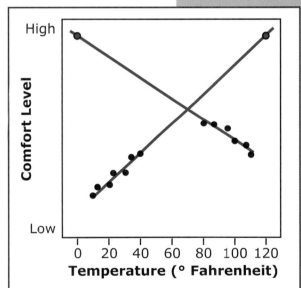

Figure 11.17 The dangers of extrapolating regression lines. This figure is interactive, with multiple parts, on the *Interactive Statistics* website and CD.

Obviously, extrapolating from either set of data alone would lead to erroneous predictions. Based on the first set of data and its regression line, we would predict that a temperature of 120 degrees would be extremely comfortable, whereas based on the second dataset and regression line, we would predict a high amount of comfort for a temperature of 0 degrees.

The lesson to be learned here is that we can only assess the **form, direction,** and **strength** of a relationship for the range of X values studied. Outside this range, the relationship between X and Y could be completely different.

Interactive
Page 287

11.5 Using Correlation and Regression in Research

The most straightforward use of **correlation** and **regression** statistics is for prediction, as described in **Section 11.4.1.** Here's an example:

> In the past, Springfield State College's policy has been that any prospective student who is deemed qualified is offered admission. This year, however, budget cuts have forced the college to adopt a goal of having 500 incoming students in the upcoming year.
>
> Jim, the head of admissions at S.S.C., has been charged with meeting this goal. The problem Jim faces is that he cannot simply admit 500 students, because he knows that some of those admitted will turn down S.S.C.'s offer and go to college elsewhere.
>
> To overcome this difficulty, Jim looks at previous years' admissions records and calculates the regression equation relating the number of students who actually enrolled each year (X) with the number of students who were offered admission (Y). The equation comes out to be:
>
> $$Y = 124 + 5.3X$$

Jim plugs 500 into the X slot of this equation and calculates that he will need to admit $Y = 124 + 5.3(500) = 2774$ applicants to meet his goal.

Note that when using a regression analysis solely for prediction purposes, it doesn't matter at all whether or not there is a causal **relationship** between X and Y. In this example, the number of students who enroll *cannot* be a causal factor of the number of students who are admitted—in fact, the causal relationship clearly goes in the other direction. But as long as two **variables** are related, one can be used to predict the value of the other.

Also note that Jim doesn't need to calculate r in order to make his prediction. However, the values of r and r^2 are very important in assessing how confident he can be in his prediction (see **Section 11.4.2**). If r is .95, for example, his prediction will probably be pretty accurate, since 90% of the variance in class size is accounted for by the number of students admitted, and vice versa.

Do not extrapolate a regression line outside the original range of X values studied.

Using regression statistics to make predictions

Interactive
Page 288

Beyond their value in assessing the accuracy of predictions, **correlations** can also be valuable tools for testing theories. Consider the following:

Using correlation statistics to test theories

> St. Louis Cardinals baseball pitcher Bob Gibson was famous for being an intimidating presence on the mound. Gibson was also an extremely proficient pitcher, posting an impressive 2.91 earned run average (ERA) for his career. (ERA measures the mean number of runs given up by a pitcher per nine innings.)
>
> Many baseball players, coaches, and fans believe that his ability to intimidate batters was part of the reason for Gibson's success, and many feel that an intimidating presence contributes to the success of other pitchers as well. To test this theory, we might ask present-day baseball hitters to provide intimidation ratings (on a scale from 1, meaning not intimidating at all, to 7, meaning most intimidating) for a randomly selected group of current pitchers. Suppose we collected average intimidation ratings for 20 pitchers and paired these data with the pitchers' ERAs from last season:

Intimidation	ERA	Intimidation	ERA
3.1	5.05	6.6	3.00
3.6	3.72	2.3	4.71
3.3	3.62	5.4	3.19
2.9	4.55	6.8	2.39
1.8	2.96	6.0	2.19
3.2	5.59	2.5	5.49
1.1	4.67	2.1	4.78
2.5	4.29	3.1	4.90
3.4	3.63	3.8	2.58
6.0	4.43	5.3	4.75

The Pearson correlation for these data is −.53. The fact that this correlation is negative (that is, greater intimidation ratings corresponded to fewer runs allowed) and relatively strong provides initial support for the theory that the ability to intimidate batters helps pitchers to be successful.

A strong correlation can support (but not prove) a theory.

However, there are two important caveats to consider. First is the warning from **Section 11.3.2,** that correlation alone can never prove causation. It is possible, for example, that batters find some pitchers intimidating, at least in part, because they are successful—the opposite causal **relationship** to the one we proposed. It is also possible that a lurking variable is responsible for the intimidation–

ERA correlation. In **Section 11.6** we will briefly consider how a technique called multiple **regression/correlation** can sometimes provide clearer evidence about causality.

The second caveat comes from the fact that we only assessed 20 pitchers. While the relationship between intimidation and ERA is undeniably strong for this sample, we do not know for sure whether the same relationship would hold for the entire population of all baseball pitchers.

Correlation statistics based on samples do not necessarily generalize to entire populations.

To address this concern, we turn to our old statistical friend, the **hypothesis test.**

11.5.1 Hypothesis Testing with *r*

To test hypotheses involving *r*, we follow the same series of steps first outlined in **Section 5.3** (see also the **Hypothesis Testing Steps** Activity).

Step 1 in the **hypothesis** testing procedure for **correlations** is to establish **null** and **alternative hypotheses** for the population correlation ρ (this is the Greek letter *rho,* the analog to *r* in the Greek alphabet).

Establishing hypotheses

Hypothesis tests are carried out because we seek evidence that our data show a *significant* effect. But recall from **Section 5.2** that since we cannot definitively prove that an effect has occurred, we set up a straw-man null hypothesis that states that there was *no* effect, and we try to prove this hypothesis wrong.

What value for ρ should we choose for our null hypothesis? To take a concrete example, consider the baseball pitcher data from the previous section. We sought to show that there was a **relationship** between intimidation ratings and earned run average. What value of ρ would be proposed by a critic who believed there was no such relationship?

11.5.1 Hypothesis Testing with *r*

H_0 for **hypothesis** tests with *r* almost always propose that ρ, the population **correlation,** is equal to .00—that is, that there is no **relationship** at all between the two variables in question.

There are three possible alternative hypotheses:

Possible alternative hypotheses

- For a bidirectional hypothesis test, H_a states that $\rho \neq 0$.

- For a directional hypothesis test, H_a can state either that $\rho > 0$ or that $\rho < 0$.

The cautions noted in **Section 5.4** and **Section 6.3.1** regarding **directional** hypothesis tests with z and t apply equally strongly here. **Bidirectional** tests are most often used when evaluating the **significance** of correlational data.

Step 2 in the hypothesis testing procedure is to collect and describe the sample data. For correlational studies, the relevant descriptive statistic is the Pearson correlation r for our sample of subjects. We also need to know the size n of the sample of subjects we've tested.

For our baseball pitcher study, we have data from $n = 20$ subjects, whose intimidation–ERA correlation is $r = -.53$.

Describing the sample data

Step 3 is to calculate the **test statistic.** There are a number of equivalent ways to do this. We will use the following formula relating r to the **t statistic:**

$$t = \frac{r\sqrt{n-2}}{\sqrt{1-r^2}}$$

Calculating the t statistic for a correlation

The correlation coefficient r has $n - 2$ **degrees of freedom,** so a t statistic based on r also has $n - 2$ df.* For the baseball pitcher data, we calculate that:

$$t = \frac{.54\sqrt{20-2}}{\sqrt{1-.54^2}} = 2.72$$

with $20 - 2 = 18$ degrees of freedom.

The final step in the hypothesis testing procedure, **step 4,** is to evaluate the hypotheses in light of the test statistic. To do this, we determine the **probability value** p of drawing a t statistic as large as we found from the distribution of t statistics with $n - 2$ degrees of freedom.

Evaluating the hypotheses

In practical terms, this means entering t and df in the **t Distribution Area** Calculation Tool, choosing whether the test is directional or bidirectional, and clicking the "calculate" button in the Tool (or using the t distribution table described in the Box on statistical tables on p. 89). For our example, $p = .014$ for a bidirectional test (confirm this value yourself using the Calculation Tool).

We interpret the p-value for hypothesis tests with correlations in exactly the same way as we have with z tests, t tests, and ANOVAs. A small p-value constitutes evidence that H_0 is probably false. In the behavioral sciences, the general convention is to declare an effect

---

*Why $n - 2$? Here's a quick and dirty explanation: You need two standard deviations (one for the X variable and one for the Y variable) to calculate r. Each standard deviation entails a loss of one degree of freedom, leaving us with $n - 2$ df.

"**significant**" if $p < .05$. Thus given the even smaller p-value for our baseball pitcher data (.014), we can make a fairly strong claim that intimidation and pitcher performance (as indicated by ERA) are significantly related.

Rather than testing whether or not two variables have a significant correlation with each other, we could instead test whether or not the **regression line** for the two variables is significantly sloped. It turns out that the p-value in a hypothesis test for r is exactly the same as the p-value in a hypothesis test for b. Thus if r can be declared significantly different than 0, so can b.

If r is significantly different than 0, b is also significantly different than 0.

Let's sum up our baseball pitcher example:

> Does the ability to intimidate batters contribute to the success of baseball pitchers? To test this theory, we asked hitters to provide intimidation ratings for a randomly selected group of 20 pitchers, and found that intimidation ratings correlated fairly strongly with the pitchers' earned run averages (ERAs): $r = -.53$. From this **correlation,** we calculated a **t statistic** of 2.72, with 18 df. The likelihood of obtaining a t statistic this small by chance alone is only **p** = .014, so we conclude that there is a significant **relationship** between intimidation and ERA for the entire population of baseball pitchers.

Batters can be intimidating to pitchers, too. Suppose we conduct a second study in which we have pitchers rate how intimidated they are by batters, and we look at the correlation between these ratings and the batters' batting averages (BAs), one measure of hitting prowess. We include a much larger sample in this study, 200 batters, and find a correlation of $r = .29$.

Calculate t and the p-value for this second study before going on to the next page. Then consider the following question: Should we consider the relationship between intimidation and batting ability to be stronger or weaker than the relationship between intimidation and pitching ability?

Here's the formula for calculating t from r again:

$$t = \frac{r\sqrt{n-2}}{\sqrt{1-r^2}}$$

Determining that the **correlation** in a sample of subjects generalizes to the entire **population** is an important step in arguing for a theory. But we remind you again (as we did in **Section 5.3.2**) that **descriptive statistics,** not **inferential statistics,** tell you what the data really mean.

This point is especially important to keep in mind when dealing with r and p, because many people have a tendency to interpret a small p as indicative of a strong correlation. It is the correlation coefficient r, *not* the **probability value** p, that tells you how strong a **relationship** is.

For our baseball data, the correlation coefficient for intimidation and ERA ($r = -.53$) is much higher than the correlation between intimidation and BA ($r = .29$; remember that it is the absolute value of r that we're interested in here). This means that the relationship between intimidation and pitching ability is stronger than that between intimidation and batting ability.

However, because we had a much larger sample in the batting study ($n = 200$) than the pitching study ($n = 20$), the ***t* statistic** is larger for the former ($t = 3.00$) than the latter ($t = 2.72$). Furthermore, the t distribution for 198 **degrees of freedom** is closer to the **normal distribution** than the t distribution for 18 df. These two factors combine to produce a lower p-value for the batting data ($p = .003$) than the pitching data ($p = .014$).

In fact, if the **sample size** is large enough, nearly any correlation will be statistically significant. For example, with 3500 subjects, an r of .05 is significant at the $p < .005$ level! In such a relationship, the X variable would account for only 0.25% of the **variance** in Y (that is, $r^2 = .0025$), so we would hardly call this a strong correlation.

The general lesson here is the same one we've hammered home a number of times before: The fact that a result is highly **significant** (i.e., that it has a very small p-value) means that there is very little chance that there is no effect at all. But it doesn't necessarily mean that the effect is large.

Since the value of t is completely determined by the values of r and n (and df is always equal to $n - 2$), researchers often simply report r, n, and p when describing a correlational result. This practice is fine, as long as it is remembered that r is really a descriptive, not an inferential, statistic. Thus to report the results of our intimidation studies, we might write:

> "The correlation between intimidation ratings and ERA for pitchers was strong and significant, $r = -.53$, $p = .014$ with $n = 20$. The relationship between intimidation and BA for batters was weaker, $r = .29$, but was also significant, $p = .003$ with $n = 200$."

The correlation coefficient r, not the p-value, is the proper gauge of relationship strengths.

With a large enough sample size, nearly any correlation will be statistically significant.

Exercise 11.6 and Exercise 11.7 take you through the steps of conducting a hypothesis test for a correlation analysis and expand on our discussion of how to interpret the results of such hypothesis tests.

11.5.2 Validity and Reliability

In addition to prediction and theory testing, there are two additional uses for **correlation** that we need to cover: testing **validity** and testing **reliability.**

Validity refers to the question of whether or not two different variables are measuring the same thing. For example, the Beck Depression Inventory (BDI) is a widely used psychological measure of clinical depression (scores on the BDI can range from 0 to 63; a score higher than 30 is considered to be indicative of severe depression). But the BDI is copyrighted, and the company holding the copyright charges researchers a certain amount of money per subject to administer it.

Suppose you're planning a massive study of depression, and want to avoid the costs associated with using the BDI for all of your subjects. You might develop your own depression measure and administer it along with the BDI to a small subset of people from your planned subject pool. If the correlation between scores on your measure (the X variable) and the BDI scores (the Y variable) for the pilot subjects is high (say, $r > .8$), you can make a reasonable claim that your test is a valid measure of depression (or at least a valid measure of those aspects of depression the BDI measures).

Reliability refers to the question of how consistent a measurement procedure is across different administrations. The simplest measure of the reliability of a behavioral science test is to simply give the test twice to the same subjects. If the test is reliable, there should be a very high correlation between the scores from the first time the test was administered (this is the X variable) and the scores from the second administration of the test (the Y variable).

The concepts of validity and reliability are easy to confuse, since both are concerned with the "quality" of a test. To keep them straight, remember that validity involves two different tests (the new measure and the previously established measure), whereas reliability involves two administrations of the same test.

11.6 Multiple Regression/Correlation

A student of the game of baseball might make the following objection to our study of intimidation and pitching ability from **Section 11.5:** Although it is true that intimidating pitchers like Bob Gibson

Glossary Term: validity

Glossary Term: reliability

are very effective, these pitchers are intimidating primarily because they throw the ball very fast. And it may be the speed of a pitcher's pitches, not intimidation per se, that leads to ow ERAs.

This new argument amounts to a proposition that there is a **lurking variable** (pitch speed) in the causal model relating intimidation and ERA. To test this proposition, we can collect data for the new variable and apply a statistical technique called **multiple regression/correlation** (MRC). The "multiple" part of the technique's name comes from the fact that we have multiple X variables that all help explain the variance in Y.

The new variable we need for the analysis we'd like to do here should indicate our 20 pitchers' velocities. Luckily, pitch speeds are regularly measured these days, so we could easily add this variable to our dataset even after the other two variables were collected. In the table below we've added average pitch speeds to our previous data on intimidation and ERA.

Intimidation	Pitch Speed	ERA	Intimidation	Pitch Speed	ERA
3.1	78.7	5.05	6.6	88.2	3.00
3.6	72.2	3.72	2.3	76.5	4.71
3.3	77.4	3.62	5.4	89.0	3.19
2.9	76.8	4.55	6.8	85.5	2.39
1.8	75.4	2.96	6.0	91.4	2.19
3.2	78.3	5.59	2.5	73.2	5.49
1.1	72.3	4.67	2.1	76.4	4.78
2.5	72.2	4.29	3.1	77.9	4.90
3.4	78.1	3.63	3.8	80.4	2.58
6.0	84.0	4.43	5.3	82.7	4.75

To ease discussion, we'll refer to these **variables** below by the abbreviations INTIM (intimidation), SPEED (pitch speed), and ERA (earned run average). The first thing we can look at are the bivariate **correlations** (that is, the r values) for each variable pairing:

Variable Pair	r
INTIM–ERA	−.53
SPEED–ERA	−.56
INTIM–SPEED	.88

Note that the INTIM–ERA correlation ($r = -.53$) is the one we were considering previously—the strong negative **relationship** between intimidation and ERA. The SPEED–ERA correlation indicates that there is a slightly stronger correlation ($r = -.56$) between pitch speed and pitching ability: the faster a pitcher pitches, the fewer runs he tends to give up. Finally, the INTIM-SPEED correlation is also strong (and positive, $r = .88$). As suspected, harder throwers are more intimidating to batters.

This set of bivariate correlations is consistent with any of the causal models presented in Figure 11.18:

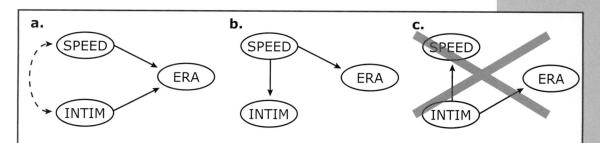

Figure 11.18 Three possible models for the causal relationships between baseball pitchers' INTIMidation, average pitch SPEED, and ERA (earned run average).

a. Both SPEED and INTIM may have direct effects on ERA.

b. INTIM may have no direct effect on ERA; instead, INTIM may simply be a side–effect of SPEED, which does have a direct effect on ERA.

c. Based on the statistics alone, it is possible that INTIM could be the cause of both SPEED and ERA. However, it seems extremely unlikely that being intimidating could cause a pitcher to throw faster. Therefore, we've crossed this diagram out in the figure.

Causal models involving three variables

There are actually even more causal models that we could consider, but we'll limit ourselves to (a) and (b) in the discussion below. On the next page, we'll get a glimpse of how an MRC analysis can help distinguish between models (a) and (b).

Interactive
Page 295

A simple MRC analysis of our pitcher data, including the variables SPEED, INTIM, and ERA, could proceed as follows. In the first stage of the analysis, we enter the SPEED variable into the **regression** equation, and calculate a regression weight b and a Y-intercept a that we could use to predict a pitcher's ERA from his pitch speed (b comes out to $-.088$, and $a = 7.22$). We enter SPEED into the equation first because it cannot be caused by any other variable under consideration. (Of course, there are other variables, such as arm strength, height, and weight, which may be causes of pitch speed, but we don't have data for any of these variables.)

At this point, the multiple regression equation looks just like the simple (bivariate) regression equation we learned about in **Section 11.4:**

$$ERA = 7.22 + (-.088)(SPEED)$$

We can also calculate r^2, the proportion of variance in ERA accounted for by SPEED. For our data, $r^2 = .315$.

In the second stage of the **MRC** analysis, we enter our other X variable, INTIM, into the regression equation. We recalculate the regression weight for SPEED, and add a second weight for INTIM. (Don't worry—you will NOT be expected to know how to calculate multiple regression equations. Indeed, few humans actually know how to perform these calculations, since computers are so much better at doing them.) Now the regression equation looks like this:

$$ERA = 5.59 + (-.063)(SPEED) + (-.097)(INTIM)$$

Given this new equation, we can calculate R^2, the proportion of **variance** in ERA accounted for by SPEED and INTIM taken together. As it turns out, R^2 is .323, almost identical to r^2 for the initial regression equation.

The conclusion we can draw from this analysis is that INTIM is adding almost no predictive power for ERA over and above the predictive power provided by SPEED. In other words, the analysis supports the causal model in Figure 11.19b over the model in Figure 11.19a: The effect of intimidation is nothing but a side-effect of pitch speed.

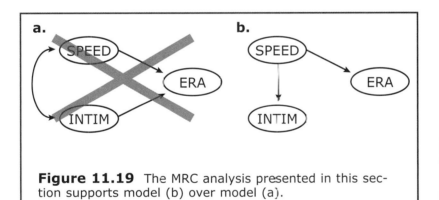

Figure 11.19 The MRC analysis presented in this section supports model (b) over model (a).

Interactive
Page 296

11.6.1 The General Linear Model

If MRC is too simplistic for your tastes, there are a host of more complicated statistical analyses you could try to master. These analyses, in addition to the ones we've learned about already, can all be considered variations on a kind of über-theory of statistics called the General Linear Model (GLM).

Glossary Term: GLM

The GLM seeks to assess the **relationship** between any number of X **variables** with any number of Y variables. Different variations of the GLM arise from different combinations for the number and types of X and Y variables. There are essentially three types of variables:

1. **Discrete** variables, which can take two or more qualitatively different values (e.g., religion: Catholic, Protestant, Muslim, Hindu, etc.).

2. **Dichotomous** variables, a special class of discrete variables that can take only two possible values (e.g., sex: male or female).

3. **Continuous** variables, which can take any numeric value within a certain range (e.g., age).

On the next page, we'll summarize how most of the analyses we've learned about up to now fit into the GLM framework. Then we'll very briefly describe some more advanced analyses. The goal here will not be to teach you how to do these analyses, but to let you know what they are so you won't be completely mystified if you see them mentioned in a research article.

Interactive Page 297

Here's how the analyses we've learned about in this course, as well as the more advanced analyses we promised to let you in on, fit into the GLM framework.

The simplest analyses have one *X* **variable** and one *Y* variable:

- When the *X* and *Y* variables are both **continuous,** we have **bivariate regression/correlation,** the subject of most of this chapter.

- When the *X* variable is **dichotomous** and the *Y* variable is continuous, we have a ***t* test** (**Section 8.3**). If the *X* variable is coded into 0's and 1's (e.g., males are coded as 0 and females as 1), we can analyze the same data via the bivariate correlation coefficient *r*. This analysis is called a **point-biserial correlation,** and the significance level of a point-biserial correlation will always be identical to that of the independent-samples *t* test for the same dataset. Thus an independent-samples *t* test can be considered a special case of a bivariate correlation analysis.

- When the *X* variable is **discrete** and the *Y* variable continuous, we have an **ANOVA** (**Chapter 10**).

- When both *X* and *Y* are discrete, we have a **chi-square test,** which we'll cover in the next chapter (**Chapter 12**). For the special case where *X* and *Y* are both dichotomous, we have a **sign test,** which we'll also cover in the next chapter (**Section 12.3**).

The next level of complexity arises when we add multiple *X* variables.

- If all the *X*'s are discrete and the *Y* variable is continuous, we have **multiple-factor ANOVA** (**Section 10.7**), while if the *X*'s

Glossary terms: discrete, dichotomous, continuous

are continuous, we have **multiple regression/correlation (MRC)**. Just as we can recode dichotomous variables into 0's and 1's to analyze them using bivariate regression/correlation, we can recode discrete variables with more than two levels and analyze the data using multiple regression/correlation. Thus ANOVA can be considered a special case of MRC.

- If we have a combination of discrete and continuous X variables with a continuous Y variable, we can use **analysis of covariance (ANCOVA),** or we can use an equivalent MRC analysis.

- If the Y variable is discrete and the X variables are continuous, we can use **discriminant function analysis** or **logistic regression.**

- If the X variables and the Y variable are all discrete, we can use **multiway frequency analysis** (chi-square and sign tests are special cases of multiway frequency analyses).

The most complex analyses have multiple X variables and multiple Y variables. These get a little hairy.

- The most general form of this type of analysis is called **canonical correlation,** in which multiple continuous X variables explain variance in multiple continuous Y variables.

- If all the X variables are discrete and the Y variables are continuous, we have **multivariate analysis of variance (MANOVA).** ANOVA can be considered a special form of MANOVA, and MANOVA, in turn, can be considered a special form of canonical correlation. **Profile analysis** is a special subclass of MANOVA when all the Y variables are measured on the same scale.

- If some of the X variables are discrete and some continuous, we can use **multivariate analysis of covariance (MANCOVA),** of which ANCOVA is a special form. Canonical correlation is the grand-daddy of MANCOVA.

Finally, there are two more procedures, called **principle components analysis** and **factor analysis,** that we should mention in this discussion. In these analyses, we start with a large set of continuous variables, with no distinction made between X's and Y's. We assume that there is a relatively small number of unmeasured variables underlying these measured variables. These unmeasured variables are the principle components, or factors, and the goal of the analyses is to determine what these components or factors are.

11.7 Chapter Summary/Review

This Chapter Summary/Review is interactive on the _Interactive Statistics_ website and CD. Also be sure to go through all the **Review Exercises** for this chapter.

- **Scatterplots** graphically illustrate the relationship (covariability) between two variables (**Section 11.1**), which can be characterized by three factors (**Section 11.2**):

  ◦ Form—the shape of the **relationship** in the scatterplot, which can generally be described as **linear** or **curvilinear**

  ◦ Direction—positive, meaning that the values of both variables increase together; or negative, meaning that the value of one variable increases as the other decreases

  ◦ Degree—the extent to which the variables are related, from completely unrelated to perfectly related (i.e., one variable's value is completely predictable on the basis of the other variable's value).

- The Pearson **correlation** r measures the direction and degree of linear relationship. One way to calculate r is to average the products of the z-scores for every pair of datapoints: $r = \Sigma z_X z_Y / n$ (**Section 11.3**). The value of r can vary from −1.00 (indicating a perfect negative relationship) through .00 (indicating no relationship at all) to +1.00 (indicating a perfect positive relationship).

- Variable pairs with correlations greater than about .60 or less than about −.60 will be considered strong in nearly any context, while r values between −.05 and .05 almost always indicate that two variables are not meaningfully related at all. But even very strong correlations cannot prove that one variable has a causal relationship to the other—instead, the causal relationship could be reversed, or a third lurking variable could be the source of the correlation (**Section 11.3.2**).

- A linear relationship can be compactly described by a linear regression line of the form $Y = a + bX$. The a term represents the Y intercept of the line, and is calculated as $\bar{Y} - b\bar{X}$. The b term, which is usually of more interest to researchers, represents the slope of the line and can be calculated from the r value: $b = r(s_Y / s_X)$ (**Section 11.4**).

- Unless the relationship between two variables is absolutely perfect, a **regression line** will not perfectly predict values of Y given new values of X. The accuracy of prediction can be assessed with the discrepancy between predicted Y values and $\hat{Y}$ for the datapoints on which the regression equation was originally based (**Section 11.4.2**). Or, we can simply take the square of r. Thus r^2, which can vary from .00 to 1.00, reveals the proportion of the **variance** in Y that can be accounted for by variance in X (**Section 11.4.2**).

- **Outliers** can have large impacts on the values of r and b. If an outlier can be clearly identified, the relationship between two variables is usually more accurately described by removing it (**Section 11.4.3**). It is important not to extrapolate a regression equation beyond the range of values on which the equation was calculated (**Section 11.4.4**).

- Correlation/regression analyses are sometimes conducted in order to predict new Y values given new values of X (**Section 11.5**). In other circumstances, r values are used to test theories that predict that two variables should be related. The null **hypothesis** for this purpose states that the **population** correlation is 0 (i.e., that the variables are not related at all). A t statistic, which has $n - 2$ df can be calculated from r via the following formula:

$$t = \frac{r\sqrt{n - 2}}{\sqrt{1 - r^2}}$$

- The **p-value** associated with the calculated **test statistic** can be looked up and assessed in exactly the same way as in any other t-test (**Section 11.5.1**). It is important to remember that r, not p, is the best indicator of the strength of the relationship between two variables. With enough subjects, even a very small correlation can produce a very small p-value. Correlation analyses can also be used to assess the validity and/or reliability of an experimental measure (**Section 11.5.2**).

- A technique called multiple regression/correlation (**MRC**) can be used to assess the influence of two or more X variables on a single Y variable. MRC and other types of analyses based on the General Linear Model (GLM) are much too complicated to be discussed in detail in this textbook, but an overview of these techniques is given in **Section 11.6** and **Section 11.6.1**.

Chapter 12
Inference for Categorical Data: Chi-Square Tests

12.1 Categorical Data

Consider the following research project:

> Fred, a sociology researcher, is hired by the Springfield City Council to write a report on the characteristics of the people who use the city's bus system. One of the central questions Fred is asked to address is whether members of minority ethnic groups are disproportionately likely to use the buses.
>
> As part of his study, Fred asks a random sample of 80 city bus riders to fill out a short questionnaire.

One of the questions probes respondents about their ethnicity. Fred finds that 42 of the respondents were white, 24 were African-American, 12 were Hispanic, and 2 identified themselves as some other ethnic origin.

Does Fred's study support the hypothesis that members of minority groups are more likely than whites to ride Springfield buses? The first thing to note when addressing this question is that, like the research questions considered in earlier chapters, the city council is asking Fred to generalize from a **sample** (the 80 bus riders he surveyed) to a larger **population** (all riders of Springfield buses). Therefore, Fred will need an **inferential statistic** to help interpret the study.

The second thing to note is that Fred's data are unlike any that we've considered since **Section 2.2.5.** What we have here is a **categorical** dataset, in which each subject is placed into one category or another. For the past 9 chapters we've been concerned exclusively with **interval data,** where each subject contributes a numeric score to the dataset.

We cannot compute the mean of a categorical **variable.** And since the various forms of **z tests, t tests,** and **ANOVAs** all test **hypotheses** about **means,** we cannot use any of the inferential statistics covered in preceding chapters to analyze Fred's ethnicity data.

The *z*, *t*, and *F* tests cannot be used with categorical data.

However, we can assess the **frequency distribution** of a sample of categorical data (see the table below), and we can then construct and test hypotheses about the population distribution from which the sample was drawn. This chapter will describe the process for such hypothesis tests, which use a new **statistic** called χ^2 ("chi-square"; the first syllable rhymes with "pie"). **Section 12.2** shows how to form **null hypotheses** for chi-square tests and calculate χ^2, and walks through the hypothesis testing procedure for the most general form of χ^2 test, the **"goodness-of-fit" test.**

Frequency distribution of Fred's data:

	Ethnicity			
	White	**African-American**	**Hispanic**	**Other**
f	42	24	12	2

In the rest of the chapter, we will see how χ^2 can be used to test hypotheses in a wide variety of research situations, including **related-samples designs** (**Section 12.3**), studies with two or more **independent samples** (**Section 12.4**), studies in which the relationship between two categorical variables is assessed (**Section 12.5**), and opinion polls (**Section 12.6**).

Interactive
Page 301

12.2 Testing One Sample: The Chi-Square Goodness-of-Fit Test

Are members of minority groups disproportionately represented in Springfield's bus ridership? Just looking at the raw data from Fred's study, illustrated by the relative frequency bar graph in Figure 12.1a, it might appear that the answer to this question is "no": 42/80—greater than 50%—of the riders in Fred's **sample** were white.

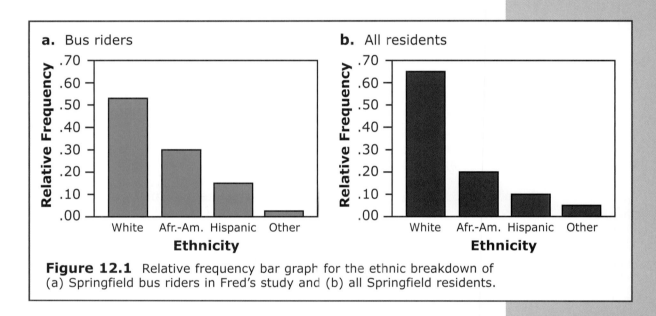

Figure 12.1 Relative frequency bar graph for the ethnic breakdown of (a) Springfield bus riders in Fred's study and (b) all Springfield residents.

But suppose the last U.S. census showed that in the city of Springfield as a whole, 65% of residents are white, 20% are African-American, 10% are Hispanic, and 5% are some other ethnicity. Comparing the bar graph in Figure 12.1b, which illustrates these percentages, with Figure 12.1a makes it clear that the **distribution** of ethnicities in Fred's sample of bus riders is quite different from the distribution of ethnicities in Springfield's general **population.**

However, it is always possible that the discrepancies between Fred's observations and the general population are due solely to sampling error (i.e., chance). To decide whether or not to believe this explanation, we turn to our old friend the hypothesis test. More specifically, we will use a test called the chi-square **goodness-of-fit test.**

Glossary Term: goodness-of-fit test

We will go through the same four steps for our chi-square tests as we have used in previous chapters. **Step 1** is to formulate **null** and **alternative hypotheses.** For Fred's data, H_0 will state that bus rid-

Steps for hypothesis testing with chi-square

ers really have the same ethnicity distribution as the general population. The H_a states that the two distributions are not the same. (Like omnibus ANOVAs, chi-square tests do not tell us how an experimental population differs from the comparison population; they only tell us that *something* is different.)

Step 2 of the **hypothesis** testing procedure is to collect and describe the sample data. As noted on the previous page, we cannot calculate summary statistics for **categorical** variables, but we can describe their distributions in a frequency table, repeated below for Fred's data:

	White	**African-American**	**Hispanic**	**Other**
f	42	24	12	2

Ethnicity

The χ^2 test statistic (the calculation of which is **step 3** of the hypothesis testing procedure) is designed to compare this set of **observed frequencies** (f_o's) to the expected frequencies predicted by the null hypothesis. (Terminology note: we will refer to each square in a frequency distribution table as a cell.) Thus f_o for the leftmost cell in the table above is 42.)

Given that 65% of all Springfield residents are white, how many of Fred's 80 bus riders should we expect to be white if our null hypothesis is correct and all ethnicities are equally likely to ride the bus?

Interactive Page 302

To calculate the **expected frequencies** for the **null hypothesis** in a **chi-square test,** we simply multiply the proportions predicted by H_0 for each **cell** by the **sample size** n. For example, we would expect to see $.65 \times 80 = 52$ of Fred's respondents to be white. The complete f_e calculations for Fred's study are shown in the second row of the table below, just below the f_o (observed frequency) values.

	White	**African-American**	**Hispanic**	**Other**
f_o	42	24	12	2
f_e	.65(80) = 52	.20(80) = 16	.10(80) = 8	.05(80) = 4
$f_o - f_e$	−10	8	4	−2
$(f_o - f_e)^2$	100	64	16	4
$(f_o - f_e)^2 / f_e$	1.92	4.00	2.00	1.00
$\Sigma((f_o - f_e)^2 / f_e) = \chi^2 = \mathbf{8.92}$				

Ethnicity

The other rows in the table above illustrate how χ^2 is calculated: For each cell in the **frequency distribution** table, we subtract f_e from

Glossary Terms:
observed frequencies,
expected frequencies,
cell

Calculating chi-square

f_o, square this difference, and divide by f_e. Summing across all cells in the table, we get χ^2, 8.92 in this case. The formula is:

$$\chi^2 = \Sigma \frac{(f_o - f_e)^2}{f_e}$$

Like the t and F statistics discussed in **Chapter 6** and **Chapter 10**, distributions of all possible χ^2 values can be mathematically defined, with the exact shape of each χ^2 distribution dependent on the **degrees of freedom** involved in calculating the test statistic. For a chi-square **goodness-of-fit test,** χ^2 has $C - 1$ df, where C is the number of columns in the frequency table.

Figure 12.2 shows the χ^2 distributions for 1, 3, and 10 df. As you can see, these are reminiscent of F distributions, in that the χ^2 distributions are bounded on the left by 0 (since the numerator of the χ^2 formula is a squared value, negative values are impossible) and positively skewed.

Chi-square distributions

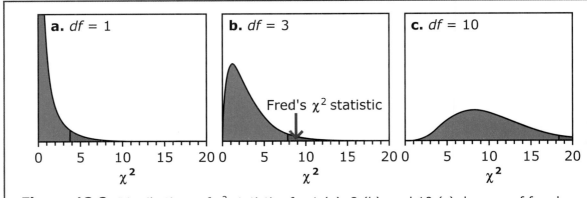

Figure 12.2   Distributions of χ^2 statistics for 1 (a), 3 (b), and 10 (c) degrees of freedom.

Step 4 of the hypothesis testing procedure demands that we evaluate H_0 by comparing the χ^2 statistic calculated from our data to the appropriate χ^2 distribution. We can thereby assess the likelihood of randomly drawing a **sample** that produces a χ^2 value as large as the one we observed. If this probability value is low enough, we conclude that H_0 is probably incorrect.

The arrow in Figure 12.2b shows the location of Fred's χ^2 value in the χ^2 distribution for $4 - 1 = 3$ df. The **Chi-Square Distribution Area** Calculation Tool will provide the exact **probability value** associated with any combination of χ^2 and df. The use of such Tools should be old hat to you by now, so see if you can look up the p-value for Fred's study yourself:

Calculation Tool: Chi-Square Distribution Area

(If you have to find the p-value for a χ^2 statistic and don't have access to a computer, you may need to use a statistical table. The procedure for looking up p-values in a χ^2 Distribution Table is exactly the same as that for looking up p-values for t statistics, and is described near Figure 3 in the **Box** on statistical tables on p. 89.)

Box: Statistical Tables

According to the **Chi-Square Distribution Area** Calculation Tool, the probability of randomly drawing a **sample** that produces a χ^2 value of 8.92 with 3 **df** is only .03. Since this **p-value** is less than .05, the traditional **alpha level** used in most behavioral science research, Fred can claim a statistically **significant** difference in ethnicity distributions between bus riders and the general **population.**

As with other inferential statistics, the end result of a chi-square test is a p-value.

And that's all there is to the **chi-square goodness-of-fit test**! To make sure you understand the procedure, let's go through another example:

> Five candidates are running in the Democratic primary for Springfield's state senate district. *The Springfield Post* conducts a poll in which 250 registered Democrats in the city are asked who they will vote for. The percentages of respondents who say they will vote for each candidate are:

Candidate	Support
Able	26%
Burns	18%
Carter	20%
Dodd	16%
Evans	20%

Does this poll provide sufficient evidence to claim that any one or more of the candidates will receive more votes than any of the other candidates? This question can be answered with a **hypothesis test** in which H_0 states that in the entire population, all candidates have equal support. This null hypothesis corresponds to **expected frequencies** of $(1 / 5) \times 250 = 50$ for each table **cell.** The chi-square calculations thus look like this:

	Candidate				
	Able	**Burns**	**Carter**	**Dodd**	**Evans**
f_o	.26 (250) = 65	.18 (250) = 45	.20 (250) = 50	.16 (250) = 40	.20 (250) = 50
f_e	.20 (250) = 50	.20 (250) = 50	.20 (250) = 50	.20 (250) = 50	.20 (250) = 50
$f_o - f_e$	15	−5	0	−10	0
$(f_o - f_e)^2$	225	25	0	100	0
$(f_o - f_e)^2 / f_e$	4.5	0.5	0	2.0	0
$\Sigma((f_o - f_e)^2 / f_e) = \chi^2 = \mathbf{7.0}$					

With 5 – 1 = 4 *df*, a χ^2 value of 7.0 corresponds to a *p*-value of .14, so no significant differences can be claimed amongst the five candidates.

This example makes especially clear the foolhardiness of "accepting" the **null hypothesis** in a chi-square (or any other) hypothesis test. The chances that the five candidates all have exactly equivalent support are practically nil. In other words, *someone* is bound to win the election, so we *know* the null hypothesis is false. The fact that H_0 could not be rejected should only be taken to mean that the poll does not have enough power to reliably predict who will win.

Interactive
Page 304

12.2.1 A Closer Look at the Chi-Square Statistic

It's not at all obvious, but the calculation of a χ^2 **test statistic** actually has quite a lot in common with the calculation of a *t* test statistic. The basic formula for *t* is:

$$t = \frac{\bar{X} - \mu}{s_{\bar{X}}}$$

The numerator of this formula measures the discrepancy between the observed data (summarized by the **sample mean** $\bar{X}$) and the value predicted by the **null hypothesis** (summarized by the **population mean** μ according to H_0).

The denominator of the *t* formula (the **estimated standard error,** $s_{\bar{X}}$) provides an indication of how much variation between the observed and predicted values we should expect by chance. If $s_{\bar{X}}$ is large, random factors alone could produce a large difference between $\bar{X}$ and μ, so $\bar{X} - \mu$ must be *very* large to produce a large *t* value. But if the denominator of the *t* formula is small, even a moderate difference between $\bar{X}$ and μ will produce a large *t* value, indicating that something other than chance was at work in the experiment.

Now look again at the χ^2 formula:

$$\chi^2 = \Sigma \frac{(f_o - f_e)^2}{f_e}$$

Just like with the *t* formula, the numerator here assesses the difference between the data we observed (f_o) and the values predicted by H_0 (f_e). Positive and negative difference scores are both equally indicative of an error in the null hypothesis, so we square the numerator of the χ^2 formula to force all the summed values to be positive.

The denominator of the χ^2 formula (f_e) also parallels the *t* formula, by scaling χ^2 to the amount of variation we should expect by chance. The best way to see why is through a quick example.

Suppose we ask 50 people whether they prefer to drink Pepsi or Coca-Cola. If 20 more people choose Pepsi than choose Coke in this

Never claim that the null hypothesis is true.

Parallels between the chi-square formula and the *t* formula

relatively small **sample,** it would represent a pretty strong preference. Now suppose instead that we polled 5000 people. In this survey, it would not be surprising at all to find 20 more people choosing Pepsi, even if both sodas are actually equally preferred.

However, if the chi-square statistic were calculated by simply adding the squared differences between f_o and f_e, we would get the same values for χ^2 in both of these surveys: for the first, we would calculate $(35 - 25)^2 + (15 - 25)^2 = 100 + 100 = 200$; for the second, we would calculate $(2510 - 2500)^2 + (2490 - 2500)^2 = 100 + 100 = 200$.

Putting f_e in the denominator of the χ^2 formula solves this problem. In the first poll, we divide by 25, whereas in the second we divide by 2500. Thus χ^2 comes out to 8.0 in the former (resulting in a p-value of .005) and 0.08 in the latter ($p = .78$).

12.2.2 Assumptions for Chi-Square Tests

Before exploring **chi-square tests** further, we should note three **assumptions** that must be fulfilled in order for any test using the χ^2 statistic to be valid:

1. As with every other inferential statistic we've discussed, the subjects whose data are to be analyzed must be drawn at random from the **population** under investigation (see **Section 6.4.1**). This assumption is violated if some subjects' responses might be influenced by other subjects' responses.

   Say, for example, that you're at a PTA meeting with 40 other parents. The principal of the school gets up and says that, due to budget cuts, either the physical education, music, or art program at the school is going to have to be eliminated. She then polls the audience, asking each parent to raise their hand to indicate which program they think should go.

   While the principal may be able to learn something from this "study," she cannot treat the data as a random **sample** of all the parents in her school. Parents who are undecided may be swayed by seeing the choices that other parents make. Therefore, it would not be valid to conduct a chi-square test on these data. (Also, the parents who choose to attend the PTA meeting are not necessarily representative of all the parents in the school.)

2. Every subject in the study must be placed in only one **cell** of the **frequency distribution** table. For example, suppose the principal follows up her PTA survey with a questionnaire mailed out to a truly random sample of her school's parents. The first 25 parents who respond check one option, stating a preference to cut physical education, music, or art. But in the 26th response, the parent checks both music and art. The principal *cannot* code this

Subjects must be randomly sampled from the population in question.

Each subject must be placed in only one cell of the frequency table.

response by adding one to both the music cell and the art cell, because this would violate the assumption for chi-square tests.

Unless this parent provided some other information (perhaps an explanatory note) indicating which program he would most prefer to cut, the principal should probably throw out the response and not count it at all in the data. (In scientific research, an investigator should always note and explain such data anomalies.)

3. The final requirement for chi-square tests is that the **expected frequencies** must not be too small. Since expected frequencies are calculated by multiplying the sample size by the proportion predicted by the **null hypothesis** for each cell, this assumption is similar to the requirement that n be greater than 15 for t tests to be valid when the underlying population is not normally distributed (**Section 6.4.2**).

> The expected frequencies must average 5 or greater, with none below 1.

So how small is "too small?" Different statisticians give different answers to this question, but the one we will use is this: *Chi-square tests should only be used when the average of the expected frequencies is 5 or greater, and when no cell has an expected frequency of less than 1.* So if you have two cells in your table and the expected frequencies are 6 and 4, you're OK (but just barely). If you have four cells and the expected frequencies are 3, 2, 5, and 6 (averaging out to 4), you shouldn't do a chi-square test on the data.

You should go through **Exercise 12.1** to make sure you understand the calculation and interpretation of goodness-of-fit tests before going on to our discussion of other χ^2 tests in the rest of the chapter.

Interactive Page 306

12.3 Testing Two Related Samples: The Sign Test

The **goodness-of-fit test** is usually used in situations in which a single **sample** of subjects is tested. However, a variation on this test, called the **sign test,** can be used in experimental designs with two **related samples.** Consider the following:

> Glossary Term: sign test

> Lilly, a biological psychologist, is investigating the function of a small section of the rat brain which is believed to be analogous to a particular part of the human brain. She conducts an experiment in which 16 rats are trained to run through a maze as quickly as possible. After a week of training, Lilly surgically removes the section of the rats' brains that she is interested in, then retests them on the maze. The table below shows the rats' performance (their times in seconds) on the last trial before surgery and the first trial after surgery, along with the post-test minus pre-test difference scores:

Subject	Pre-surgery	Post-surgery	Difference
1	15.9	33.0	−17.1
2	18.5	38.0	−19.5
3	20.9	—	—
4	23.2	37.7	−14.5
5	22.1	27.3	−5.2
6	20.6	23.7	−3.1
7	19.8	—	—
8	21.6	18.1	3.5
9	23.2	27.3	−4.1
10	23.4	41.4	−18.0
11	22.8	21.9	0.9
12	19.9	27.0	−7.1
13	19.4	23.9	−4.5
14	23.7	25.4	−1.7
15	19.3	—	—
16	22.6	29.9	−7.3

As indicated in the table, three rats (numbers 3, 7, and 15) were unable to make it through the maze at all after the surgery. These rats' performance clearly suffered, but since we have no numerical score for them, we cannot calculate a mean score for all 16 rats, and we can't use a **related-samples *t* test** to analyze the results.

What we can do, however, is divide the 16 rats into two categories: those who went through the maze faster after the surgery than before the surgery, and those who were slower after the surgery than before the surgery. Rats 8 and 11 improved (they were faster post-surgery than they were pre-surgery), while the rest of the rats, including the ones who were not able to complete the maze at all, showed a decline in performance. In table form, we have:

Categorizing data from related samples as positive or negative difference scores

Performance Improved	Performance Declined
2	14

We can now conduct a goodness-of-fit test on the frequency data in this table. All we need is a set of f_e's. What values should we expect to see in these two **cells** if the surgery actually had no effect on the rats' maze-running abilities?

Interactive Page 307

If 16 rats are tested before and after surgically removing a part of their brains, and if the removal actually has no effect whatsoever on their maze-running abilities, then we would expect 8 of the rats to do slightly better after the surgery, and the other 8 to do slightly worse.

The null hypothesis in a sign test always predicts an equal number of improvers and decliners.

(You might think that the null hypothesis predicts that each rat will do exactly the same before and after the surgery. But if we could get measurements of the rats' maze-running times that were accurate to .00000001 seconds, surely no two times would be identical. So H_0 predicts that *on average,* the rats will be identically fast before and after surgery, but that some individual rats will be slightly faster and some slightly slower.)

Armed with these f_e's, along with the f_o's of 2 and 14 from the previous page, we can go on to compute the χ^2 statistic:

$$\chi^2 = (2 - 8)^2 / 8 + (14 - 8)^2 / 8 = 36/8 + 36/8 = 9.0$$

The *p*-value associated with this χ^2 value (with 1 *df*) is .003, so on the basis of this **sign test,** Lilly can soundly reject the hypothesis that the surgery had no effect on the rats' maze-running abilities.

The sign test can also be used as an alternative to the **related-samples *t* test** when **assumptions** for using the *t* statistic are violated (see **Section 8.4**). For example, if the **distributions** of scores in an experiment are severely **skewed,** a *t* test may not be appropriate. But the chi-square test can be used as long as **_n_** is greater than 10 (if *n* is smaller than 10, the **expected frequencies** for the two **cells** in the sign-test table will be less than 5, violating an assumption of the χ^2 statistic).

Note, however, that the sign test is generally less powerful than the related-samples *t* test. In other words, for an experimental manipulation that has a real effect, we are more likely to be able to reject H_0 with a *t* test than with a sign test.

Sign tests are less powerful than *t* tests.

Many textbooks present the sign test as a version of the binomial test, which we will discuss in **Section 12.6.** The sign test is also formally equivalent to a statistic called the Φ ("phi") coefficient, a version of the Pearson **correlation** that can be used with **categorical** data.

Interactive Page 308

Before we leave the **sign test,** we should discuss one tricky aspect of this test: what to do when a tie occurs. Suppose we ask 13 randomly selected political science professors to rate on a 1–9 scale how effective they thought Ronald Reagan and Bill Clinton were as presidents. Since we suspect the ratings will not be normally distributed, we want to use a sign test to analyze the data. The subjects provide the following scores:

Clinton	2	3	3	4	3	4	6	6	6	4	3	8	5
Reagan	5	6	5	6	5	5	7	7	7	4	3	6	2
Difference	−3	−3	−2	−2	−2	−1	−1	−1	−1	0	0	+2	+3

Dividing the pairs of observations into negative and positive **difference scores** (where a negative score means the professor thought Reagan was the more effective president, and a positive difference score means the professor thought Clinton was more effective), we find 9 of the former and 2 of the latter. But what do we do with the 2 professors who gave equal ratings to the two presidents (i.e., the difference scores of 0)?

When a professor gives the same rating to Reagan and Clinton, it really means that their opinions are too similar for us to be able to measure the difference using our crude 9-point scale. It is very unlikely that anyone's opinion of the two presidents is *exactly* the same. The problem is that we don't know which way the people with 0 difference scores are leaning.

Excluding 0 difference scores in a sign test

The **liberal** solution to this problem (that's "liberal" in the statistical sense, not the political sense!) is to reason that if we can't tell whether a person should really be classified as a positive or negative difference score, we should simply exclude that person from the analysis. This approach leads to the following table:

− Difference	+ Difference
2	9

For this table, the **expected frequencies** are 5.5 for both **cells.** Although there were 13 subjects originally, we threw two out, so the f_e's are both $(.5)(11) = 5.5$. The χ^2 statistic is thus 4.45 with 1 *df*, and $p = .035$. By analyzing the data this way, we can make the claim that significantly more political science professors think Reagan was a more effective president than Clinton.

Splitting 0 difference scores evenly between improvers and decliners in a sign test

The **conservative** approach is to assume that half the people with 0 difference scores really prefer Reagan, and half really prefer Clinton. Thus we divide these subjects equally between the two cells in our sign test table (if there were an odd number of 0 difference scores, we would throw out one of them and divide the rest equally between the two cells):

− Difference	+ Difference
3	10

Here, the expected frequencies are 6.5 for both cells, $\chi^2(1) = 3.77$, and $p = .052$. Analyzing the data this way, we cannot claim to be able to distinguish whether political science professors think Reagan or Clinton was more effective.

This is one of those situations in which the presumed objectivity of statistical analysis breaks down. Neither of the two approaches (excluding 0's or dividing them equally) is universally accepted as "right," so every individual researcher must make a subjective decision about which approach they feel is most appropriate. And, as we've seen in this example, this subjective decision can sometimes determine the supposedly objective decision for whether or not to reject the **null hypothesis.**

Exercise 12.2 asks you to do a sign test yourself, and notes some additional subtleties to keep in mind when interpreting sign tests.

Interactive
Page 309

12.4 Testing Independent Samples: The Test for Independence

Now that we've seen how to use the **goodness-of-fit test** as an analog to single-sample designs and the **sign test** as an analog to **related-samples designs,** let's look at how we can use the χ^2 statistic to test **hypotheses** in experimental designs with two or more **samples** of subjects. For example:

Chi-square tests involving multiple independent samples

> Harold, a developmental psychologist, is investigating how styles of play differ for 3-year-old boys and girls. In his experiment, he brings each child into a room with three play stations set up. One station has a set of blocks, one has a collection of toy cars, and the third features a dollhouse. The child is told that he or she can play at one and only one station. His or her choice is recorded, the child is allowed to play for 15 minutes, and then he or she leaves and the next child is brought in. The children's choices are as follows:

Boys:	Cars,	Cars,	Blocks,	Dolls,	Cars,	Cars,	Cars,	Dolls,
	Cars,	Blocks,	Dolls,	Dolls,	Cars,	Cars,	Dolls,	Cars,
	Cars,	Blocks,	Blocks,	Cars,	Cars,	Cars		
Girls:	Dolls,	Dolls,	Dolls,	Dolls,	Cars,	Cars,	Dolls,	Dolls,
	Blocks,	Dolls,	Blocks,	Dolls,	Dolls,	Dolls,	Cars,	Cars,
	Dolls,	Cars,	Blocks,	Dolls				

Harold has not collected any numerical data in this experiment, so he cannot use a **t test** to assess the differences between boys' and girls' toy choices. However, he can assess the **frequency distributions** of the choices made by the two samples. To do so, he should organize his data in a 2 (row) × 3 (column) table, which will look like this:

	Chosen Toy		
	Blocks	**Cars**	**Dolls**
Boys	4	13	5
Girls	3	5	12

As the relative frequency bar graph in Figure 12.3 shows, the **distributions** of toy choices did vary between Harold's **sample** of boys and his sample of girls. (Bar graphs are essential to understanding the results of any experiment in which the data are divided into three or more categories.) But is the variation in distributions greater than we would expect if chance processes alone were responsible for the differences between the two samples?

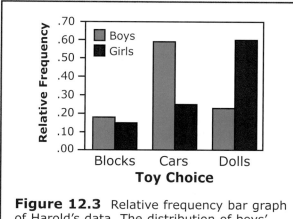

Figure 12.3 Relative frequency bar graph of Harold's data. The distribution of boys' toy choices is shown by the green bars and the distribution of girls' choices by the purple bars.

To answer this question, we need to test the **null hypothesis** that the two population distributions (the distribution of toy choices for all boys and girls) are identical. In other words, this H_0 says that toy choices are **independent** of (i.e., unrelated to) the sex of the child. For this reason, the **hypothesis test** is called a **test for independence.**

Note that H_0 in a test for independence does *not* assert that all three toys were chosen with equal probability. Instead, it holds that however likely it is for boys to choose the blocks (for example), girls should choose the blocks just as often. This makes calculating the **expected frequencies** a bit trickier than in a **goodness-of-fit test.**

The first thing we need to do to calculate the f_e's is to compute the column frequencies and row frequencies for our **observed frequency** table. A column frequency (abbreviated f_c) is just what it sounds like: the total frequency count for a column in the table, collapsed over all the rows. Likewise, a row frequency (f_r) is the total frequency count for a row collapsed over all the columns.

Thus the column count for block choices (the first column in Harold's table) is 4 (the number of boys who chose the blocks) plus 3 (the

Glossary Term: test for independence

The null hypothesis in a test for independence predicts equivalent category distributions for all populations.

Column and row frequencies

number of girls who chose the blocks) = 7. Filling in the rest of the column frequencies and the row frequencies, we end up with the following table:

Observed Frequencies:

	Blocks	Cars	Dolls	f_r
Boys	4	13	5	22
Girls	3	5	12	20
f_c	7	18	17	42

We can now use the column and row frequencies to compute the f_e for the top-left **cell** in the table by going through the following two logical steps:

1. If 3-year-old boys and girls really do choose toys in the same way, then our best estimate for the proportion of children who choose blocks is the column frequency f_c for this toy choice (7) divided by the total number of children in the study N (42) = .167. (Note that you can calculate N in three different ways: by counting all the subjects in the study, by summing all the column frequencies, or by summing all the row frequencies. All three methods should arrive at the same value.)

2. The row frequency f_r for boys is 22. If the null hypothesis is right, .167 of these boys should choose the blocks. Thus f_e for the top-left cell is .167(22) = 3.67.

Putting these two steps together, we arrive at a simple formula for calculating f_e's in a test of independence:

$$f_e = (f_c \, / \, N) \times f_r = f_c f_r \, / \, N$$

Calculating expected frequencies in a test for independence

Applying this formula to the bottom-left cell of Harold's table, we find that f_e = (7)(20) / 42 = 3.33 for girls choosing blocks. Filling in the rest of the expected frequencies, we get:

Expected Frequencies:

	Blocks	Cars	Dolls	f_r
Boys	3.67	9.43	8.90	22
Girls	3.33	8.57	8.10	20
f_c	7	18	17	42

(Note that the column and row frequencies for the f_e table are identical to those of the f_o table. This should always be the case, so adding up the f_c's and f_r's for your expected frequency table is a good way to check your f_e calculations.)

The calculation of χ^2 is exactly the same for the test for independence we want to do here as it was for the **goodness-of-fit test**: for each **cell** in the table, we subtract f_e from f_o, square this difference, and divide by f_e. Then we sum these values over all cells to get χ^2. Here's the formula again:

Chi-square is calculated identically in goodness-of-fit tests and tests for independence

$$\chi_2 = \Sigma \frac{(f_o - f_e)^2}{f_e}$$

For Harold's data, we have:

$$\chi^2 = ((4 - 3.67)^2 / 3.67) + ((13 - 9.43)^2 / 9.43)$$
$$+ ((5 - 8.90)^2 / 8.90) + ((3 - 3.33)^2 / 3.33)$$
$$+ ((5 - 8.57)^2 / 8.57) + ((12 - 8.10)^2 / 8.10) = \textbf{6.50}$$

To find the **probability** value associated with this χ^2 statistic, we need to determine the **degrees of freedom** for the test. In the goodness-of-fit test, *df* was equal to the number of cells in the table minus one. Since Harold's table has 6 cells, you might think that *df* for his hypothesis test is 5.

Unfortunately, it's not this simple.

The *df* value for a chi-square test is determined by the number of expected frequencies that are "free to vary." The only restriction on f_e variation in a test for a table with one row is that the f_e's must add up to the total number of subjects in the experiment *N*. For example, in Fred's study on Springfield bus ridership (**Section 12.1**), there were four cells and *N* was 80. So if we chose expected frequencies of 10, 10, and 10 for the first three columns, f_e for the fourth column would have to be 80 − 10 − 10 − 10 = 50. Or, if we chose expected frequencies of 25, 30, and 5 for the first, third, and fourth columns, f_e for the second cell would have to be 80 − 25 − 30 − 5 = 20. The f_e's for all but one of the columns are free to vary, so *df* = C − 1.

In a chi-square test on a table with multiple rows, though, the situation is different. Here, we have an extra restriction on the amount that they can vary: the f_e's for each column have to add up to f_c for the column, and the f_e's for each row have to add up to f_r for the row.

	Blocks	Cars	Dolls	f_r
Boys	5	10	7	22
Girls	2	8	10	20
f_c	7	18	17	42

Figure 12.4 If we choose f_e values for the red-shaded cells in Harold's data table, the f_e's for the blue-shaded cells are determined by the row and column frequencies. This figure is interactive on the *Interactive Statistics* website and CD.

In Harold's study, for example, we could choose expected frequencies of 5 and 10 for the top-left and top-middle cells (shaded red in Figure 12.4). After freely choosing these two f_e's, however, the other four (shaded blue in the figure) are locked:

- For the top-right cell, f_e must be 7 so that the top row frequency equals 22.

- For the bottom-left cell, f_e must be 2 so that the left column frequency equals 7.

- For the bottom-middle cell, f_e must be 8 so that the middle column frequency equals 18.

- And for the bottom-right cell, f_e must be 10 so that the right column frequency equals 17 and the bottom row frequency equals 20.

Restrictions on expected frequency choices in a test for independence

On the *Interactive Statistics* website and CD you can click on the table in Figure 12.4 to randomly choose and set the f_e values for a new pair of cells, then click again to fill in the values for the rest of the cells given these new choices.

As Figure 12.4 hopefully makes clear, the number of expected frequencies that are free to vary is always equal to the number of columns $C - 1$ times the number of rows $R - 1 = 2 \times 1 = 2$. This rule holds in tests for independence on tables with any number of columns and rows:

Calculating *df* in a test for independence

$$df = (C - 1)(R - 1)$$

**Interactive
Page 312**

Now that we know χ^2 (6.50) and *df* (2), we can find the **p-value** for Harold's experiment. Consulting the **Chi-Square Distribution Area Calculation Tool**, we find that $p = .039$. As anyone who has actually observed 3-year-olds would expect, there are **significant** differences between the toy choices boys and girls make.

The graph of Harold's results (Figure 12.3, reproduced below) indicates that boys are more likely than girls to choose the cars or the blocks, while girls are much more likely than boys to choose the

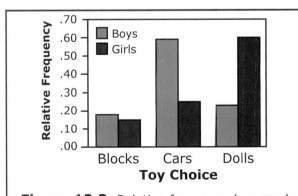

Figure 12.3 Relative frequency bar graph of Harold's data. The distribution of boys' toy choices is shown by the green bars and the distribution of girls' choices by the purple bars.

dollhouse. Unfortunately, there is no chi-square analog to **post hoc tests** (**Section 10.5.2**), so we cannot tell on the basis of these data whether any of these categorical "pairwise comparisons" are significant. The significant χ^2 statistic tells us only that something about the two toy choice **distributions** is different.

Interactive
Page 313

12.4.1 The Median Test

Now consider the following example:

> Springfield State College's Psychology Department requires all majors to take a course in statistics, but does not dictate when they take the course. Dr. Tan, a member of the department's curriculum committee, has a theory that students who take statistics earlier in their college careers learn more (and therefore get better grades) in the rest of their psychology courses.
>
> To test this theory, Dr. Tan draws a random sample of 31 psychology majors from the previous year's graduating class and sorts them into those who took statistics in their Freshman, Sophomore, Junior, and Senior years. He then calculates each student's psychology grade point average (that is, the average grade for all the psychology classes the student took at S.S.C.). The data look like this:

Students who took statistics as:			
Freshmen	**Sophomores**	**Juniors**	**Seniors**
4.0	4.0	4.0	3.8
	4.0	3.7	3.6
	4.0	3.3	3.3
	3.6	3.2	2.9
	3.5	3.1	2.8
	3.3	3.0	2.5
	3.3	2.9	2.4
	2.7	2.7	2.3
		2.6	1.7
		2.6	1.2
		2.6	
		2.3	

The first problem Dr. Tan faces in analyzing these data is that his **sample size** for students who took statistics as freshmen is only 1.

Little can be learned from a **sample** with $n = 1$, so Dr. Tan's best course of action is probably to combine the freshmen and sophomores. (Alternatively, he could drop the category of freshmen altogether, but it is usually preferable to find a way to use all your data in a study rather than exclude some data points.)

Now, since Dr. Tan has numeric scores (GPAs) for each of his three groups, he could compute means and sums of squares for each group and conduct an independent-samples **ANOVA** on the data. The problem with this analysis strategy is that, as the **histogram** in Figure 12.5 shows, the distribution of scores for freshmen/sophomore statistics-takers is negatively **skewed** (a number of these students had perfect GPAs of 4.0) and much less variable than the distribution for senior statistics-takers. In fact, the **sample variance** for seniors (.66) is more than three times the variance for freshmen/sophomores (.20).

Thus two of the **assumptions** for ANOVA (**Section 10.6**) have been violated: The population of freshmen/ sophomore scores does not appear to be **normally** distributed (and the sample size is not large enough for the Central Limit Theorem to compensate for this non-normality), and the variances in the conditions do not appear to be equal.

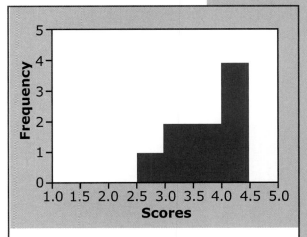

Figure 12.5 Histogram of the GPAs in Dr. Tan's dataset for students who took statistics as freshmen/sophomores.

But as we saw in **Section 12.2.2**, the only assumptions for a **chi-square test** are that subjects must be drawn at random from the population and placed into only one **cell** of the table, and that the **expected frequencies** for the χ^2 statistic must not be too small. Therefore, χ^2 is an example of a **nonparametric statistic** (see **Section 8.4.1**), and can be used as a substitute for t and F tests when population **parameter** assumptions are violated. (Indeed, we already discussed in **Section 12.3** how the sign test can be used as an alternative to a **related-samples t test** when the assumptions of that test are violated.)

For Dr. Tan to conduct a χ^2 test, he will need to transform his numeric data into **categorical data.** One way to do this is through a procedure called the **median test.** The first step in a median test is to combine all the data in all conditions and find the overall **median.** To review (see **Section 3.4.1**), the median is the middle score in a dataset with an odd number of scores, or the mean of the two middle scores of a dataset with an even number of scores.

Glossary Term: median test

Combining across all four classes, the **median** of Dr. Tan's scores is 3.1. To construct the frequency table for the **median test,** we include:

- One row for each condition in the study. For our example, we will have three rows, one for freshmen/sophomores, one for juniors, and one for seniors.

- Two columns: one for subjects who were below the overall median and one for subjects above the overall median.

Here are Dr. Tan's data again; the scores that fall above the median are printed on a blue background and the scores below the median appear on a red background:

Students who took statistics as:		
Freshmen/ Sophomores	Juniors	Seniors
4.0	4.0	3.8
4.0	3.7	3.6
4.0	3.3	3.3
4.0	3.2	2.9
3.6	3.1	2.8
3.5	3.0	2.5
3.3	2.9	2.4
3.3	2.7	2.3
2.7	2.6	1.7
	2.6	1.2
	2.6	
	2.3	

Thus we construct the following table:

Observed Frequencies:

	Students who took statistics as:			
	Freshmen/ Sophomores	Juniors	Seniors	f_r
Above Median	8	4.5	3	15.5
Below Median	1	7.5	7	15.5
f_c	9	12	10	31

Note that any score that equals the median exactly (there is one such score in Dr. Tan's dataset; it is printed on a green background in the table above) is divided evenly, with .5 placed in the "Above Median" row and .5 in the "Below Median" row. That's why we have frequencies of 4.5 and 7.5 in the "Juniors" column.

Categorizing data-points from independent samples as above or below the median

To complete the median test, we conduct a chi-square **test for independence** on this table. The f_e for the top-left **cell** is 9 (f_c for the left column) times 15.5 (f_r for the top row) / 31 (N) = 4.5. The following table is filled in with the rest of the f_e's:

Once the data have been categorized, a median test is identical to a test for independence.

Expected Frequencies:

	Students who took statistics as:			
	Freshmen/ Sophomores	Juniors	Seniors	f_r
Above Median	4.5	6	5	15.5
Below Median	4.5	6	5	15.5
f_c	9	12	10	31

We can now calculate that $\chi^2 = 7.79$, with $(C - 1)(R - 1) = (1)(2) = 2$ df, so $p = .020$. The median test supports Dr. Tan's hypothesis that early statistics-takers get better grades in their psychology courses than students who take statistics later in their college careers. (Note, however, that Dr. Tan *cannot* claim on the basis of these data that taking statistics early caused the students to do better in later classes—it might be, for example, that students who are naturally more gifted tend to take statistics earlier.)

Just as the **sign test** can be used instead of a **related-samples *t* test,** the median test can be substituted for an **independent-samples *t* test** in any experimental situation. The median test can also be used instead of a one-way independent-samples **ANOVA.** However, like the sign test, the median test is almost always less powerful than a *t* or *F* test on the same data. So if you have a choice, use the **parametric inferential statistic.**

Median tests are less powerful than *t* and *F* tests.

Exercise 12.3 will test your ability to conduct a basic chi-square **test for independence,** and **Exercise 12.4** walks you through the process of doing a median test.

12.5 Testing Relationships between Categorical Variables

Now let's consider the following research project:

> Dr. Fletcher, a family practice doctor working at a clinic in a low-income neighborhood, saw 100 patients in the past month. For each patient, she noted whether or not the patient smoked cigarettes regularly and whether or not the patient should be classified as overweight. She constructs the following table of her observations:

Observed Frequencies:

	Overweight	Not Overweight	f_r
Smokers	29	19	48
Nonsmokers	11	41	52
f_c	40	60	100

We could treat Dr. Fletcher's observations as data regarding the body weights of a **sample** of 48 smokers and a second sample of 52 nonsmokers. Alternatively, we could see them as data on the smoking habits of a sample of 40 overweight people and a sample of 60 people who are not overweight. Either way, we could use the chi-square **test for independence** to analyze the data in lieu of an independent-samples t test, since the data are **categorical** in nature.

But a better way to conceptualize these data is as a categorical X **variable** (smoking habits) and a categorical Y variable (body weight) for a *single* sample of 100 low-income clinic patients. When we frame the data this way, the relevant question becomes whether or not these two categorical variables are **correlated** with each other—that is, whether or not there is a significant **relationship** between weight and smoking. Because the data are categorical, we cannot use the Pearson correlation r (**Section 11.3**) to assess this relationship. Fortunately, though, we can use the chi-square test for independence as an analog to r for assessing relationships between categorical variables.

The row and column frequencies are already given in the table above, so we're all set to conduct the chi-square test. The table of **expected frequencies** is:

Expected Frequencies:

	Overweight	Not Overweight	f_r
Smokers	19.2	28.8	48
Nonsmokers	20.8	31.2	52
f_c	40	60	100

And the χ^2 statistic is calculated as follows:

$$\chi^2 = ((29 - 19.2)^2 / 19.2) + ((19 - 28.8)^2 / 28.8)$$
$$+ ((11 - 20.8)^2 / 20.8) + ((41 - 31.2)^2 / 31.2) = 16.03$$

For a 2 × 2 table, we only have 1 df. The p-value for $\chi^2(1) = 16.03$ is less than .0001. If Dr. Fletcher's last 100 patients can be considered a random sample of all low-income Americans, there is clearly

Correlations between two categorical variables can be assessed using the test for independence.

a strong relationship between whether or not people in this **population** smoke and whether or not they are overweight.

Again, as far as the calculation of χ^2 goes, it doesn't make any difference whether we treat a 2×2 table as two samples of data for a single variable (e.g., weight data for smokers and nonsmokers) or as two variables (e.g., smoking and weight) from a single sample. Use whichever conceptualization makes sense for any given dataset. See **Exercise 12.5** for another example of a situation in which the test for independence is used as a test for categorical relationships.

12.6 Opinion Polls and the Binomial Test

Glossary Term:
binomial test

When a chi-square test is done on a table with one row and only two columns, it is formally equivalent to another type of hypothesis test, the **binomial test.** * For example, suppose *USA Today* conducts a poll asking Americans whether or not the judicial system should ever impose the death penalty. The results might come out like this:

Yes	No	*N*
224 (56%)	176 (44%)	400

Suppose we wanted to use these poll results to test the **hypothesis** that a majority of Americans favors the death penalty. The null hypothesis would be that 50% of the **population** falls into each **cell,** so the **expected frequencies** would be 200 for both. The χ^2 value would be $((224 - 200)^2 / 200) + ((176 - 200)^2 / 200) = 5.76$, with 1 *df*, resulting in a *p*-value of .016. Thus a claim that most Americans support capital punishment would be statistically justified on the basis of these results.

USA Today might report the results of the poll described on the previous page like this: "A poll of 400 Americans found that 56% of respondents support the death penalty, with a margin of error of ± 5%." You might wonder where the "margin of error" comes from in reports like this.

---

*Actually, the exact form of the binomial test is based on the binomial distribution, which we have not covered in this textbook. However, as long as the **sample size** in a study is greater than about 20, the binomial distribution will usually look almost identical to the **normal distribution.** Hypothesis tests based on the normal approximation of the binomial distribution are formally equivalent to the chi-square test. That is, you always get exactly the same ***p*-value** with the chi-square test as you would with the normal approximation of the binomial test. So the bottom line is that the chi-square test can substitute for the binomial test whenever $n \geq 20$.

Basically, the margin of error represents a **confidence interval** (see **Chapter 9**; more specifically, a 95% confidence interval is usually used) for the reported percentages. We won't go through the exact procedure for calculating confidence intervals with binomial data, but it turns out that a good, **conservative** approximation of the 95% confidence interval for a poll result is $1 / \sqrt{n}$.

The margin of error in an opinion poll is approximately equal to 1 divided by the square root of *n* for the poll.

Thus for a poll with $n = 400$ respondents, the margin of error will always be $1 / \sqrt{400} = .05$. *USA Today*'s statement that "56% of respondents support the death penalty, with a margin of error of ± 5%" means that if we repeated the poll 100 times, the actual percentage of people saying they support the death penalty should be between 51% and 61% in 95 of these repetitions.

Alternatively, we can interpret this statement to mean that we can be 95% confident that the percentage of the entire **population** that supports the death penalty is between 51% and 61%. Note, however, that we cannot *guarantee* that the actual population percentage is within this range—we can only be 95% confident.

Moreover, the margin of error only tells us about the **variability** in poll results due to sampling error. But polls can be wrong for all sorts of reasons not related to sampling error. For example, in the 2000 U.S. presidential election, exit polls indicated that in the state of Florida, substantially more people had voted for Al Gore than for George Bush. The margins of error from these exit polls were small enough to convince the major news networks to claim that Gore had won the state.

There are many potential sources of error in an opinion poll.

But it turned out that there were many problems with the voting systems in Florida. One result of these problems was that many of the people who told pollsters they had voted for Gore failed to cast their votes properly. When the actual ballots were counted, Bush ended up with more votes, and the networks ended up with egg on their faces. The exit polls were probably accurate in their assessment of the percentage of people who *intended* to vote for Gore. But they nevertheless failed to predict the outcome of the election.

This is a great example to end this book with, because it brings us full circle back to the observations made in the first few pages: Many people inherently mistrust **statistics,** but we shouldn't blame the numbers themselves. The exit poll statistics were probably accurate guides to how people intended to vote, but the networks were overly confident in them as predictors of how people actually *did* vote.

By now, you've hopefully become an informed enough consumer of number-crunching that you will better understand when people are properly using and when they're misusing statistics.

Interactive
Page 318

12.7 Chapter Summary/Review

This Chapter Summary/Review is interactive on the *Interactive Statistics* website and CD. Also be sure to go through all the **Review Exercises** for this chapter.

- Since we can't calculate **means** for categorical data, there is no way to test **hypotheses** about such data using t or F tests. Instead, we can use the χ^2 statistic to evaluate null hypotheses that make predictions about the **population distributions** of **categorical** variables (**Section 12.1**).

- A **chi-square test** is performed by constructing tables of observed frequencies (f_o's) and expected frequencies (f_e's) for a dataset, then calculating χ^2 according to the formula below. The calculated χ^2 value is then compared to the χ^2 distribution with the appropriate **degrees of freedom** (**Section 12.2**).

$$\chi^2 = \Sigma \frac{(f_o - f_e)^2}{f_e}$$

- A **goodness-of-fit test** is used to test a single sample of categorical data against an arbitrarily selected set of **expected frequencies** and is therefore roughly equivalent to a single-sample t test for categorical data. The only restriction on the selection of f_e's is that they add up to the total number of observations in the dataset N. Therefore, the χ^2 statistic for a goodness-of-fit test has $C - 1$ degrees of freedom, where C is the number of columns in the frequency table (i.e., the number of categories into which the datapoints are divided; **Section 12.2**). Here is an example of observed and expected frequencies for a dataset, along with the χ^2 calculation:

Category	A	B	C	D	Total
f_o	10	8	2	5	25
f_e	5	5	5	5	25

χ^2= ((10 − 5) / 5) + ((8 − 5) / 5) + ((2 − 5) / 5) + ((5 − 5) / 5) = **8.6**

- Like the t statistic, χ^2 is calculated by taking a measure of the discrepancy between the observed data and the predictions made by the **null hypothesis,** and dividing by an indicator of how large we might expect this discrepancy to be if chance factors alone were causing it (**Section 12.2.1**).

- The following **assumptions** must hold for a χ^2 test to be valid: 1) subjects must be drawn at random from the population in question, 2) each subject must be placed in only one category, and 3) the average f_e value must be at least 5, with no one f_e being less than 1 (**Section 12.2.2**).

- To conduct a **sign test,** we divide datapoint pairs from two **related samples** into two categories: those where the score from sample 1 is greater than the score from sample 2, and those where the score from sample 1 is less than the score from sample 2. Ties are either excluded or divided evenly between negative and positive **difference scores.** We then conduct a chi-square goodness-of-fit test on this newly created categorical variable. The sign test thus serves as an alternative to the related-samples *t* test (**Section 12.3**).

- The chi-square **test for independence** is used to analyze datasets with two or more **independent samples** of categorical data. Expected frequencies are calculated for each cell by multiplying the column frequency for the cell by the row frequency for the cell, then dividing by *N*: $f_e = f_c f_r / N$. The χ^2 statistic is then calculated exactly as in the goodness-of-fit test and is compared to the χ^2 distribution with $(C - 1)(R - 1)$ *df,* where *R* is the number of rows in the frequency table (**Section 12.4**).

- The **median test** is a version of the test for independence that serves as an alternative to independent-samples *t* tests and one-way independent-samples ANOVAs. To create the categories in the median test, we first find the **median** of all *N* observations in the dataset (collapsed over all samples), then classify each observation as above or below the median (**Section 12.4.1**).

- Any dataset that can be analyzed using a test for independence can be conceptualized as a set of *R* **samples** of data, a set of *C* samples of data, or two categorical variables for a single sample of subjects. In the latter conceptualization, the test for independence is the categorical analog to the Pearson correlation ***r*** (**Section 12.5**).

- An opinion poll asking a yes–no question can be analyzed using a chi-square goodness-of-fit test on a two-cell table, which is roughly equivalent to a **binomial test.** When opinion researchers give a "margin of error" for a poll result, they are really giving a 95% confidence interval for the population values of the cells in the chi-square table (**Section 12.6**).

Appendix 1
Formula Reference

▼ Central Tendency

Mean

$$\text{Population: } \mu = \frac{\Sigma X}{N} \qquad \text{Sample: } \bar{X} = \frac{\Sigma X}{n}$$

Median

If *n* is odd, middle score

If *n* is even, mean of middle two scores

▼ Variability

Sum of Squares

$$SS = \Sigma(X - \mu)^2$$

Variance

$$\text{Population: } \sigma = \frac{SS}{N} \qquad \text{Sample: } s^2 = \frac{SS}{n-1}$$

Standard Deviation

$$\text{Population: } \sigma = \sqrt{\sigma^2} = \sqrt{\frac{SS}{N}}$$

$$\text{Sample: } s = \sqrt{s^2} = \sqrt{\frac{SS}{n-1}}$$

Interquartile Range

75th percentile of scores − 25th percentile of scores

▼ z-Scores

z-Score for Individual Datapoints

$$z = \frac{X - \mu}{\sigma}$$

z-Score Test Statistic

$$z = \frac{\overline{X} - \mu}{\sigma_{\overline{X}}}, \quad \text{where} \quad \sigma_{\overline{X}} = \sigma / \sqrt{n}$$

▼ t Statistics

Single-Sample t Statistic

$$t = \frac{\overline{X} - \mu}{s_{\overline{X}}}, \quad \text{where} \quad s_{\overline{X}} = s / \sqrt{n} \quad \text{and} \quad df = n - 1$$

Related-Samples t Statistic

$$t = \frac{\overline{D} - \mu_D}{s_{\overline{D}}}, \quad \text{where} \quad s_{\overline{D}} = s_D / \sqrt{n} \quad \text{and} \quad df = n - 1$$

Independent-Samples t Statistic

$$t = \frac{(\overline{X}_1 - \overline{X}_2) - (\mu_1 - \mu_2)}{s_{\overline{X}_1 - \overline{X}_2}}, \quad \text{where}$$

$$s_{\overline{X}_1 - \overline{X}_2} = s_p \sqrt{\frac{1}{n_1} + \frac{1}{n_2}}$$

$$s_p = \sqrt{\frac{(n_1 - 1)s_1^2 + (n_2 - 1)s_2^2}{(n_1 - 1) + (n_2 - 1)}}$$

$$\text{and} \quad df = n_1 + n_2 - 2$$

▼ Confidence Intervals

Single-Sample Confidence Interval

$$\overline{X} \pm t_{\alpha/2}(s_{\overline{X}}), \quad \text{where} \quad df = n - 1$$

Related-Samples Confidence Interval

$$\overline{D} \pm t_{\alpha/2}(s_{\overline{D}}), \quad \text{where} \quad df = n - 1$$

Independent-Samples Confidence Interval

$$(\overline{X}_1 - \overline{X}_2) \pm t_{\alpha/2}(s_{\overline{X}_1 - \overline{X}_2}), \text{ where } df = n_1 + n_2 - 2$$

▼ Analysis of Variance

Independent-Samples ANOVA

$$SS_T = \Sigma(X - \overline{X}_T)^2 \qquad df_T = N - 1$$

$$SS_B = \Sigma n_i(\overline{X}_1 - \overline{X}_T)^2 \qquad df_B = k - 1$$

$$SS_E = \Sigma(X - \overline{X}_i)^2 = \Sigma SS_i \quad df_E = N - k$$

$$MS_B = SS_B / df_B \qquad MS_E = SS_E / df_E$$

$$F = MS_B / MS_E$$

where k = the number of conditions,

$\overline{X}_T$ = the mean of all N scores in the dataset, and

$\overline{X}_i$ = the mean of the n_i scores in each condition.

Related-Samples ANOVA

$$SS_T = \Sigma(X - \overline{X}_T)^2 \qquad df_T = kn - 1$$

$$SS_B = n\Sigma(\overline{X}_i - \overline{X}_T)^2 \qquad df_B = k - 1$$

$$SS_W = \Sigma(X - \overline{X}_i)^2 = \Sigma SS_i \quad df_W = k(n - 1)$$

$$SS_S = k\Sigma(\overline{X} - \overline{X}_T)^2 \qquad df_S = n - 1$$

$$SS_E = SS_W - SS_S \qquad df_E = (k - 1)(n - 1)$$

$$MS_B = SS_B / df_B \qquad MS_E = SS_E / df_E$$

$$F = MS_B / MS_E$$

where k = the number of conditions,

$\overline{X}_T$ = the mean of all N scores in the dataset, and

$\overline{X}_i$ = the mean of the n_i scores in each condition.

Least Significant Differences (LSD) Post Hoc Test

Independent samples: $t = \dfrac{(\bar{X}_1 - \bar{X}_2)}{\sqrt{MS_E}\sqrt{\dfrac{1}{n_1} + \dfrac{1}{n_2}}}$

Related samples: $t = \dfrac{\bar{D}}{\sqrt{MS_E} \, / \, \sqrt{n}}$

where MS_E is taken from an ominibus ANOVA including the two groups and $df = df_E$ from the ominbus ANOVA.

▼ Correlation / Regression

Pearson Correlation r

$$r = \frac{\sum z_X z_Y}{n}$$

Linear Regression Line

$Y = a + bX$, where

$$b = r\frac{\sigma_Y}{\sigma_X} = r\frac{s_Y}{s_X} \quad \text{and} \quad a = \bar{Y} - b_{\bar{X}}$$

t Statistic for r

$$t = \frac{r\sqrt{n-2}}{\sqrt{1 - r_2}}, \quad \text{where} \quad df = n - 2$$

▼ Chi-Square Tests

χ^2 Statistic

$$\chi_2 = \sum \frac{(f_o - f_e)^2}{f_e}$$

Chi-Square Goodness-of-Fit Test

$$df = C - 1$$

Chi-Square Test for Independence

$$f_e = f_c f_r \, / \, N$$
$$df = (C - 1)(R - 1)$$

Appendix 2
Statistical Tables

In the days before computers, the only realistic way to find the *p*-value associated with a given *z*-score was to use a table. Figure 1 below shows part of the Normal Distribution Table used for this purpose (this full table and others are found beginning on page 364 and on the *Interactive Statistics* website and CD). Your professor may also ask you to use this table to look up *p*'s when you are sitting for an exam and can't access the *Interactive Statistics* Calculation Tool or another statistics program.

z	**.00**	**.01**	**.02**	**.03**	**.04**	**.05**	**.06**	**.07**	**.08**	**.09**
1.6	.0548	.0537	.0526	.0516	.0505	.0495	.0485	.0475	.0465	.0455
1.7	.0446	.0436	.0427	.0418	.0409	.0401	.0392	.0384	.0375	.0367
1.8	.0359	.0351	.0344	.0336	.0329	.0322	.0314	.0307	.0301	.0294
1.9	.0287	.0281	.0274	.0268	.0262	.0256	.0250	.0244	.0239	.0233
2.0	.0228	.0222	.0217	.0212	.0207	.0202	.0197	.0192	.0188	.0183

Figure 1 Excerpt from the Standard Normal Table. This figure is interactive and includes parts (a–d) on the *Interactive Statistics* website and CD.

To use this table, you must first find the whole number and first decimal place of your *z*-score in the column of numbers on the far left. If, for example, we were interested in a *z*-score of 1.76, we would find the row for 1.7, shaded blue in the figure.

Next, we read across the table to find the column for the second decimal place of our *z*-score (.06 in our example, the column for which is shaded red in Figure 1).

The number at the intersection of the row and column you've just located (.0392, shaded green in the figure) provides the information you need to calculate the *p*-value you're looking for. More specifically, this number shows the area under the normal distribution and to the right of this *z*-score, as illustrated in Figure 2. This area corresponds to the probability of randomly drawing a *z*-score this large or larger from a normal distribution.

The probability of randomly drawing a z-score as small as −1.76 or smaller is exactly the same: .0392. If you need the "two-tailed" probability, multiply the p-value from the table by two, thus giving you the likelihood of randomly drawing a z-score ≤ −1.76 OR ≥ +1.76.

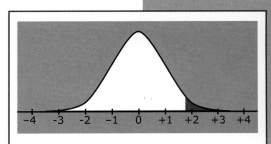

Figure 2 Area under the normal distribution for z-scores greater than or equal to 1.76.

Our Normal Distribution Table only lists z-scores up to z-scores of 3.5 or so. The reason it doesn't go beyond this is that the likelihood of randomly drawing a z-score of 3.7 or above is about .0001. That's already pretty unlikely, so there's rarely any need to be more precise than this. If you need to state the probability value for a z-score greater than 3.69, just say "p < .001."

Probability values associated with t, F, and χ^2 statistics (covered in Chapters 6, 10, and 12, respectively) can also be looked up in a table. (If you haven't made it to these chapters yet, you can stop reading now and come back to this box when you encounter these statistics later.) For these statistics, however, the situation is complicated by the fact that the exact shape of the t, F, and χ^2 distributions depends on the degrees of freedom for the test statistic in which we are interested.

This complication forces us to abridge and reformat the tables for these statistics. Figure 3 shows part of the t Distribution Table. Each row in the table represents a df value and each column represents a p-value. Suppose we calculate a t statistic of 2.19 with 15 df. To find the one-tailed p-value, go to the row of the table for 15 df (shaded blue in the figure), then read across the table until you find the two t values that straddle your calculated t. These values are 2.131 and 2.249, as highlighted in green in the figure.

df	.25	.20	.15	.10	.05	.025	.02	.01	.005
14	0.692	0.868	1.076	1.345	1.761	2.145	2.264	2.625	2.977
15	0.691	0.866	1.074	1.341	1.753	2.131	2.249	2.602	2.947
16	0.690	0.865	1.071	1.337	1.746	2.120	2.235	2.583	2.921
17	0.689	0.863	1.069	1.333	1.740	2.110	2.224	2.567	2.898
18	0.688	0.862	1.067	1.330	1.734	2.101	2.214	2.552	2.878
14	0.692	0.868	1.076	1.345	1.761	2.145	2.264	2.625	2.977
15	0.691	0.866	1.074	1.341	1.753	2.131	2.249	2.602	2.947
16	0.690	0.865	1.071	1.337	1.746	2.120	2.235	2.583	2.921
17	0.689	0.863	1.069	1.333	1.740	2.110	2.224	2.567	2.898
18	0.688	0.862	1.067	1.330	1.734	2.101	2.214	2.552	2.878

Figure 3 Excerpt from the t Distribution Table. This figure is interactive and includes parts (a–d) on the *Interactive Statistics* website and CD.

Now move up to the column headings of the table and find the p-values for the columns you just identified (for our example, .025 and .02, circled in Figure 3). The one-tailed p-value for your test statistic is somewhere between these two values. Thus for $t(15) = 2.19$, you would report that $.02 < p < .025$. Or, more succinctly, you might just say that $p < .025$. If you need a two-tailed p-value for a t test, multiply the tabled p-value by two (in our example, the two-tailed p is between .04 and .05).

19	.10	2.99	2.61	2.40	2.27	2.18	2.11	2.06
	.05	4.38	3.52	3.13	2.90	2.74	2.63	2.54
	.01	8.18	5.93	5.01	4.50	4.17	3.94	3.77
	.001	15.08	10.16	8.28	7.27	6.62	6.18	5.85
20	.10	2.97	2.59	2.38	2.25	2.16	2.09	2.04
	.05	4.35	3.49	3.10	2.87	2.73	2.60	2.51
	.01	8.10	5.85	4.94	4.43	4.10	3.87	3.70
	.001	14.82	9.95	8.10	7.10	6.46	6.02	5.69
df_E	p				df_B			
		1	2	3	4	5	6	7

Figure 4 Excerpt from the F Distribution Table. This figure is interactive and includes parts (a–e) on the *Interactive Statistics* website and CD.

The Chi-Square Distribution Table is used in exactly the same way as the t Table, since each χ^2 statistic is also associated with one df value. An F statistic, however, has two df values, one for the numerator and one for the denominator of the F ratio. Figure 4 shows part of the F table and illustrates how to use it to find the p value for $F = 2.53$ with 3 df in the numerator and 20 df in the denominator.

First find the intersection of the set of rows corresponding to the denominator df (20) and the column corresponding to the numerator df (3). Then scan the set of F values in this block to find the values that straddle your test statistic. In this case, the bracketing values are 2.38 and 3.10, which according to their row headings correspond to p-values of .10 and .05. Therefore, the p-value for $F(3, 20) = 2.53$ is less than .10 but greater than .05.

Normal Distribution Table

z	.00	.01	.02	.03	.04	.05	.06	.07	.08	.09
0.0	.5000	.4960	.4920	.4880	.4840	.4801	.4761	.4721	.4681	.4641
0.1	.4602	.4562	.4522	.4483	.4443	.4404	.4364	.4325	.4286	.4247
0.2	.4207	.4168	.4129	.4090	.4052	.4013	.3974	.3936	.3897	.3859
0.3	.3821	.3783	.3745	.3707	.3669	.3632	.3594	.3557	.3520	.3483
0.4	.3446	.3409	.3372	.3336	.3300	.3264	.3228	.3192	.3156	.3121
0.5	.3085	.3050	.3015	.2981	.2946	.2912	.2877	.2843	.2810	.2776
0.6	.2743	.2709	.2676	.2643	.2611	.2578	.2546	.2514	.2483	.2451
0.7	.2420	.2389	.2358	.2327	.2297	.2266	.2236	.2207	.2177	.2148
0.8	.2119	.2090	.2061	.2033	.2005	.1977	.1949	.1922	.1894	.1867
0.9	.1841	.1814	.1788	.1762	.1736	.1711	.1685	.1660	.1635	.1611
1.0	.1587	.1562	.1539	.1515	.1492	.1469	.1446	.1423	.1401	.1379
1.1	.1357	.1335	.1314	.1292	.1271	.1251	.1230	.1210	.1190	.1170
1.2	.1151	.1131	.1112	.1093	.1075	.1056	.1038	.1020	.1003	.0985
1.3	.0968	.0951	.0934	.0918	.0901	.0885	.0869	.0853	.0838	.0823
1.4	.0808	.0793	.0778	.0764	.0749	.0735	.0721	.0708	.0694	.0681
1.5	.0668	.0655	.0643	.0630	.0618	.0606	.0594	.0582	.0571	.0559
1.6	.0548	.0537	.0526	.0516	.0505	.0495	.0485	.0475	.0465	.0455
1.7	.0446	.0436	.0427	.0418	.0409	.0401	.0392	.0384	.0375	.0367
z	.00	.01	.02	.03	.04	.05	.06	.07	.08	.09
1.8	.0359	.0351	.0344	.0336	.0329	.0322	.0314	.0307	.0301	.0294
1.9	.0287	.0281	.0274	.0268	.0262	.0256	.0250	.0244	.0239	.0233
2.0	.0228	.0222	.0217	.0212	.0207	.0202	.0197	.0192	.0188	.0183
2.1	.0179	.0174	.0170	.0166	.0162	.0158	.0154	.0150	.0146	.0143
2.2	.0139	.0136	.0132	.0129	.0125	.0122	.0119	.0116	.0113	.0110
2.3	.0107	.0104	.0102	.0099	.0096	.0094	.0091	.0089	.0087	.0084
2.4	.0082	.0080	.0078	.0075	.0073	.0071	.0069	.0068	.0066	.0064
2.5	.0062	.0060	.0059	.0057	.0055	.0054	.0052	.0051	.0049	.0048
2.6	.0047	.0045	.0044	.0043	.0041	.0040	.0039	.0038	.0037	.0036
2.7	.0035	.0034	.0033	.0032	.0031	.0030	.0029	.0028	.0027	.0026
2.8	.0026	.0025	.0024	.0023	.0023	.0022	.0021	.0021	.0020	.0019
2.9	.0019	.0018	.0018	.0017	.0016	.0016	.0015	.0015	.0014	.0014
3.0	.0013	.0013	.0013	.0012	.0012	.0011	.0011	.0011	.0010	.0010
3.1	.0010	.0009	.0009	.0009	.0008	.0008	.0008	.0008	.0007	.0007
3.2	.0007	.0007	.0006	.0006	.0006	.0006	.0006	.0005	.0005	.0005
3.3	.0005	.0005	.0005	.0004	.0004	.0004	.0004	.0004	.0004	.0003
3.4	.0003	.0003	.0003	.0003	.0003	.0003	.0003	.0003	.0003	.0002
3.5	.0002	.0002	.0002	.0002	.0002	.0002	.0002	.0002	.0002	.0002
3.6	.0002	.0002	.0001	.0001	.0001	.0001	.0001	.0001	.0001	.0001
z	.00	.01	.02	.03	.04	.05	.06	.07	.08	.09

t Distribution Table

df	.25	.20	.15	.10	.05	.025	.02	.01	.005	.0025	.001	.0005
1	1.000	1.376	1.963	3.078	6.314	12.707	15.895	31.819	63.655	127.345	318.493	636.045
2	0.817	1.061	1.386	1.886	2.920	4.303	4.849	6.965	9.925	14.089	22.328	31.599
3	0.765	0.978	1.250	1.638	2.353	3.182	3.482	4.541	5.841	7.453	10.214	12.924
4	0.741	0.941	1.190	1.533	2.132	2.776	2.999	3.747	4.604	5.598	7.173	8.610
5	0.727	0.920	1.156	1.476	2.015	2.571	2.757	3.365	4.032	4.773	5.893	6.869
6	0.718	0.906	1.134	1.440	1.943	2.447	2.612	3.143	3.707	4.317	5.208	5.959
7	0.711	0.896	1.119	1.415	1.895	2.365	2.517	2.998	3.500	4.029	4.785	5.408
8	0.706	0.889	1.108	1.397	1.860	2.306	2.449	2.896	3.355	3.833	4.501	5.041
9	0.703	0.883	1.100	1.383	1.833	2.262	2.398	2.821	3.250	3.690	4.297	4.781
10	0.700	0.879	1.093	1.372	1.812	2.228	2.359	2.764	3.169	3.581	4.144	4.587
11	0.697	0.876	1.088	1.363	1.796	2.201	2.328	2.718	3.106	3.497	4.025	4.437
12	0.695	0.873	1.083	1.356	1.782	2.179	2.303	2.681	3.055	3.428	3.930	4.318
13	0.694	0.870	1.079	1.350	1.771	2.160	2.282	2.650	3.012	3.372	3.852	4.221
14	0.692	0.868	1.076	1.345	1.761	2.145	2.264	2.625	2.977	3.326	3.787	4.140
15	0.691	0.866	1.074	1.341	1.753	2.131	2.249	2.602	2.947	3.286	3.733	4.073
16	0.690	0.865	1.071	1.337	1.746	2.120	2.235	2.583	2.921	3.252	3.686	4.015
df	**.25**	**.20**	**.15**	**.10**	**.05**	**.025**	**.02**	**.01**	**.005**	**.0025**	**.001**	**.0005**
17	0.689	0.863	1.069	1.333	1.740	2.110	2.224	2.567	2.898	3.222	3.646	3.965
18	0.688	0.862	1.067	1.330	1.734	2.101	2.214	2.552	2.878	3.197	3.610	3.922
19	0.688	0.861	1.066	1.328	1.729	2.093	2.205	2.539	2.861	3.174	3.579	3.883
20	0.687	0.860	1.064	1.325	1.725	2.086	2.197	2.528	2.845	3.153	3.552	3.850
21	0.686	0.859	1.063	1.323	1.721	2.080	2.189	2.518	2.831	3.135	3.527	3.819
22	0.686	0.858	1.061	1.321	1.717	2.074	2.183	2.508	2.819	3.119	3.505	3.792
23	0.685	0.858	1.060	1.319	1.714	2.069	2.177	2.500	2.807	3.104	3.485	3.768
24	0.685	0.857	1.059	1.318	1.711	2.064	2.172	2.492	2.797	3.091	3.467	3.745
25	0.684	0.856	1.058	1.316	1.708	2.060	2.167	2.485	2.787	3.078	3.450	3.725
26	0.684	0.856	1.058	1.315	1.706	2.056	2.162	2.479	2.779	3.067	3.435	3.707
27	0.684	0.855	1.057	1.314	1.703	2.052	2.158	2.473	2.771	3.057	3.421	3.690
28	0.683	0.855	1.056	1.313	1.701	2.048	2.154	2.467	2.763	3.047	3.408	3.674
29	0.683	0.854	1.055	1.311	1.699	2.045	2.150	2.462	2.756	3.038	3.396	3.659
30	0.683	0.854	1.055	1.310	1.697	2.042	2.147	2.457	2.750	3.030	3.385	3.646
40	0.681	0.851	1.050	1.303	1.684	2.021	2.123	2.423	2.704	2.971	3.307	3.551
60	0.679	0.848	1.045	1.296	1.671	2.000	2.099	2.390	2.660	2.915	3.232	3.460
100	0.677	0.845	1.042	1.290	1.660	1.984	2.081	2.364	2.626	2.871	3.174	3.390
df	**.25**	**.20**	**.15**	**.10**	**.05**	**.025**	**.02**	**.01**	**.005**	**.0025**	**.001**	**.0005**

F Distribution Table

df_E	p	\multicolumn{10}{c}{df_B}
		1	**2**	**3**	**4**	**5**	**6**	**7**	**8**	**9**	**10**
1	.10	39.86	49.50	53.60	55.83	57.23	58.21	58.91	59.44	59.86	60.20
	.05	161.5	199.5	215.7	224.5	230.1	234.0	236.8	238.9	240.5	241.8
	.01	4063	4992	5404	5637	5760	5890	5890	6025	6025	6025
2	.10	8.53	9.00	9.16	9.24	9.29	9.33	9.35	9.37	9.38	9.39
	.05	18.51	19.00	19.16	19.25	19.30	19.33	19.35	19.37	19.39	19.40
	.01	98.50	99.00	99.15	99.27	99.30	99.34	99.34	99.38	99.38	99.38
	.001	997.6	997.6	997.6	997.6	997.6	997.6	997.6	997.6	997.6	997.6
3	.10	5.54	5.46	5.39	5.34	5.31	5.28	5.27	5.25	5.24	5.23
	.05	10.13	9.55	9.28	9.12	9.01	8.94	8.89	8.85	8.81	8.79
	.01	34.11	30.82	29.46	28.71	28.24	27.91	27.67	27.49	27.34	27.23
	.001	167.0	148.5	141.1	137.1	134.6	132.8	131.6	130.6	129.8	129.3
4	.10	4.54	4.32	4.19	4.11	4.05	4.01	3.98	3.95	3.94	3.92
	.05	7.71	6.94	6.59	6.39	6.26	6.16	6.09	6.04	6.00	5.96
	.01	21.20	18.00	16.69	15.98	15.52	15.21	14.98	14.80	14.66	14.55
	.001	74.15	61.24	56.18	53.44	51.71	50.53	49.66	48.99	48.48	48.05

df_E	p	\multicolumn{10}{c}{df_B}
		1	**2**	**3**	**4**	**5**	**6**	**7**	**8**	**9**	**10**
5	.10	4.06	3.78	3.62	3.52	3.45	3.40	3.37	3.34	3.32	3.30
	.05	6.61	5.79	5.41	5.19	5.05	4.95	4.88	4.82	4.77	4.74
	.01	16.26	13.27	12.06	11.39	10.97	10.67	10.46	10.29	10.16	10.05
	.001	47.18	37.12	33.20	31.08	29.75	28.83	28.16	27.65	27.24	26.92
6	.10	3.78	3.46	3.29	3.18	3.11	3.05	3.01	2.98	2.96	2.94
	.05	5.99	5.14	4.76	4.53	4.39	4.28	4.21	4.15	4.10	4.06
	.01	13.75	10.92	9.78	9.15	8.75	8.47	8.26	8.10	7.98	7.87
	.001	35.51	27.00	23.70	21.92	20.80	20.03	19.46	19.03	18.69	18.41
7	.10	3.59	3.26	3.07	2.96	2.88	2.83	2.78	2.75	2.72	2.70
	.05	5.59	4.74	4.35	4.12	3.97	3.87	3.79	3.73	3.68	3.64
	.01	12.25	9.55	8.45	7.85	7.46	7.19	6.99	6.84	6.72	6.62
	.001	29.24	21.69	18.77	17.20	16.21	15.52	15.02	14.63	14.33	14.08
8	.10	3.46	3.11	2.92	2.81	2.73	2.67	2.62	2.59	2.56	2.54
	.05	5.32	4.46	4.07	3.84	3.69	3.58	3.50	3.44	3.39	3.35
	.01	11.26	8.65	7.59	7.01	6.63	6.37	6.18	6.03	5.91	5.81
	.001	25.41	18.49	15.83	14.39	13.48	12.86	12.40	12.05	11.77	11.54

df_E	p	\multicolumn{10}{c}{df_B}
		1	**2**	**3**	**4**	**5**	**6**	**7**	**8**	**9**	**10**

F Distribution Table

df_E	p	\multicolumn{10}{c}{df_B}
		1	2	3	4	5	6	7	8	9	10
9	.10	3.36	3.01	2.81	2.69	2.61	2.55	2.51	2.47	2.44	2.42
	.05	5.12	4.26	3.86	3.63	3.48	3.37	3.29	3.23	3.18	3.14
	.01	10.56	8.02	6.99	6.42	6.06	5.80	5.61	5.47	5.35	5.26
	.001	22.86	16.39	13.90	12.56	11.71	11.13	10.70	10.37	10.11	9.89
10	.10	3.29	2.92	2.73	2.61	2.52	2.46	2.41	2.38	2.35	2.32
	.05	4.96	4.10	3.71	3.48	3.33	3.22	3.14	3.07	3.02	2.98
	.01	10.04	7.56	6.55	5.99	5.64	5.39	5.20	5.06	4.94	4.85
	.001	21.04	14.91	12.55	11.28	10.48	9.93	9.52	9.20	8.96	8.75
11	.10	3.23	2.86	2.66	2.54	2.45	2.39	2.34	2.30	2.27	2.25
	.05	4.84	3.98	3.59	3.36	3.20	3.09	3.01	2.95	2.90	2.85
	.01	9.65	7.21	6.22	5.67	5.32	5.07	4.89	4.74	4.63	4.54
	.001	19.69	13.81	11.56	10.35	9.58	9.05	8.66	8.35	8.12	7.92
12	.10	3.18	2.81	2.61	2.48	2.39	2.33	2.28	2.24	2.21	2.19
	.05	4.75	3.89	3.49	3.26	3.11	3.00	2.91	2.85	2.80	2.75
	.01	9.33	6.93	5.95	5.41	5.06	4.82	4.64	4.50	4.39	4.30
	.001	18.64	12.97	10.80	9.63	8.89	8.38	8.00	7.71	7.48	7.29

df_E	p	\multicolumn{10}{c}{df_B}
		1	2	3	4	5	6	7	8	9	10
13	.10	3.14	2.76	2.56	2.43	2.35	2.28	2.23	2.20	2.16	2.14
	.05	4.67	3.81	3.41	3.18	3.03	2.92	2.83	2.77	2.71	2.67
	.01	9.07	6.70	5.74	5.21	4.86	4.62	4.44	4.30	4.19	4.10
	.001	17.82	12.31	10.21	9.07	8.35	7.86	7.49	7.21	6.98	6.80
14	.10	3.10	2.73	2.52	2.39	2.31	2.24	2.19	2.15	2.12	2.10
	.05	4.60	3.74	3.34	3.11	2.96	2.85	2.76	2.70	2.65	2.60
	.01	8.86	6.51	5.56	5.04	4.70	4.46	4.28	4.14	4.03	3.94
	.001	17.14	11.78	9.73	8.62	7.92	7.44	7.08	6.80	6.58	6.40
15	.10	3.07	2.70	2.49	2.36	2.27	2.21	2.16	2.12	2.09	2.06
	.05	4.54	3.68	3.29	3.06	2.90	2.79	2.71	2.64	2.59	2.54
	.01	8.68	6.36	5.42	4.89	4.56	4.32	4.14	4.00	3.89	3.80
	.001	16.59	11.34	9.34	8.25	7.57	7.09	6.74	6.47	6.26	6.08
16	.10	3.05	2.67	2.46	2.33	2.24	2.18	2.13	2.09	2.06	2.03
	.05	4.49	3.63	3.24	3.01	2.85	2.74	2.66	2.59	2.54	2.49
	.01	8.53	6.23	5.29	4.77	4.44	4.20	4.03	3.89	3.78	3.69
	.001	16.12	10.97	9.01	7.94	7.27	6.81	6.46	6.19	5.98	5.81

df_E	p	\multicolumn{10}{c}{df_B}
		1	2	3	4	5	6	7	8	9	10

F Distribution Table

df_E	p	df_B									
		1	**2**	**3**	**4**	**5**	**6**	**7**	**8**	**9**	**10**
17	.10	3.03	2.64	2.44	2.31	2.22	2.15	2.10	2.06	2.03	2.00
	.05	4.45	3.59	3.20	2.96	2.81	2.70	2.61	2.55	2.49	2.45
	.01	8.40	6.11	5.19	4.67	4.34	4.10	3.93	3.79	3.68	3.59
	.001	15.72	10.66	8.73	7.68	7.02	6.56	6.22	5.96	5.75	5.58
18	.10	3.01	2.62	2.42	2.29	2.20	2.13	2.08	2.04	2.00	1.98
	.05	4.41	3.55	3.16	2.93	2.77	2.66	2.58	2.51	2.46	2.41
	.01	8.29	6.01	5.09	4.58	4.25	4.01	3.84	3.71	3.60	3.51
	.001	15.38	10.39	8.49	7.46	6.81	6.36	6.02	5.76	5.56	5.39
19	.10	2.99	2.61	2.40	2.27	2.18	2.11	2.06	2.02	1.98	1.96
	.05	4.38	3.52	3.13	2.90	2.74	2.63	2.54	2.48	2.42	2.38
	.01	8.18	5.93	5.01	4.50	4.17	3.94	3.77	3.63	3.52	3.43
	.001	15.08	10.16	8.28	7.27	6.62	6.18	5.85	5.59	5.39	5.22
20	.10	2.97	2.59	2.38	2.25	2.16	2.09	2.04	2.00	1.96	1.94
	.05	4.35	3.49	3.10	2.87	2.71	2.60	2.51	2.45	2.39	2.35
	.01	8.10	5.85	4.94	4.43	4.10	3.87	3.70	3.56	3.46	3.37
	.001	14.82	9.95	8.10	7.10	6.46	6.02	5.69	5.44	5.24	5.08

df_E	p	df_B									
		1	**2**	**3**	**4**	**5**	**6**	**7**	**8**	**9**	**10**
25	.10	2.92	2.53	2.32	2.18	2.09	2.02	1.97	1.93	1.89	1.87
	.05	4.24	3.39	2.99	2.76	2.60	2.49	2.40	2.34	2.28	2.24
	.01	7.77	5.57	4.68	4.18	3.85	3.63	3.46	3.32	3.22	3.13
	.001	13.88	9.22	7.45	6.49	5.89	5.46	5.15	4.91	4.71	4.56
30	.10	2.88	2.49	2.28	2.14	2.05	1.98	1.93	1.88	1.85	1.82
	.05	4.17	3.32	2.92	2.69	2.53	2.42	2.33	2.27	2.21	2.16
	.01	7.56	5.39	4.51	4.02	3.70	3.47	3.30	3.17	3.07	2.98
	.001	13.29	8.77	7.05	6.12	5.53	5.12	4.82	4.58	4.39	4.24
50	.10	2.81	2.41	2.20	2.06	1.97	1.90	1.84	1.80	1.76	1.73
	.05	4.03	3.18	2.79	2.56	2.40	2.29	2.20	2.13	2.07	2.03
	.01	7.17	5.06	4.20	3.72	3.41	3.19	3.02	2.89	2.78	2.70
	.001	12.22	7.96	6.34	5.46	4.90	4.51	4.22	4.00	3.82	3.67
100	.10	2.76	2.36	2.14	2.00	1.91	1.83	1.78	1.73	1.69	1.66
	.05	3.94	3.09	2.70	2.46	2.31	2.19	2.10	2.03	1.97	1.93
	.01	6.90	4.82	3.98	3.51	3.21	2.99	2.82	2.69	2.59	2.50
	.001	11.50	7.41	5.86	5.02	4.48	4.11	3.83	3.61	3.44	3.30

df_E	p	df_B									
		1	**2**	**3**	**4**	**5**	**6**	**7**	**8**	**9**	**10**

Chi-Square Distribution Table

df	.25	.20	.15	.10	.05	.025	.02	.01	.005	.0025	.001	.0005
1	1.32	1.64	2.07	2.71	3.84	5.02	5.41	6.64	7.88	9.14	10.83	12.12
2	2.77	3.22	3.79	4.61	5.99	7.38	7.82	9.21	10.60	11.98	13.82	15.20
3	4.11	4.64	5.32	6.25	7.81	9.35	9.84	11.35	12.84	14.32	16.27	17.73
4	5.39	5.99	6.74	7.78	9.49	11.14	11.67	13.28	14.86	16.42	18.47	20.00
5	6.63	7.29	8.12	9.24	11.07	12.83	13.39	15.09	16.75	18.39	20.52	22.11
6	7.84	8.56	9.45	10.64	12.59	14.45	15.03	16.81	18.55	20.25	22.46	24.11
7	9.04	9.80	10.75	12.02	14.07	16.01	15.62	18.48	20.28	22.04	24.32	26.02
8	10.22	11.03	12.03	13.36	15.51	17.53	13.17	20.09	21.96	23.77	26.12	27.87
9	11.39	12.24	13.29	14.68	16.92	19.02	19.68	21.67	23.59	25.46	27.88	29.67
10	12.55	13.44	14.53	15.99	18.31	20.48	21.16	23.21	25.19	27.11	29.59	31.42
11	13.70	14.63	15.77	17.27	19.67	21.92	22.62	24.72	26.76	28.73	31.27	33.14
df	.25	.20	.15	.10	.05	.025	.02	.01	.005	.0025	.001	.0005
12	14.85	15.81	16.99	18.55	21.03	23.34	24.05	26.22	28.30	30.32	32.91	34.82
13	15.98	16.99	18.20	19.81	22.36	24.73	25.47	27.69	29.82	31.89	34.53	36.48
14	17.12	18.15	19.41	21.07	23.69	26.12	26.87	29.14	31.32	33.43	36.12	38.11
15	18.24	19.31	20.60	22.31	24.99	27.49	28.26	30.58	32.80	34.95	37.70	39.72
16	19.37	20.47	21.79	23.54	26.30	28.85	29.63	32.00	34.27	36.46	39.25	41.31
17	20.49	21.62	22.98	24.77	27.59	30.19	30.99	33.41	35.72	37.95	40.79	42.88
18	21.61	22.76	24.16	25.99	28.87	31.53	32.35	34.80	37.16	39.42	42.31	44.44
19	22.72	23.90	25.33	27.20	30.14	32.85	33.69	36.19	38.58	40.89	43.82	45.98
20	23.83	25.04	26.50	28.41	31.41	34.17	35.02	37.56	40.00	42.33	45.32	47.50
30	34.80	36.25	37.99	40.25	43.78	46.98	47.96	50.89	53.68	56.33	59.70	62.16
50	56.33	58.17	60.34	63.16	67.51	71.43	72.60	76.16	79.50	82.66	86.66	89.57
df	.25	.20	.15	.10	.05	.025	.02	.01	.005	.0025	.001	.0005

Glossary

alpha level   The alpha (α) level is chosen by the experimenter as a guideline for what should be considered a statistically significant effect. The null hypothesis is rejected if an experiment produces a probability value lower than alpha. The most common alpha level in the social sciences is .05.

alternative hypothesis   In the hypothesis testing procedure, the alternative hypothesis stands in opposition to the null hypothesis and claims that the experimental treatment had an effect. Formally, the alternative hypothesis proposes that the mean of the experimental population is not equal to, is greater than, or is less than the mean of the general population (i.e., the population of subjects who don't receive the experimental treatment). Denoted by the symbol H_a.

analysis of variance   Analysis of variance (ANOVA) is a statistical analysis approach in which variance statistics are used to test hypotheses involving two or more means.

assumption   An assumption is a condition that must be met for a statistical procedure to result in valid conclusions.

beta   The Greek letter beta (β) is used to represent the probability of a Type II error–the likelihood that a false null hypothesis will fail to be rejected.

between-condition variance   Between-condition variance is the ANOVA term for variability due to differences between experimental conditions in a research study—that is, variability due to the effects of one or more independent variables. Between-condition variance is measured by MS_B, which forms the numerator for the F statistic.

bias   A sample statistic is considered biased if it consistently under- or overestimates the value of its corresponding population parameter. For example, σ for a sample is a biased estimator of the true σ for the population from which the sample is drawn, so we need to alter the formula for sample standard deviation s to correct for this bias.

bidirectional test   In a bidirectional hypothesis test, the alternative hypothesis states that the experimental mean is not equal to the mean proposed by the null hypothesis. Thus effects of the experimental manipulation in either direction (an increase or a decrease in mean compared to the general population) are considered in this type of test. Also called a two-tailed test, because the resulting probability value includes the area in both tails of the test-statistic distribution.

bimodal   A bimodal dataset has two distinct peaks in its frequency distribution.

binomial test   The binomial test can be used to assess a categorical dataset in which there are exactly two categories. The null hypothesis of the test specifies the proportion of datapoints for each category (e.g., 50%–50%). The binomial test is formally equivalent to a chi-square test for a dataset with two categories.

bivariate outlier   A bivariate outlier has reasonable X and Y values when the values are considered separately (i.e., the X

and *Y* values separately are not univariate outliers), but when the two values are taken together the datapoint falls outside the range of other datapoints in the relationship's scatterplot.

carry-over effect   A carry-over effect occurs in a repeated-measures design when the experience of completing one of the conditions (e.g., the pre-test) affects subjects' performance in another condition (e.g., the post-test).

categorical variable   A categorical (or discrete) variable is a collection of categories. Categorical data is usually generated by placing each subject in a study into one of a group of pre-defined categories, for example male/female or Catholic/Protestant/Muslim. Compare to *continuous variable*.

cell   A cell refers to one particular frequency in a frequency distribution table.

Central Limit Theorem   The Central Limit Theorem states that for any population with mean μ and standard deviation σ, the distribution of means of samples of size *n* will approach a normal distribution with mean μ and standard deviation $\sigma / \sqrt{n}$ as *n* approaches infinity.

central tendency   A measure of central tendency describes a typical or average member of a population. The most common measure of central tendency is the mean.

chi-square test   The chi-square (χ^2) test is used to test hypotheses about frequency distributions.

confidence interval   A confidence interval is an interval estimate for a population parameter that is specially designed to reflect a certain level of confidence (i.e., 95%) that the parameter actually falls in the specified interval, given a set of sample data.

conservative   A conservative statistician tends to be cautious when making claims on the basis of experimental results, and/or to only use statistical procedures that are the most widely accepted by the scientific community. Compare to *liberal*.

continuous variable   A continuous (or interval) variable can take on any of a range of values, and different values of a continuous variable can be meaningfully compared to each other quantitatively (i.e., we can say that one value is greater than, less than, or equal to another value). Compare to *categorical variable*.

control condition   A control condition in an experiment is a condition that serves as a baseline for assessing effects in an experimental condition.

correlation   The Pearson correlation (usually shortened to just "correlation," and represented by the symbol *r*) is the most common statistic used to measure the relationship between two variables. It is calculated as:

$$r = \frac{\sum z_X z_Y}{n}$$

covariability   Two variables covary to the extent that the variability in one variable goes along with the variability in the other variable. For example, height and weight covary to a great extent in humans, since short people tend to weigh less and tall people tend to weigh more. See also *related*, which captures essentially the same concept.

criterion   A criterion is a standard a researcher uses to compare other values to. For example, the alpha level in an experiment serves as a criterion to compare probability values to.

critical value   The term "critical value" refers to a test-statistic value that must be surpassed if the *p*-value for an experimental result is to be lower than the alpha level chosen for the hypothesis test. For example, in a bidirectional hypothesis test using the *z* statistic, *z* must be greater than 1.96 or less than −1.96 for the probability value to be lower than .05; thus 1.96 is the critical value for this hypothesis test.

cumulative frequency   The cumulative frequency for an interval in a frequency distribution is the number of scores less than or equal to the upper limit of the interval. For example, if the first interval in a table has 5 scores and the second interval has 8 scores, the cumulative frequency for the second interval is 13.

cumulative frequency distribution polygon   A cumulative frequency distribution polygon graphs the cumulative frequencies of each interval in a frequency distribution.

cumulative relative frequency   The cumulative relative frequency for an interval in a frequency distribution table is calculated by dividing the cumulative frequency for the interval by the total number of scores in the dataset.

curvilinear   The datapoints in a curvilinear relationship form a curved line in a scatterplot of the relationship. The Pearson correlation does not provide an accurate measurement of the degree of curvilinear relationships.

degree   The degree or strength of the relationship between two variables refers to the extent to which the two variables covary: from a perfect relationship, in which the value of one variable perfectly predicts the value of the other variable, to no relationship, in which the value of one variable has no power at all to predict the value of the other variable.

degrees of freedom   The degrees of freedom (df) for a statistic refers to the number of scores that are "free to vary" in the calculation of the statistic. The practical significance of the term is that the distributions of many test statistics have different shapes depending on the degrees of freedom for the statistic.

density curve   A density curve is simply a smoothed-out frequency distribution polygon. It is usually used to represent an idealized distribution of an entire population, rather than a distribution of a single sample.

dependent variable   A dependent variable in a research study is a variable whose values may be determined, at least in part, by changes to one or more independent variables. In other words, a dependent variable is a set of scores collected by the researcher which may reveal the effect(s) of one or more experimental manipulations.

descriptive statistic   A descriptive statistic describes some aspect of a set of numbers (as opposed to inferential statistics, which help researchers to interpret what the numbers mean).

deviation score   A deviation score (or difference score) is the difference between two numbers–often a raw score (X) minus the mean (μ or $\overline{X}$). The related-samples t test is used to test whether a sample of D scores differs from a value (usually 0) specified by the null hypothesis.

df_B   In ANOVA terminology, df_B is the degrees of freedom between conditions.

df_E   In ANOVA terminology, df_E is the degrees of freedom for experimental error.

df_S   In ANOVA terminology, df_S is the degrees of freedom between subjects.

df_T   In ANOVA terminology, df_T is the total degrees of freedom for the dataset.

df_W   In ANOVA terminology, df_W is the degrees of freedom within subjects.

dichotomous   A dichotomous variable is a discrete variable that has exactly two possible values, for example male/female.

direction   The direction of a relationship can be either positive, in which case an increase in the value of one variable corresponds to an increase in the value of the other variable, or negative, in which case an increase in the value of one variable corresponds to a decrease in the value of the other variable.

directional test   In a directional hypothesis test, the alternative hypothesis states that the experimental mean is either greater than or less than the mean proposed by the null hypothesis. Thus effects of the experimental treatment in only one direction are considered in this type of test. Also called a one-tailed test, because the resulting p-value includes the area in only one tail of the test-statistic distribution.

distribution   A frequency distribution is a graph or table of the number of individual scores that fall in each of a set of categories for a dataset or population. We often use the word "distribution" to refer to the population of scores themselves (as opposed to just the table or graph of the scores).

distribution of sample means   The distribution of the means of all possible samples of a particular size from a population. Often referred to as a sampling distribution.

effect size   The effect size of an experiment is a measure of the impact of the experimental manipulation. The simplest measure of effect size is obtained by taking the mean of the scores of the experimental subjects and subtracting some standard value.

error variance   Error variance is the ANOVA term for variability due to "background noise" in a research study–that is, variability that cannot be accounted for by any independent variable. Error variance is measured by MS_E, which forms the denominator for the F statistic.

estimated standard error   The estimated standard error ($s_{\bar{x}}$ is an estimate of the standard deviation of a sampling distribution, calculated by substituting the sample standard deviation s for the population standard deviation σ in the standard error formula: $s_{\bar{x}} = s / \sqrt{n}$

expected frequencies   The expected frequencies in a χ^2 goodness-of-fit test or χ^2 test for independence are the values predicted by the null hypothesis for each cell in the frequency distribution table. See also *observed frequencies.*

experimental condition   In an experiment with two or more conditions, an experimental condition is one in which some experimental treatment is applied to the subjects. Results from experimental conditions are usually compared to results from control conditions.

experimental treatment   An experimental treatment (or manipulation) is the factor that differentiates the experimental population from the wider general population in a research study.

exploratory data analysis   Exploratory data analysis is the process of searching a set of data for effects that the study was not specifically designed to uncover.

factor   In ANOVA terminology, a factor is an independent variable to be analyzed in the ANOVA. A factor is said to have some number of levels.

first quartile   The 25th percentile is also called the first quartile.

form   The form of a relationship between two variables refers to the shape formed by the points in the scatterplot of the relationship. For example, many relationships can best be described as linear in form.

frequency distribution polygon   A frequency distribution polygon provides an alternative way to graph a frequency distribution; it is constructed by placing dots at the centers of the top sides of each of the bars in a histogram of the distribution, then connecting the dots.

F statistic   The F statistic is a test statistic used in analysis of variance to compare the means of two or more different conditions. It is calculated by dividing MS_B by MS_E.

goodness-of-fit test   A χ^2 goodness-of-fit test is a general-purpose hypothesis testing procedure for categorical data that are organized in a frequency distribution table. The null hypothesis in a goodness-of-fit test specifies an arbitrary set of expected frequencies for the cells of the table.

histogram   A histogram graphs the frequency distribution for a group of scores.

hypothesis   Generally, a hypothesis is an educated guess about something. In inferential statistics, a hypothesis is a formal prediction about the outcome of an experiment. The hypothesis testing procedure pits a null hypothesis against an alternative hypothesis.

independent   Two event outcomes are independent if the probability of one does not depend on the probability of the other. Two samples of subjects in an experiment are independent if the subjects in one sample do not influence in any way the subjects in the other sample.

independent-samples design   In an independent-samples research design, two or more samples of scores are taken from independent sets of individuals and the samples are compared. See also *independent-samples* t *test* and *related-samples design.*

independent-samples *t* test   An independent-samples t test is used in experimental designs in which two unrelated groups of subjects participate in different experimental conditions. The t statistic is used to evaluate whether or not the means of the two groups differ significantly from each other.

independent variable   An independent variable in a research study is a variable whose values are controlled or manipulated, at least to some extent, by the researcher. Most research studies are done with the hope that an independent variable will have some measurable effect on one or more dependent variables. Compare *dependent variable.*

inferential statistic   An inferential statistic is designed to help a researcher draw conclusions about a population from the descriptive statistics calculated for a random sample of the population.

interaction   An interaction occurs in an experiment when the outcome of one aspect of the experiment is affected by the outcome of another aspect. For example, suppose groups of men and women in an experiment take a pre-test, then undergo some training, then take a post-test. If the women's performance improves on the post-test compared to the pre-test while the men's scores decline, we say that the effect of training *interacts* with gender.

interquartile range   The interquartile range (IQR), a resistant measure of variability, is calculated as the 75th percentile of a dataset minus the 25th percentile of the dataset.

interval estimate   An interval estimate for a population parameter is a range of values within which we guess the parameter falls. Compare to *confidence interval* and *point estimate.*

level   In ANOVA terminology, a level is a value that a factor can assume. For example, a study might include an "age" factor that has three levels: young (15–30 years old), middle-age (31–50 years old), and elderly (51–80 years old).

liberal   A liberal statistician tends to make strong claims on the basis of experimental results, and/or to use statistical procedures that might not be accepted by more conservative members of the scientific community.

linear   The datapoints of a linear relationship form a straight line in a scatterplot of the relationship. The Pearson correlation measures the strength of a linear relationship.

LSD   In statistics, LSD stands for Fisher's least significant differences method for doing post hoc pairwise comparisons.

lurking variable   A lurking variable in correlation analysis is an unmeasured variable that has some effect on the measured relationship between variables X and Y.

main effect   In ANOVA terminology, a main effect is the effect of one factor when all other factors are ignored. For example, suppose groups of men and women in an experiment take a pre-test, then undergo some training, then take a post-test. If we average across the pre-test and post-test, we can assess the main effect of gender. See also *interaction.*

matched-pairs design   In a matched-pairs experimental design, subjects are paired off based on some pre-experimental measure(s), then one member of each pair is placed in one condition and the other member in a second condition. The data are then analyzed as if each pair of scores came from a single subject, using the related-samples t test.

mean   The mean is the most-used measure of central tendency, calculated by adding all the scores in question together and dividing by the number of scores. See *population mean* and *sample mean* for definitional formulas.

mean squared deviation   Mean squared deviation (MS) is the ANOVA term for sample variance.

median   The middle score in a distribution is called the median, and it is used as a resistant measure of central tendency. If the distribution has an even number of scores, the median is formally defined as the mean of the two middle scores.

median test   A median test is a special form of the χ^2 test for independence that serves as a nonparametric alternative to the independent-samples t test or independent-samples ANOVA.

mixed-factors design   An experiment with a mixed-factors design includes at least two independent variables, with at least one factor manipulated between subjects (i.e., as in an independent-samples design) and at least one factor manipulated within subjects (i.e., as in a related-samples design).

mode   The mode is the most common score in a dataset. We can also speak of the modal interval in a frequency distribution.

MS_B In ANOVA terminology, MS_B is the mean squared deviation between conditions, a measure of the between-condition variance. The MS_B is used as the numerator for the F statistic.

MS_E In ANOVA terminology, MS_E is the mean squared deviation for experimental error, a measure of the error variance. The MS_E is used as the denominator for the F statistic.

multimodal A multimodal dataset has two or more distinct peaks in its frequency distribution (if the dataset has exactly two peaks, it is usually called a bimodal distribution).

multiple regression/correlation Multiple regression/correlation (MRC) is a statistical analysis in which the relationships between three or more variables are assessed all at once.

nonparametric statistic "Nonparametric" statistics do not assume anything about underlying population parameters. As such, they can be valuable alternatives to parametric statistics (such as t) when parameter assumptions may be violated.

normal A normal distribution has a classic "bell-shaped" curve, shown below, which can be precisely defined by a complex mathematical formula.

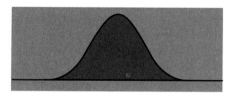

Scores of a remarkable number of diverse populations turn out to be normally distributed.

null hypothesis In the hypothesis testing procedure, the null hypothesis claims that the experimental manipulation has no effect at all. More formally, the null hypothesis usually proposes that the mean of the experimental population is identical to the mean of the general population (i.e., the population of subjects who don't receive the experimental treatment). Denoted by the symbol H_0.

observed frequencies The observed frequencies in a χ^2 goodness-of-fit test or χ^2 test for independence are the actual observed values in each cell of the frequency distribution table. See also *expected frequencies.*

omnibus An omnibus ANOVA assesses whether or not any condition mean differed from any other condition mean. Omnibus tests must be followed by pairwise comparisons or some other post-hoc test to determine exactly which means significantly differ from which other means.

outlier An outlier is a single score in a dataset that is so atypical (i.e., so much higher or lower than the rest of the scores in the dataset) that it deserves special consideration when calculating descriptive and inferential statistics for the dataset.

pairwise comparison A pairwise comparison assesses the statistical significance of the difference between one pair of means in a multicondition experiment. Pairwise comparisons are usually done as post hoc tests after an initial assessment of the data using an omnibus ANOVA.

parameter A parameter is a statistic that is based on data from every member of a population. Parameters are symbolized by Greek letters (such as μ for the population mean and σ for the population standard deviation) to distinguish them from statistics based on data from samples.

parametric statistic A parametric statistic requires certain assumptions about population parameters to be met in order for conclusions based on the statistic to be valid. Compare to *nonparametric statistic.*

percentile The Pth percentile of a dataset is the score in the dataset's distribution that is greater than P% of the other scores in the distribution and less than $(1 - P)$% of the other scores.

percentile rank The percentile rank of a score X in a dataset is the percentage of scores in the dataset that is less than or equal to X.

pooled variance The pooled variance is the weighted mean of the variances of two samples. It is used as the basis for calculating the estimated standard error for a independent-samples t test.

population   In statistical jargon, a population is the set of all individuals of interest in a particular study. The difference between a population and a sample is sometimes tricky. If you are working with the data for every single individual you'd like to draw conclusions about, you're working with a population. If you'd like to generalize from the individuals you're directly studying to a larger group, you've got a sample.

population mean   The mean of a population of scores, symbolized . Calculated by summing the scores (X) and dividing by the population size (N):

$$\mu = \frac{\Sigma X}{N}$$

population standard deviation   The standard deviation of a population of scores, symbolized σ. Calculated as follows:

$$\sigma = \sqrt{\frac{SS}{N}} = \sqrt{\frac{\Sigma(X - \overline{X})^2}{N}}$$

See also *sum of squares, population variance, sample standard deviation.*

population variance   The variance of a population of scores, symbolized σ^2. Calculated as follows:
$\sigma^2 = SS / N = \Sigma(X - \mu)^2 / N$
See also *sum of squares, population standard deviation, sample variance.*

power   Statistical power is the probability of rejecting the null hypothesis on the basis of experimental data, if the null hypothesis is actually false. A "powerful" experiment has a very good chance of finding an effect if in fact an effect exists.

power analysis   Power analysis is a process by which a researcher analyzes the power of a planned experiment by estimating the effect size and standard deviation she expects to find. If a power analysis reveals that there is a poor chance of rejecting the null hypothesis, the experiment may be scrapped or more subjects may be included.

point estimate   A point estimate of a population parameter is a single value that serves as an estimate for the value of the parameter. For example, a sample mean can be used as a point estimate of the mean of the underlying population. Compare to *interval estimate.*

post hoc test   A post hoc test is conducted after another hypothesis test (usually an omnibus ANOVA) indicates that further hypothesis tests are justified. The most common type of post hoc tests are pairwise comparisons.

probability   The probability, or likelihood, of a possible outcome A, $p(A)$, for a random event is the proportion of times the outcome will occur over a large number of trials of the event.

probability value   The term "probability value" (usually shortened to *p*-value) is usually used to indicate the probability of obtaining a given value of a test statistic by chance alone. For example, if $z = 2.0$, the *p*-value is .046, meaning that there is a 4.6% chance of randomly drawing a z-score this large from a normally distributed population.

random event   A random event is an event whose outcome is uncertain, such as whether a flipped coin will come up with the "heads" or "tails" side up. Over the long term, the probabilities of the various outcomes of a random event are usually predictable.

range   The range of a dataset is the difference between the highest and lowest score.

regression line   The regression line for a pair of variables X and Y runs through the points in the scatterplot of the two variables in such a way that the average distance from all points to the line is minimized.

related   In statistics, two variables are said to be related if knowing the value of one variable helps to predict the value of the other variable. The degree of relationship for two variables is typically measured with a statistic called the Pearson correlation.

related-samples design   In a related-samples research design, two or more samples of scores are taken from the same set of individuals, or from matched sets of individuals, and the samples are compared. See also *related-samples t test* and *independent-samples design.*

related-samples *t* test   A related-samples *t* test is used in experimental designs in which each subject, or each pair of matched subjects, participates in two experimental conditions. One member of each pair of scores is subtracted from the other to produce a difference score, and the *t* test helps determine whether the mean of the difference scores is significantly different than 0.

relative frequency   The percentage of the total scores from a dataset that falls in one particular interval is called the relative frequency for that interval.

relative frequency distribution histogram   Bars in a relative frequency distribution histogram represent the relative frequency of each interval in the graph, as opposed to the absolute number of scores in each interval that is shown on a regular histogram.

reliability   In correlation analyses, reliability refers to the consistency of a variable in measuring an underlying property.

repeated-measures design   In a repeated measures experimental design, each subject participates in two or more experimental conditions. For exactly two conditions, the data are usually analyzed via the related-samples *t* test.

representative   A sample is said to be representative of a population if the central tendency and variability of the scores in the sample are similar to the central tendency and variability in the population as a whole.

resistant   A resistant statistic is not overly influenced by outliers. The median and interquartile range are prototypical resistant statistics.

robust   A test statistic is said to be robust if conclusions based on the test statistic are generally valid even if the assumptions behind the test statistic are violated to some extent. Sometimes, test statistics are robust to violations of some assumptions but very sensitive to violations of other assumptions.

sample   A sample is a set of individuals selected from a population that is being studied in an experiment. Usually, subjects are selected randomly from the population to help ensure that the sample is representative of the population.

sample mean   The mean of a sample of scores, symbolized $\overline{X}$. Calculated by summing the scores and dividing by the sample size:

$$\overline{X} = \frac{\Sigma X}{n}$$

sample size   The sample size is the number of individuals in a sample taken from some population, abbrevi-

ated *n*. Usually, *n* is the number of subjects in an experiment. (Note that capital *N* represents the number of individuals in an entire population.)

sample space   The sample space for an random event is the set of all possible outcomes of the event.

sample standard deviation   The standard deviation of a sample of scores, symbolized *s*. Calculated as follows:

$$s = \sqrt{\frac{SS}{n-1}} = \sqrt{\frac{\Sigma(X - \overline{X})^2}{n-1}}$$

See also *sum of squares, sample variance, population standard deviation.*

sample variance   The variance of a sample of scores, symbolized s^2. Calculated as follows:
$$s^2 = SS / (n - 1)$$
See also *sum of squares, sample standard deviation, population variance.*

sampling distribution   A sampling distribution is a distribution of statistics that reflects all possible samples of a given size *n* selected from a population. The most commonly used sampling distribution is of sample means, so "sampling distribution" is often used as a synonym for "distribution of sample means."

scatterplot   A scatterplot graphs a relationship between two variables *X* and *Y* by plotting a single point for each pair of values.

sign test   A sign test is a special form of chi-square test for frequency distributions of discrete variables that have exactly two categories (e.g., male/female).

single-sample *t* test   A single-sample *t* test is used to help decide whether or not a single mean in an experiment is significantly different from some arbitrary comparison value.

skewed   A skewed distribution is one in which scores are very common on one side and gradually taper off on the other side. For example, in the figure below the distribution is positively skewed:

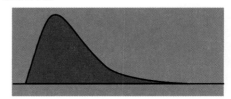

SS_B In ANOVA terminology, SS_B is the sum of squared deviations between conditions.

SS_E In ANOVA terminology, SS_E is the sum of squared deviations for experimental error.

SS_S In ANOVA terminology, SS_S is the sum of squared deviations between subjects.

SS_T In ANOVA terminology, SS_T is the total sum of squared deviations for the dataset.

SS_W In ANOVA terminology, SS_W is the sum of squared deviations within subjects.

standard deviation The standard deviation, calculated as the square root of the variance of a sample or population, is the most commonly used measure of the variability of a distribution. The term comes from the notion that we are measuring the typical (standard) difference (deviation) from the mean of the sample or population. See *sample variance* and *population variance* for formulas.

standard error The standard deviation of a distribution of sample means with sample size n is called the standard error of $\overline{X}$, usually shortened to simply "standard error" and symbolized by $\sigma_{\overline{X}}$. The Central Limit Theorem tells us that the standard error is equal to $\sigma/\sqrt{n}$. If we don't know σ when we're evaluating an experiment, we can substitute s for σ to calculate the estimated standard error.

standardize To standardize a number is to transform it in some way that puts the number into context.

statistic A single number that summarizes some aspect of a set of other numbers. For example, the mean summarizes the central tendency of a set of numbers.

statistically significant An experimental result is said to be statistically significant if the p-value for the test statistic calculated for the experiment is less than the alpha level set by the experimenter. This term is sometimes shortened to "significant," but "significance" should not be equated with "importance."

stem and leaf diagram Stem and leaf diagrams combine elements of frequency distribution histograms and tables. They are rather infrequently used in modern statistics, since computers can generate graphical histograms so easily.

sum of squares The term "sum of squares" (*SS*) is short for "sum of squared deviations," a component in the formulas for variance and standard deviation:

$$SS = \Sigma(X - \mu)^2$$

(To calculate *SS* for a sample, substitute $\overline{X}$ for μ.)

symmetrical A symmetrical distribution is one that can be divided in half at the mean, resulting in two halves that are mirror images of each other. The classic symmetrical distribution is the normal curve:

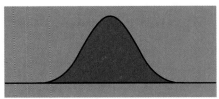

tail A distribution is said to have a tail if it tapers off at one end (often distributions have two tails, one on the left and one on the right side). In the figure below the left tail is highlighted in red:

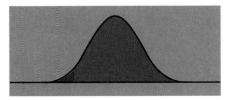

t distribution A t distribution is a distribution of all possible t statistics derived from samples of a certain size. The t distributions for different degrees of freedom differ slightly in shape: the fewer degrees of freedom, the "flatter" the t distribution. As the degrees of freedom approach infinity, the t distribution resembles a normal distribution more and more. Note that the shape of the t distribution does *not* depend on the mean or standard deviation of a particular sample or population.

test for independence A χ^2 test for independence is a hypothesis testing procedure for assessing the relationship between categorical variables organized in a frequency distribution table. The null hypothesis in a test for independence specifies that the two variables are not related at all.

test statistic   A test statistic is an inferential statistic used to evaluate hypotheses about experimental results.

third quartile   The 75th percentile is also called the third quartile.

transform   To transform a number means to change it by applying a regular mathematical operation. For example, we transform a raw score into a z-score by subtracting the mean and dividing by the standard deviation of the population the score came from.

trimmed mean   If outliers are removed from a dataset before calculating the mean and standard deviation of the dataset, we may call those statistics the trimmed mean and trimmed standard deviation.

t statistic   The t statistic is a test statistic used to compare the mean of a single condition to a standard value (single-sample t test), or to compare the means of two different conditions to each other (related-samples t test or independent-samples t test). The basic formula for t is:

$$t - \frac{\overline{X} - \mu}{s_{\overline{X}}}$$

where μ is the population mean specified by the null hypothesis and $s_{\overline{X}}$ is the estimated standard error of the sampling distribution.

Type I error   A Type I error occurs when the hypothesis testing procedure leads a researcher to reject the null hypothesis (and claim that the experimental treatment had an effect), when the null hypothesis is actually true (the treatment actually had no effect).

Type II error   A Type II error occurs when the hypothesis testing procedure leads a researcher to "accept" the null hypothesis (and claim that the experimental treatment had no effect), when the null hypothesis is actually false (the treatment actually did have an effect).

uniform   A uniformly distributed dataset has no distinct peaks in its frequency distribution (compare to unimodal and bimodal distributions).

unimodal   A unimodal dataset has a single peak in its frequency distribution.

validity   In correlation analyses, validity refers to the extent to which one variable measures the same underlying property as another variable.

variable   Formally, a quantity, usually denoted by a symbol such as X, that may assume any one of a set of values. Informally, a set of data collected in an experiment.

variability   The variability of a collection of scores is the degree to which the scores are spread out around the center of the distribution of scores. The most common measure of variability is the standard deviation.

variance   The variance is one measure of the variability in a set of scores. The standard deviation, which is the square root of the variance, is more commonly used to describe a distribution's variability.

z-score   A z-score is an individual score from a population that has been standardized by subtracting the mean and dividing by the standard deviation of the population:

$$z = \frac{X - \mu}{\sigma}$$

Thus a z-score of 0 is equal to the mean of the population, a z-score of -1 is exactly one standard deviation below the mean, and a z-score of 2.3 is 2.3 standard deviations above the mean.

Index